Physiology in the American Context
1850–1940

S. R. Geiger, *Publications Manager and Executive Editor*

B. B. Rauner, *Production Manager*

L. S. Chambers, M. J. Carnahan, G. I. Armstrong, S. P. Mann, *Editorial Staff*

D. L. Witt, *Indexer*

Physiology in the American Context

1850–1940

Edited by
Gerald L. Geison

AMERICAN PHYSIOLOGICAL SOCIETY
Bethesda, Maryland

Library of Congress Catalog Card Number 86-32286
International Standard Book Number 0-683-03446-4
Printed in the United States of America by
Waverly Press, Inc., Baltimore, Maryland 21202
Distributed for the American Physiological Society by
The Williams & Wilkins Company, Baltimore, Maryland 21202

Preface

This book had its origins in a conference held in mid-January 1986 at the National Library of Medicine in Bethesda, Maryland, with financial support from the American Physiological Society. John Parascandola of the National Library of Medicine generously offered its facilities and his help with the local arrangements. Toby Appel, historian/archivist for the American Physiological Society, in addition to contributing a chapter to the volume, oversaw the local arrangements for the conference and contributed in a host of other ways.

The American Physiological Society deserves special thanks, not only because of its financial support for the conference or the contributions of Dr. Toby Appel, but also because in its role as publisher, the Society has done all that an editor might wish.

No less important, the Society made no attempt whatever to influence the content or tone of the chapters, even though the volume was conceived and scheduled with an eye toward its centennial celebration this year. If this volume serves in part to mark that important occasion, it is by no means a centennial volume of the usual sort. For long stretches, the American Physiological Society is entirely absent from these pages; it is in fact the focus of only one of the chapters (Appel's). The other authors and myself as editor appreciate the Society's willingness to serve as publisher of a historical volume in which it is not the central object of concern.

My own research in this field also received an early boost through a small but fertile grant from the American Physiological Society in 1981. I am grateful to Orr Reynolds, past Executive Secretary of the Society, for helping me to secure that grant and for his encouragement ever since. The data gathered with the support of that small grant helped lay the basis for a more ambitious proposal to the National Library of Medicine, which very generously supported my research on the history of American physiology from 1983 through 1986. To the extent that the production of this volume depended on my own time and research, it was supported in part by National Library of Medicine Grant LM-03936; as program officer for that grant, Jeanne Brand was consistently helpful and supportive.

One crucial benefit of the National Library of Medicine grant was that it allowed me to employ a series of talented part-time research assistants, all of them graduate students in the Program in History of Science at Princeton University. They included, sequentially, Pauline Carpenter [now Dear], Peter Dear, John Carson, Tom Broman, George Anastoplo, and Andrea Rusnock. They compiled much of the data for my discussion of "domestic elites" in chapter V. In particular, George Anastoplo provided valuable insights into research productivity at the institutional level and taught me a lot about the history of physiology at the University of Chicago. He and Andrea Rusnock contributed in major ways to several of the tables and figures in chapter V.

Finally, I am indebted to Faye Angelozzi and Vanessa Noya for their skillful assistance with clerical and organizational tasks.

GERALD L. GEISON

Contents

Toward a History of American Physiology

GERALD L. GEISON

The history of physiology in the United States is a vast and mostly uncharted domain. Even as the American Physiological Society (APS) enters its centennial year, we know much less about the history of physiology in this country than we do about such other disciplines as astronomy, chemistry, geology, and physics, or even such related fields as biology, biochemistry, and genetics.[1] This volume seeks to fill part of that gap. It provides a sort of base camp from which further expeditions into the history of American physiology can be launched. As the first book-length study of the subject, it is bound to set part of the agenda for further research. It is therefore important to specify both what the volume tries to accomplish and what it leaves undone.

The volume title has been chosen with some care. This is a history of physiology in the American context from ca. 1850 to 1940, where "context" is to be understood mainly as a shorthand way of saying "social, institutional, and cultural context." This is not the place to look for any extended or detailed account of the history of physiological research or ideas in the United States. Except for allusions to the cultural appeal of laboratory science or scientific medicine in the United States, or occasional attention to the general intellectual style of physiology pursued in particular institutional contexts, the conceptual and technical side of the story is largely confined to portions of the essays in Part 3, especially Louise Marshall's chapter on the history of American neurophysiology.

This lack of attention to the intellectual history of American physiology is unfortunate, for it is of course the conceptual and technical side of any scientific discipline that ultimately distinguishes it from other activities and its practitioners from other folk. Yet when these conceptual and technical histories come to be written, this book should serve as a useful introduction to the social and institutional contexts in which American physiological research can be situated. Meanwhile, the focus here on the social and institutional history of the discipline has at least one major virtue: it gives the book an unusual unity of theme and approach. In this introduction, I summarize the leading concerns in each of the fifteen chapters that follow, suggesting along the way my rationale for dividing the volume into three parts. I end by emphasizing the extent to which many of these apparently disparate chapters in fact share common themes. At whatever unfortunate cost to the conceptual and technical history of American physiology, this book has a coherence that is especially welcome in a collective volume that opens up a new field of study.

PART 1: SOCIAL AND INSTITUTIONAL HISTORY

The five chapters in Part 1 are unified by their concern with aspects of the social and institutional history of physiology in the United States and (in one case) Canada. Robert G. Frank, Jr. emphasizes the extent to which the research ideal entered American physiology from Germany after ca. 1860. Through statistical analyses, Frank shows that the majority of the early leaders in American physiology traveled to Germany (above all to Ludwig's laboratory in Leipzig) for advanced training, especially before 1885, when the percentage began to drop. From about 1900 until the outbreak of World War I in 1914, the percentage of American research physiologists who traveled abroad stabilized at 15%–20% of the cohort but even then a disproportionate number of them later attained distinction. From Frank's lively accounts of American physiologists abroad, it is easy to see why they enjoyed the experience, even if they did bring back a partly illusory vision of the German system that was then further transformed in the process of adapting it to local circumstances.

W. Bruce Fye identifies the full range of personal, institutional, and social factors that frustrated or promoted the aspirations of would-be professional physiologists in the United States from ca. 1850 to 1900. He does so partly by surveying the careers of four early leaders in American physiology (John Call Dalton, Jr., S. Weir Mitchell, Henry P. Bowditch, and H. Newell Martin), the last two of whom were finally able to establish a foothold for professional physiology in the United States after ca. 1875. In accounting for their success and the more general emergence of physiology as an independent discipline, Fye gives special attention to the role of the broader American medical reform movement with its emphasis on full-time academic careers and basic scientific research. Because this reform agenda was expensive, private wealth and institutional endowments played a crucial part in the story, including the early success of Bowditch at Harvard and Martin at Johns Hopkins. By ca. 1900, Fye insists, the medical reform movement enjoyed substantial support in the United States, thanks in no small part to the spectacular achievements in bacteriology after 1880, and the professional pursuit of physiology and the other preclinical medical sciences in America was henceforth assured.

The chapters by Alejandro C. Laszlo and Sandra McRae focus more sharply on the rise and fate of experimental physiology at three universities—Harvard and Columbia (Laszlo) and the University of Toronto (McRae). Both essays develop their stories in the context of medical reform at these universities, with Laszlo giving an unusual amount of attention to the role of university presidents (Seth Low and Nicholas M. Butler at Columbia and especially Charles Eliot at Harvard). Discussing the period from ca. 1870 to 1920, Laszlo contrasts the success of the medical style of physiology pursued at Harvard under Bowditch and his successor Walter B. Cannon with the relative failure and ultimate defeat of the biological program that John G. Curtis and Frederic S. Lee tried to establish at Columbia.

McRae shows that the University of Toronto developed a vigorous tradition in physiological research under A. B. Macallum from the 1880s onward, long before the discovery of insulin in 1922 made Toronto famous. She also contends that medical and biological interests were well integrated at Toronto from ca. 1890 until 1903, when the Department of Physiology was transferred from the biological building to the new medical building. Her survey of the careers of leading Toronto graduates and faculty reveals a pattern whereby the best students often went to the United States for further training (and often remained there), while the senior posts at Toronto itself were often filled by biomedical scientists from elsewhere in the British Empire. In the end, McRae suggests, Canadian physiologists and biologists developed a fertile and somewhat distinctive style of scientific research by combining a dominant British influence with strong ties to the United States, though at some cost to Canadians' awareness of their own achievements.

My own chapter is divided into two parts. In the first part, I show that the international ascendancy of American physiology can be traced quite precisely to World War I. I briefly discuss a few of the factors that account for the new international standing of American physiology, including especially the number and quality of new laboratories in this country. In the second part, I identify the leading research physiologists and major research centers in the United States up to ca. 1940 through a quantitative study of publication patterns in the *American Journal of Physiology*. Using that data and a sort of "time-lapse photography," I then construct a set of social maps of the discipline from 1880 to 1940. Finally, I emphasize how very few institutions and indeed how very few men it took to lead the United States onto center stage in the international world of physiology, although I also suggest that these American physiological elites were the products of broader social and institutional forces.

PART 2: PHYSIOLOGY IN RELATION TO OTHER FIELDS

Part 2 of the volume, which contains nearly half of the chapters, draws attention to the perennial problem of how physiology is to be defined in relation to other fields, both pure and applied. As Toby A. Appel shows in her essay, the question of the place physiology should occupy vis-à-vis biology, the biochemical sciences, and clinical medicine played a pivotal role in the very founding of the APS in 1887. The catalyst was the International Medical Congress of 1887, which met in the United States that year under the organizational control of the American Medical Association and ultimately of rank-and-file medical practitioners throughout the country. In response, elite and reform-minded elements in the American medical community organized their own alternative Congress of American Physicians and Surgeons, made up of constituent medical specialty societies. In order to participate in the elite Congress, professional physiologists formed their own specialty society, the APS. At the same time, they became affiliated with a new biological specialty society, the American Society of Naturalists. The physiologists met alternatively and sometimes a bit uneasily with the (medical) Congress and the Naturalists until 1912, when the APS became one of the charter societies in the new Federation of American Societies for Experimental Biology. By then, the Society had largely resolved its identity crisis in favor of affiliation with other preclinical biomedical disciplines.

However, the chapters by Jane Maienschein and Philip J. Pauly show that not all experimental biologists were willing to concede the label "physiology" to the medically oriented conception of the field that dominated the APS. In complementary ways, Maienschein and Pauly discuss the antagonism of several leading American biologists toward medical physiology (i.e., vertebrate organ physiology) and their efforts to articulate an alternative conception of the field that went variously by the names "biological physiology," "physiological morphology," or "general physiology."

Maienschein begins by tracing the American roots of biological physiology to the biology department at Johns Hopkins (founded in 1876), where early graduate students were exposed simultaneously to physiology broadly conceived as part of biology (through H. Newell Martin) and to morphology (through W. K. Brooks), with no sharp disciplinary boundaries being drawn. However, she argues, this broad vision was largely lost by 1893, when Martin's resignation and the opening of the Hopkins medical school combined to encourage a more medically oriented conception of physiology there. Maienschein then turns her attention to Charles Otis Whitman and his efforts to institutionalize a biological conception of physiology at the University of Chicago and (in the summers) at the Marine Biological Laboratory in Woods Hole, Massachusetts. Whitman was particularly eager to show the advantages of combining traditional physiological and embryological approaches in the investigation of the fundamental problems of development and differentiation. Insofar as Jacques Loeb's important research could be placed in this framework, Whitman promoted his career and secured a place for him at Chicago. Maienschein suggests that a promising Chicago School of Biology coalesced around Whitman in the 1890s but lost much of its vitality within a decade as the University failed to nourish it. Even then, however, Chicago remained a relative haven for biological physiology as it became increasingly marginal throughout the United States.

Pauly focuses on Loeb rather than Whitman and on general rather than biological physiology. He moves beyond Hopkins and Chicago to other institutions, notably Harvard, and carries the story into the 1930s. What emerges is a tale of personal idiosyncracy and arrogance, conceptual confusion, and institutional fragility. Loeb's own notions of general physiology shifted continually as he moved from one research problem to another and from one institution to another. Other scientists who claimed title to general physiology promoted still other versions of the specialty, including the study of the simplest organisms, the properties common to all, the organism as a whole, or even colloid chemistry. The only unifying features were hostility toward medical physiology and a search for some basic understanding of the problem of life, usually in physical and chemical terms. Surprisingly, these amorphous and even competing notions of general physiology enjoyed some favor during the 1920s, when the Rockefeller Foundation funded several versions simultaneously. During that decade, a few general physiologists, including the insufferable W. J. Crozier and his gifted disciples at Harvard, carved out a small niche for themselves in the interstices between physiology, biochemistry, biology, and physical chemistry. Even those few opportunities were soon lost in the face of persistent conceptual confusion and personal arrogance, which Pauly compares with that of the molecular biologists of the 1950s. By the time general physiology did begin to stabilize as a discipline in the 1940s it had narrowed its scope to bioenergetics and the physical chemistry of membrane phenomena and bore little resemblance to the once grandiose visions of Loeb. For a while, however, general physiology had provided an institutional framework for the training and research of a number of creative individualistic scientists.

Russell C. Maulitz's chapter directs our attention to another discipline that once sought to incorporate the term "physiology" into its conception of itself: pathology. He reminds us that the great German cellular pathologist Rudolf Virchow promoted the idea of a unified pathophysiology, which some American pathologists enthusiastically embraced. At the turn of this century, reform-minded physiologists and pathologists seemed united behind that idea, just as they were then joined in a single section of the American Medical Association. However, two sets of forces soon began to drive the pathologists away from the experimental model associated with physiology. First, pathologists themselves became increasingly preoccupied elsewhere: with hospital posts and an associated morphological (rather than experimental) model of pathology and with the flourishing new subdiscipline of bacteriology, a daughter specialty that threatened to dominate its parent science. Second, a new group of crusading clinical investigators had emerged by World War I with an appealing program that usurped much of what had been called experimental pathology. Thereafter, regular pathologists were largely relegated to the sidelines as physiologists, biochemists, and the new clinical investigators struggled for hegemony over the broad territory of experimental medicine in the United States.

Diana E. Long's essay concerns the relations between physiology and endocrinology or sex research more generally. She offers a statistical analysis of the disciplinary affiliations and identities of a group of thirty-one sex researchers who published articles in the *American Journal of Physiology* between 1923 and 1947 (with funding from the Rockefeller Foundation through the Committee for Research in Problems of Sex). By examining their pattern of publication in various journals and their disciplinary self-definition in *American Men of Science*, Long investigates the extent to which this group of sex researchers considered themselves physiologists and thought of their research as physiology. Although barely one-third identified themselves primarily as "physiologists," two-thirds belonged to the APS and Long argues that they gained important benefits from their association with the prestigious discipline, its Society, and its journal, especially in the early part of the period she covers. Indeed, she suggests that "physiology formed the stable base on which sex research was built" even as she emphasizes the extent to which the field was a striking example of interdisciplinary success in American science, fueled by the larger social goal of reforming attitudes toward sex.

The chapters by Richard Gillespie and Joel D. Howell in Part 2 take us directly into the realm of applied physiology. Richard Gillespie examines the possibilities that physiologists saw in the concept of industrial fatigue during World War I and briefly thereafter. The concept, which gained currency in the wake of production shortfalls during the war, was almost as loose and labile as general physiology, embracing (among other things) muscular fatigue, metabolic and respiratory changes in bodily fluids and tissues, or general decline in physiological and mental performance under the stress of long working hours. Its diversity of meanings briefly gave industrial fatigue its power; it held out the hope of applying physiological principles to the entire range of problems in the work place. For physiologists, it offered an opportunity to escape from the constraints of teaching medical students and to do so in the service of a social problem of great national significance. If industrial fatigue owed its early appeal to its breadth and versatility, however, its conceptual and institutional base was correspondingly shallow. By the mid-1920s, it was clear that physiologists were having trouble transferring physiological principles from the laboratory to the work place, and they now found themselves in competition with psychologists and physicians for a place in industry. Moreover, the government funding that underwrote research in the field declined after the war as the state again withdrew from industrial problems. Insofar as experimental research on industrial fatigue survived after the 1920s, it was an attenuated version that became increasingly remote from the work place.

Joel D. Howell's chapter explores the relationship between basic physiological research and medical practice in the case of heart disease. Specifically, he asks when and how the electrocardiogram (ECG) and pressure tracings of the cardiac chambers came to be seen as useful to practicing physicians, whether for purposes of therapy, diagnosis, or conceptualizing heart disease. A physician as well as historian, Howell refuses to take at face value the initial self-serving claims of research physiologists or scientific clinicians as to the medical utility of these techniques, both of which were developed in physiological laboratories in the early years of this century. Instead, he investigates actual medical practice as reflected in clinical case records from the Massachusetts General Hospital, medical licensing examinations, and his own studies of the use of medical technology in hospital patient records. He insists that despite the claims of early enthusiasts, neither the ECG nor pressure curves had much therapeutic utility before the 1950s. Also, neither technique was routinely used for diagnostic purposes much before then, despite the potential value of the ECG in the diagnosis of myocardial infarction after about 1920. For much of this century, both techniques were used instead mainly as aids in conceptualizing heart disease, partly because they produced graphic records that accorded so well with the ideology of scientific medicine. Howell does not dispute the current diagnostic or therapeutic utility of either technique, especially the now-ubiquitous ECG, but he does insist that we should analyze utilitarian claims for basic research critically and should recognize the other goals that may lie behind such claims. The neglected history of typical medical practice can thus yield new insights into the relationship between basic physiological knowledge and its application.

PART 3: INSTRUMENTS, MATERIALS, AND TECHNIQUES

Part 3 of the volume consists of three chapters that explore the crucial but neglected topic of laboratory instruments, materials, and techniques in the history of American physiology. With the aid of several striking illustrations, Merriley Borell provides a full and palpable sense of the laboratory arrangements and instruments that were available to the first generation of professional physiologists in the United States. She focuses on the Harvard Physiological Laboratory from 1871 to 1906, giving special attention to the role of William Townsend Porter, who has long been cast in the shadow of his distinguished colleague W. B. Cannon. At Harvard, Porter introduced a novel version of laboratory instruction in which all beginning students gained hands-on experience with physiological apparatus and techniques even before they heard the related lectures. Porter's pedagogical innovations required him to redesign existing laboratory instruments and to invent still others that were inexpensive and durable enough for large classes of introductory students. To supply these needs, he established the highly successful Harvard Apparatus Company, based on the "American system of manufacture," i.e., mass production through standardized, interchangeable parts. By 1906 the company was exporting laboratory equipment to more than 200 laboratories, including more than 160 in this country alone. As Borell emphasizes, Porter's innovations in laboratory teaching and in the design and manufacture of laboratory instruments played a significant part in the larger social and conceptual process by which physiology emerged as an independent discipline in the United States.

Adele Clarke's essay brings to light the virtually unstudied problem of how physiologists and other experimental biologists secured animals and animal materials for research during the first half of this century. Stressing the radically different needs for research materials that accompanied the shift from morphological to physiological approaches, she offers a case study of the situation in mammalian reproductive research in the United States from ca. 1900 to 1940. Like other experimental biologists, reproductive scientists now sought large quantities of live animals or fresh animal materials, a task that was challenging, time-consuming, and expensive. At first, they went mainly to local farms and slaughterhouses (for pig and cow embryos and ovaries or for hormone-rich cow's urine and mare's blood), though cats and dogs were secured from medical school supplies. To collect human materials, they arranged with local physicians to receive discarded embryos, ova, and ovaries from miscarriages, tubal pregnancies, and hysterectomies. They often exchanged materials and information about their acquisition and maintenance through formal and informal networks. In time, new infrastructural arrangements arose to meet the need, including commercial biological supply firms, but most important by far were new on-site laboratory colonies, especially of rats, mice, and guinea pigs, which were thus transformed into "laboratory animals." More expensive and less common were opossum and primate colonies.

As Clarke emphasizes, the acquisition and maintenance of animal materials for reproductive research was by no means a merely technical or organizational problem. It also affected the pace, direction, and even the content of research. In particular, once scientists mastered the tasks of acquiring and maintaining a particular animal, they tended to use that animal in all their research and often bequeathed it to their students or colleagues at the same site. Such entrenched commitments to one research material could limit the range of problems investigated and of results achieved. The ways in which the direction and content of physiological research were shaped by the use and availability of particular animals or other materials clearly deserves further investigation.

Louise Marshall's essay on the history of American neurophysiology is divided into two parts. After noting the extent of British influence on American neurophysiologists, the first part briefly surveys the institutional history of the subdiscipline in the United States, including the early role of the APS, the emergence of research units at Harvard, Johns Hopkins, the University of Pennsylvania, the University of Michigan, Washington University in St. Louis, the University of Chicago, and Northwestern, and the stimulus provided by the informal but influential group known as the "Axonologists." Even in this part of her chapter, Marshall often alludes to the particular concepts, instruments, and techniques associated with leading individuals or centers. In the second part, that set of issues becomes the thematic focus. In particular, Marshall emphasizes the extent to which several of the major concepts and research programs in American neurophysiology during the 1930s depended on new or improved devices for amplifying and recording biological signals, notably the cathode-ray oscilloscope, the electroencephalograph, and the microelectrode. The important role of the stereotaxic instrument in localizing centers of neurophysiologic activity is also described. Indeed, Marshall suggests that instruments or techniques could themselves be a major cohesive force in research units, citing the role of the cathode-ray oscilloscope in Joseph Erlanger's group at Washington University, the stereotaxic instrument in Stephen W. Ranson's Institute of Neurology at Northwestern, and the microelectrode in Ralph Gerard's laboratory at the University of Chicago. Finally, she raises the question of whether one can detect a distinctive American style of neurophysiology in a combination of technical ingenuity and the relatively democratic tone of research units in the subdiscipline in the United States.

UNIFYING FEATURES OF THE VOLUME

The division of the volume into three parts serves the useful purpose of emphasizing the distinctive unity of the chapters within each part. However, it should not be allowed to obscure the unifying features of the volume as a whole. Without pretending that every chapter can be easily or equally accommodated within the following scheme, let me draw attention to a few of these common themes.

In the first place, all fifteen chapters discuss the social and institutional history of American physiology in some sense and to some degree. As a whole, the volume is dominated by one central theme: the emergence, maturation, and ascendance of physiology as an independent discipline in the United States. Even in Part 2, which may sometimes seem to focus on other fields, the common thread is the relation between those fields and the discipline of physiology. Part 3, for all of its special and novel attention to instruments, materials, and techniques, also links those particular issues with the general rise of experimental physiology and the subsequent emergence of reproductive research and neurophysiology as subdisciplines in the United States.

In addition, many of the chapters share a set of more specific concerns within this broader theme: (1) the competition between physiology and other fields, especially clinical medicine, for limited resources and a secure institutional base; (2) the crucial role of the American medical reform movement in the emergence and ascendance of physiology as an independent discipline; and (3) the equally crucial role of large-scale private funding, partly from wealthy individuals or local elites but especially from the large national philanthropies, the Carnegie and Rockefeller Foundations, which often enough supplied not only the money but also the leadership and vision required to bring about major reforms in American medical education and research.

W. Bruce Fye's chapter surveys these themes in American physiology for the period from 1850 to 1900 and several of the succeeding chapters provide a more focused account of their manifestations in particular institutional contexts. For good reasons, Harvard, Johns Hopkins, Columbia, the University of Chicago, and Washington University receive special and repeated attention, while McRae's essay on the University of Toronto extends the range of case studies to Canada. Several other departments are discussed at least incidentally and my own chapter (chapter V) offers a set of social maps of the leading individuals and institutions in the discipline throughout North America from ca. 1880 to 1940.

This volume thus provides a newly firm sense of the basic contours of the social and institutional history of physiology in the United States from ca. 1850 to 1940. We now have a more systematic understanding of who practiced physiology when and where in this country, under what conditions, and in what relations with their colleagues in neighboring or competing disciplines. We now have a much sharper picture of the leading American physiologists and research centers up to World War II. We now have a new appreciation of the extent to which local institutional contexts, especially the presence or absence of strong medical schools, shaped the rôle and configuration of physiology departments. We also now have a convenient means of access to a wide range of hitherto scattered and underused sources, including notably archival sources. At the least, we now have a solid base from which to begin further research into the history of physiology in the United States.

NOTE

1. For a thoroughly up-to-date survey of the state of scholarship in the history of American science, see Sally Gregory Kohlstedt and Margaret W. Rossiter, eds., *Historical Writing on American Science*, 2nd ser. (Osiris, 1985), vol. 1. The volume includes essays on the history of geology, astronomy, chemistry, biology, physics, and the social sciences; "physiology" does not even appear in the index. Nor does physiology find any place in Alexandra Oleson and John Voss, eds., *The Organization of Knowledge in Modern America, 1860–1920* (Baltimore, MD: Johns Hopkins Univ. Press, 1979).

I

American Physiologists in German Laboratories, 1865–1914

ROBERT G. FRANK, JR.

Anniversaries have a way of eliciting from those who celebrate them an epitome, a summary judgment, of the past. The banquet celebrating the fiftieth anniversary of the American Physiological Society (APS), held in 1938, was no exception. Warren Plimpton Lombard, eighty-two years old and the retired professor of physiology at the University of Michigan, spoke to the assembled Society and engaged in a bit of reminiscence:

> In '85 I came back from Ludwig's laboratory in Leipzig; like any other young man I was full of aspiration; I wanted to teach physiology and above all I wanted to get into a good laboratory where I could do some research work, because in Ludwig's laboratory I caught the disease.[1]

The "disease" of which he spoke was the primary reason that the APS was founded two years thereafter, in 1887.

The aims of the Society included, of course, the more general ones of promoting the advancement of physiology as a science and of facilitating personal contacts among physiologists. However, it intended to go beyond mere collegiality. The Society's major focus was, and continued to be, the encouragement, recognition, and communication of research. In admitting members it allowed entrance only to those who had "conducted and published an original research in Physiology or Histology (including Pathology and experimental Therapeutics and experimental research in Hygiene)," or to those who have "promoted and encouraged physiological research."[2] The very breadth of this definition of "physiology" served to emphasize that the process of investigation, and the publication of original results, was more important than a strict delineation of subject matter. Moreover, the most desirable research was experimental in character. Therefore, from its very inception, as Toby Appel has explained in chapter VI in this book, the major activity of the Society at its annual meetings in various cities was not to sponsor review lectures or to pass resolutions, but to provide a forum from which its members could present research results, and a format that allowed them to view research facilities.

Such an investigatory and experimental ideal—laboratory ideal, if you will—came into American physiology during the 1870s, 1880s, and 1890s from precisely the source that Lombard suggested in his remarks (Germany) and was carried back to America via a process that the Michigan physiologist himself exemplified, that is, by extensive foreign study and a research apprenticeship in a physiological laboratory. Lombard was neither alone nor merely prejudiced in tracing the new physiology in America to its Teutonic roots. William Henry Howell, whose training and career were purely American, used his more systematic retrospective on the first quarter century of the Society's existence to emphasize that the key American laboratories in Boston, New Haven, and Baltimore were inspired by, and organized along the lines of, the leading physiological institutes in Germany.[3] Later chroniclers have rung changes on this theme enunciated by the founding generations of American physiology.

The affirmation of a fact, and its wide acceptance, enlightens us little about its character and significance. Therefore, I examine more precisely the historical phenomenon of Americans studying physiology abroad, especially in German laboratories, during the period from the end of the Civil War to the outbreak of World War I. First, I sketch briefly the conception of physiology and its institutional framework, as it existed in America about the middle of the nineteenth century, and then explore some possible sources of influence upon it, especially from Parisian physiology. Second, I delineate some of the patterns of travel that took Americans to German laboratories in search of advanced training in physiological subjects. Who and how many went, and where did they go? Whence did they come, when did they go, and what characteristics did these sojourners have? Third, I describe what kinds of universities, facilities, and institutions they found east of the Rhine. Fourth, I explore what kinds of motivations, practices, and attitudes they found to be unique to the German laboratories, and that constituted the particular strength of German physiology. Finally, I assess what advantages, concepts, and ideals these sojourners brought back with them—the models according to which they hoped to reconstruct and reorient American physiology.

In an inquiry like this it is useful to define "physiologists" no less broadly than they did themselves. Such a category certainly includes those (like Lombard, Henry Pickering Bowditch, Frederic Schiller Lee, or Graham Lusk) who would later list their *Fach* as "Physiology" in biographical compendia such as *American Men of Science.* It also includes physiological and biological chemists, such as Russell Henry Chittenden and Lafayette B. Mendel, and pharmacologists like John Jacob Abel and Reid Hunt. The boundary between physiology and anatomy, especially histology, was subject to constant incursions from both sides; my sojourners therefore include men like Henry Herbert Donaldson and Franklin Paine Mall, medical scientists who were trained in physiology but who ended up as professors of anatomy. On the flank of physiology exposed to the behavioral disciplines, I also bring in men such as William James and G. Stanley Hall, who worked in physiology before finding careers of a more psychological bent. Finally, one must include a number of academic physicians, like William Henry Welch, William Osler, or Lewellys Barker, whose pathological or clinical interests led them to seek laboratory training in physiology. All of the foregoing thought of themselves as having physiological interests; all were elected members of the APS.

My picture of Americans abroad is drawn in part from the obvious sources: printed biographies and autobiographies of individual scientists. I have attempted, however, to obviate some of the inevitable distortions that come from trading, as the historian must, in the particular, by also dealing with biography en masse. Two different populations of physiologists, a smaller one with a focus ca. 1890 and a larger one centered ca. 1920, serve to put the idiosyncrasies of individual biography into a context that is, if not fully rigorous, at least better defined. For qualitative detail and fine structure, one must always have recourse to witnesses. I have tried, wherever possible, to use letters written back from German laboratories and autobiographical reminiscences of scientific travels. The evidence is rich, but dispersed; I can use here only some first fruits of archival diggings.

THE NATIVE LANDSCAPE AND FRENCH INFLUENCE

Physiology in America, both as a subject of teaching and an object of interest, had a history that long predated the German sojourners and such organizational efforts as the APS.[4] That very history underscores the difference between the United States and Europe in the development of physiology as a science in the first two-thirds of the nineteenth century.

The most prominent feature of American physiology before the Civil War was its lack of an institutional base fitted for investigative activity. Every one of the medical schools in the United States, scores of them by the 1840s and 1850s, taught the subject,[5] but they did so in an entirely didactic way. A "Professor," whose subject was usually designated as "Anatomy and Physiology," or "The Institutes of Medicine," gave twenty to thirty lectures on physiology to the beginning students. If they were lucky these same students would have received some practical instruction in anatomy given over a cadaver, but there was no attempt to provide the same kind of practical and hands-on training in physiology. There were few physiological experiments that were thought pedagogically useful for students to do, and there was no assigned space similar to the dissecting room in which such exercises could be carried out. In this respect the professor had no better facilities than his students: no laboratory space and no money for animals, apparatus, and supplies. If such a man wished to do investigations, he had to do so in private space, often at home, and at his own expense. As William Williams Keen recalled the scene of his collaborative work with Silas Weir Mitchell on poisons in the 1860s, "the front office was full of patients and the back office of rattlesnakes and guinea-pigs."[6]

The very nature of the American medical school mitigated against any changes in this institutional ambiance. All of them, even those with a university or collegiate connection, were run as "professional" schools, that is, by practicing professionals, and for the purpose of training succeeding generations of practitioners. The man who lectured on anatomy or physiology saw a daily docket of patients just as did his confreres who lectured on diagnosis, surgery, or obstetrics, but there was one major difference in the outcome. While the activities of a clinician outside the lecture room were providing him with an income, they also enriched his teaching activities. In an opposite way, laboratory activity in physiology not only stole hours from treating paying patients, it made one appear odd and rather less trustworthy to one's professional colleagues. "Older men," Mitchell recalled, "not only disapproved of" any divorce between practice and research, "but also considered the active pursuit of physiological knowledge in the laboratory as inimical at least to worldly success in the practice of medicine." Samuel Jackson, professor of the Institutes of Medicine at Pennsylvania, told Mitchell, "You will lose a patient for every experiment you make in the laboratory."[7]

Because it was considered quite proper that a medical school both reflect and serve a city, or at most the surrounding region, the faculty of that institution was necessarily drawn from the local professional community. Bostonians staffed Harvard, New Yorkers ran the College of Physicians and Surgeons, and Philadelphians would not have thought of importing a New Englander, no matter how scientifically eminent, to teach at Pennsylvania or Jefferson. Not only would such a chair have failed to yield a living income, but such a "foreigner" would by definition lack the social and professional prominence that was the sine qua non of faculty membership in a medical college. Abraham Flexner and his fellow reformers of a later period have led us to believe that the small, freestanding, and private medical college resisted change because it was proprietary. Investigatory subjects such as physiology had difficulty within this ambiance, not because of the money-making intentions of the faculty—few schools produced wealth for their professors in any case—but because of the way in which the intellectual agenda and organizational framework was set by the local medical community. Such schools failed to provide a home for the medical sciences, not because they were too proprietary but because they were too professional.

Within this overall picture of American medical sciences, including physiology, defined in limited, didactic, and local ways, there might have been one possible external source of innovation: France. From the late eighteenth century until well into the 1840s, France—and here one means almost exclusively Paris—was not only the center of clinical medicine but also of medicine's basic sciences, such as pathology and most especially physiology. While German physiology before the 1840s explored the lushly seductive natural world conceived by the Romantics, postrevolutionary physiology in France took an empirical and experimental tack. Bichat, Magendie, Legallois, Flourons, Claude Bernard, and Charles Edouard Brown-Séquard were successive luminaries in the Paris firmament from the late 1790s to the 1860s.[8]

Moreover, Paris was a medical world to which American physicians had easy access. In the period after the end of the Napoleonic wars in 1815, an increasing number of the well-heeled and better-trained American medical graduates added a year or more in the Paris hospitals to round out their clinical educations. In doing so a certain number of them came in contact with leading French physiologists, especially Bernard. On paper, at least, it looks as if a modest generation of American physician/physiologists was so trained in the 1850s and 1860s. John Call Dalton, Mitchell, Austin Flint, Stanford Emerson Chaillé, and Bowditch are all cited in the standard biographical sources as being students of Bernard. The natural implication is that they worked with him, learned his methods, became oriented to his lines of research, and brought back to the United States some kind of Parisian ideal and research momentum.

On closer inspection, however, the usefulness of such a category, and its implications, becomes suspect. Take Mitchell as an example. The time he actually spent in Paris was from late November 1850 to early June 1851. In the middle of his sojourn he was seriously ill with a varioloid disease that made it almost impossible to work. While healthy, his primary occupation in the French capital was taking courses in surgery, with some minor courses in auscultation and ophthalmology. Much later, in the autobiography written at the end of his life, he recalled: "I took courses designed for surgical training, but liked better the lessons of Bernard in physiology and Robin in microscopy."[9] This, plus another remark of late maturity acknowledging Bernard's skills as an educator,[10] has become transmuted into a shorthand characterization of Mitchell's training such as one finds in the *Dictionary of Scientific Biography*: "He then spent a year in Europe attending the lectures of Claude Bernard and the microscopist Philippe Robin."[11] Dalton's case was similar. We know that in 1850 he visited Paris and attended Bernard's lectures, an experience that had a great effect on him. Mitchell expressed the admiration he and Dalton shared by once recalling a favorite anecdote. While in Paris Dalton once said in answer to a question, "I think it must be so"; to which Bernard replied, "Think! Why think when you can experiment?"[12] Yet despite such warm feelings, there is no evidence that the relationship was one of colleague and co-worker rather than simply lecturer and auditor.

Moreover, one cannot discern in the postsojourn activities of Bernard's "students" any clearly marked intentions to change the direction of American physiology. Dalton, at least in the 1850s, was content to surprise his colleagues with the innovation of using vivisectional demonstrations, a la Bernard, to illustrate his didactic lectures in medical school. In the course of such demonstrations, he repeated many of the experimental investigations recently carried out in Europe, especially on the action of the pancreatic juice, the production of sugar by the liver, and the effects on body temperature of sectioning sympathetic nerves. This in turn led Dalton to do experiments on gastric digestion, bile, the cerebellum, and the corpus striatum. These research interludes were carried out in private quarters, however, without support from the medical schools where he taught, without collaborators or students, and with a momentum that was interrupted first by military service in the Civil War and later by administrative responsibilities at the College of Physicians and Surgeons in New York. Dalton's life was essentially that of a teacher.[13]

Mitchell, easily as brilliant, followed a similarly well-worn path. Upon his return he modeled his life on that of his father: general practice, with some part-time teaching and an occasional investigation squeezed into the interstices. When he did try, unsuccessfully, to obtain a chair in physiology, first at

Pennsylvania in 1863 and then at Jefferson in 1868, it was still seemingly not with the intention of giving up his practice; nor would it have been financially possible. When defeated in both these sorties by physicians better connected in the local medical community, he did not attempt to get a position outside of Philadelphia.[14] This was similar with Flint in New York and Chaillé in New Orleans; each did some physiology, but in the mode and context of local medical practice and traditional physician–medical school relations.

Why did French physiology fail to exert a significant influence over its American counterpart? One can perhaps see some of the reasons in the unfolding of Bowditch's visit to Paris, from September 1868 to April 1869.[15] In the first place, arrangements for laboratory facilities, while better than those in America, were still rather haphazard. Bowditch had come to Paris not only to walk the wards in the usual way but especially to study with Brown-Séquard. This mercurial, transatlantic bird of passage had taught at Harvard Medical School while Bowditch was an undergraduate there.[16] He had promised the young Bostonian that he would open a laboratory in Paris by the end of October, a prospect that elated young Bowditch when he wrote home soon after his arrival in the capital:

> [Brown-Séquard's] object is to study physiology in general (not particularly the nervous system) & to teach young men to make discoveries which he says is as easy as picking berries in the fields in Summer & is by no means the lot of a privileged few as is very generally supposed in America.
>
> Those who study with him he says will be rather his assistants than his pupils. He does not intend to charge anything as he thinks that in that way he will be able to get more work done for science.[17]

However, within a few weeks such grand plans had fallen through,[18] and Bowditch had to cast around for an alternative: "I hear that there is a probability that Bernard will deliver his course of lectures this Winter. If he takes pupils I shall try to study with him."[19]

Bowditch did work his way into Bernard's circle, but he did not find even that very satisfactory. His first extensive contacts in Bernard's laboratory were not with the master, but with his "preparateur," Louis-Antoine Ranvier. Ranvier did, however, let Bowditch and his friend John Collins ("Coll") Warren work in his own private laboratory whenever they wished, although Ranvier himself could only be there from 12 P.M. to 4 P.M.[20] The young Frenchman and the two Americans hit it off well, went out eating and drinking together, and "wandering about in the suburbs of Paris hunting frogs and salamanders for physiological experiments."[21] Bernard himself was more distant, concerned with his own investigations and lectures, and taken up with French politics.[22] Moreover, he was in the midst of translating chairs from the College de France to the Jardin des Plants. Unfortunately, Ranvier's metier did not fit Bowditch's emerging plans:

> The only trouble is that Ranvier is more of a histologist than a physiologist, so that we are rather left to our own resources in physiology & as the laboratory is in very bad order we do not get on so fast as I could desire. When Bernard begins his lectures, which he is advertised to do early in January, I hope the laboratory will be put in order & things will go on more regularly.[23]

When the lectures finally began, they were hardly the immersion that Bowditch had hoped for. "Bernard lectures twice a week & always after the lecture & often before he spends some time in the laboratory doing experiments & talking physiology." Rather weakly, Bowditch concluded, "This is very instructive."[24] As the winter wore on his letters reflected a growing admiration for Bernard as an experimentalist and physiological thinker,[25] but frustration at his own inability to profit from the proximity, due to French customs and arrangements. "Bernard's laboratory, where I work nearly every day," he wrote to his old Harvard teacher, Jeffries Wyman, "is dark, poorly provided with apparatus, and with no proper arrangement for pupils. . . . The Frenchmen acknowledge the great advantage that the Germans have over them in this respect."[26]

In summary, Bowditch's experience reflected both the advantages and disadvantages of Paris as a scientific center. It had brilliant individuals in haphazard facilities and was ill financed, especially in comparison to the money spent on culture and the military. Supplies and animals—vide Bowditch's frogs—were expected to be conjured up by the investigator's ingenuity. Laboratory experimentation, insofar as it concerned students, was largely for the purpose of demonstration. The overall umbrella for the medical sciences was the system of hospitals and learned academies, rather than universities. Neither hospitals, with their clinical concerns, nor academies, with their isolation from daily teaching and their emphasis on formal, almost set-piece, lectures, were well adapted to the goal of bringing master and pupil together to carry out investigations.[27] Indeed, the French seem to have had only a dim sense of the laboratory as a teaching venue. Moreover, the social life of the capital tended to separate teachers and students. With the single exception of the young bachelor, Ranvier, Bowditch never in his nine months in Paris got to see a French scientist work at close range or to break down the barrier that separated the scientific and social spheres. Finally, at least to this one visiting American, the situation was compounded by a Gallic style that did not value order and precision:

> I hear the same story all round. The government lavishes millions on the army but leaves science to starve & freeze. However, I don't think that scientific men make the most of the opportunities which they have. Bernard's laboratory, for instance, might be made much more useful than it is if they would take a little more pains to keep it in order. None of Bernard's assistants seem to have the bump of order developed in the slightest degree. All the drawers seem to contain a little of everything.[28]

The Bostonian was ready for a change.

TRAVELERS BEYOND THE RHINE

Bowditch was, at one and the same time, both the last of the old travelers and one of the first of the new. In the dreary late winter of 1869 he found that "I have been devoting myself lately almost entirely to the purely scientific part of my profession, which certainly has much greater attractions for me than the more practical part."[29] He began to think about seeking the necessary training beyond the Rhine. He read a pamphlet entitled *De la réforme des études médicales par les laboratoires*, "by a Dr. Lorain in which the writer compares the advantages for studying medical science in Germany & in France much to the disadvantage of the latter country."[30] A few weeks later a family friend, Dr. Ellis, came to town and the two visited hospitals together. Bowditch wrote to his "Uncle Doctor," Bowditch, that "Dr. Ellis sends his best regards to all the family. He wants me to tell you that he finds Paris nowhere compared to Germany."[31] To his mother, young Bowditch wrote that Ellis "draws comparisons between the advantages of studying in France and in Germany which are not very favorable to the former country and which make me quite anxious to get to work in Germany."[32] Bowditch's plans were crystallized when he met that enthusiastic ambassador for physiology, Willy Kühne. He wrote to Uncle Doctor from Paris in April 1869:

> I have met Prof. Kühne here in Paris. He was Virchow's assistant for a long while you know. He is now teaching Physiology on his own account at Amsterdam. I asked his advice as to the best course for getting a thorough physiological education. He advised me to go first to Bonn to study Histology with Max Schultze for a few months & then in November to begin with Ludwig at Leipzic. He thinks I ought to spend a full year there so as to get the whole course & then divide another year between Helmholtz & Virchow. This marks out rather an extensive course but I don't know that I can do better than to follow it. At any rate, I shall probably start for Germany.[33]

Which he did, the following month.

In the following quarter century many other Americans bent on careers in the physiological sciences did exactly the same thing. The fact is well known, but the magnitude and timing of the movement have not been clearly defined. If we examine the founding cluster of men who organized the APS in 1887 and add to those the members who were elected in the first five years, this "founding group" totals forty-five (Table 1). Thirty-one of the forty-four identifiable physiologists, 72 percent of the membership of the Society, had been educated in Europe. Only two of these, Henry Newell Martin and Samuel James Meltzer, were Europeans by birth and training. The rest were Americans, the overwhelming proportion of whom went to Europe after getting their initial medical and/or scientific training in the United States or Canada. Of the thirteen who stayed at home, seven of them (largely in the younger age cohort) had been educated in physiological laboratories, at Harvard and especially Johns Hopkins, which had been explicitly founded on the European model.

Moreover, it was not just in Europe generally that these men had caught the "disease" of research, but in Germany particularly. Of the thirty-one European-trained members, twenty-eight had had their apprenticeships as physiologists either partially, or completely, in German universities. This was 64 percent of the entire Society. Even more specifically, almost 40 percent of the Society had, in the previous two decades, passed through a single German university: Leipzig. The attraction of that city was, of course, the complex of hospitals and institutes that had been built there in the 1850s and 1860s, especially the magnificent Physiologisches Institut under the direction of Carl Ludwig. The list of Americans who studied with him, and who were later elected to the Society in its first five years, reads like a who's who of the physiological sciences in the 1890s: Bowditch, Hall, Ott, Martin, Warren, Welch, Beyer, Minot, Smith, Lombard, Edward Tyson Reichert, Henry Sewall, Abel, Lee, Hare, and Mall. The APS, at least at its foundation, might well have been nicknamed the *Verein alter Leipziger Studenten.*

Yet American physiology was not to remain a preserve of those who had studied in Germany, as one can see clearly by looking at a rather broader, later, and less exclusive sample of practitioners in the physiological sciences. In the first three editions of James McKeen Cattell's *American Men of Science* (1906, 1910, 1921), there are 412 individuals who declared their fields to be within the broad domain of physiology/biochemistry.[34] In contrast with the early members of the APS, only 105 (25.5 percent) had had some part of their training abroad. Clearly, by the end of World War I, North America was training its own physiological scientists, without nearly such a need to go abroad. Such a pattern continued and accelerated well into the 1920s and 1930s. Of the 175 members elected into the Society in the five years before its semicentennial (1933–1937) only 14 percent had any foreign training. Take away from this number the physiologists who were foreign-born emigrés, and one is left with less than 6 percent of this five-year cohort that felt the need for European study and research experience.

Despite this growth in the domestic production of scientists, however, the most respected and prestigious physiologists and biochemists of the pre-1920 sciences were drawn quite disproportionally from among those who had sought additional training in Europe. One measure of such prestige is that a scientist was designated a leader in his discipline by a vote of his peers in that field. In each edition of *American Men of Science*, Cattell tagged the names of such leaders with an asterisk. One can divide the population of physiological scientists into travelers and nontravelers (Table 2), and then examine the ratio of individuals with asterisks to those without in each of the two moieties, broken down by discipline.

These patterns, translated into words, tell the tale. A physiologist was almost three times as likely to end up in the elite if he had had foreign training. In the more traditional and medically related subject of physiological chemistry, only one of the elite had not had his European *Wanderjahre.* Biochemistry, being younger, less closely linked to physiology, less dependent on medical patronage, and less likely to confine its research topics to mammalian or even vertebrate material, had a more distinctly indigenous quality about it. In pharmacology the advantage of foreign study was not so pronounced as in physiology and physiological chemistry, but it still doubled one's chances of ending up in the elite.

The chronological clusterings of these European travels show some very interesting and consistent patterns. Here, as before, it is useful to compare those who went abroad with those who stayed home. Moreover, the comparison can be performed both for the physiological sciences, broadly defined, and for physiology proper. The results are embodied in Figures 1 and 2, which plot the number of physiologists embarking on their careers in each quinquennium and divides the cohort between those who went abroad and those who stayed home. Three distinct phases can be seen.

In the first phase, from the 1860s to the mid-1880s, the majority of the

Table 1
American Physiologists in European Laboratories
Training of members elected to the American Physiological Society in first five years of its existence, 1887–1891

Name	Birth Date	Degree(s)	Period Abroad	Place(s)
John Call Dalton	1825	M.D. Harvard, 1847	1850	Paris
Silas Weir Mitchell	1829	M.D. Jefferson, 1850	1850–1851	Paris
George Miller Sternberg	1838	M.D. Columbia P & S, 1860		
George Lincoln Goodale	1839	M.D. Harvard, 1863		
Henry Pickering Bowditch	1840	M.D. Harvard, 1868	1868–1871	Paris, Bonn, Leipzig, Munich
Horatio Curtis Wood	1841	M.D. Pennsylvania, 1862		
William James	1842	M.D. Harvard, 1869	1867–1868	Berlin
William Powell Wilson	1844	Sc.D. Tübingen, 1880	ca. 1878–1880	Tübingen
Granville Stanley Hall	1844	Ph.D. Harvard, 1878	1869–1870	Berlin, Bonn
			1871–1872	Berlin, Heidelberg
			1878–1880	Berlin, Leipzig
John Green Curtis	1844	M.D. Columbia P & S, 1870		
Henry Cadwalader Chapman	1845	M.D. Pennsylvania, 1867	ca. 1869–1872	London, Paris, Berlin, Vienna
Thomas Wesley Mills	1847	M.B. McGill, 1878	ca. 1879–1880	London
			ca. 1882–1883	London
Isaac Ott	1847	M.D. Pennsylvania, 1869	ca. 1871–1873	Leipzig, Berlin, Würzburg
Henry Newell Martin*	1848	M.B., B.S. London, 1872	1864–1876	London, Cambridge, Leipzig
Joseph Weatherhead Warren	1849	M.D. Bonn, 1880	1871–1880	Leipzig, Berlin, Bonn
William Osler	1849	M.B. McGill, 1872	1872–1874	London, Berlin, Vienna
William Henry Welch	1850	M.D. Columbia P & S, 1875	1876–1878	Strassburg, Leipzig, Breslau, Vienna
			1884–1885	Munich, Leipzig, Göttingen, Berlin
Henry Gustav Beyer	1850	M.D. Bellevue, 1876	1880–1881	Leipzig
Victor Clarence Vaughan	1851	Ph.D. Michigan, 1876	1888	Berlin, Munich
		M.D. Michigan, 1878		
Samuel James Meltzer*	1851	M.D. Berlin, 1882	1879–1882	Berlin
Charles Sedgwick Minot	1852	Sc.D. Harvard, 1878	1873–1876	Leipzig, Würzburg, Paris

Albert Philson Brubaker	1852	M.D. Jefferson, 1874		
Robert Meade Smith	1854	M.D. Pennsylvania, ca. 1880	ca. 1880–1884	Leipzig
William Thomas Councilman	1854	M.D. Maryland, 1878	ca. 1880–1884	Vienna, Leipzig
Warren Plimpton Lombard	1855	M.D. Harvard, 1881	1882–1885	Leipzig
William Thompson Sedgwick	1855	Ph.D. Johns Hopkins, 1881		
Edward Tyson Reichert	1855	M.D. Pennsylvania, 1879	1882–1885	Berlin, Leipzig, Geneva
Henry Sewall	1855	Ph.D. Johns Hopkins, 1879	1879–1880	Leipzig, Heidelberg, Cambridge
John Marshall	1855	M.D. Pennsylvania, 1878	1878–1882	Göttingen, Tübingen (partial years?)
		M.D. Göttingen, 1879	1886	Christiana [Oslo] (partial year?)
		Nat.Sc.D. Tübingen, 1882		
Russell Henry Chittenden	1856	Ph.D. Yale, 1880	1878–1879	Strassburg, Heidelberg
William Gilman Thompson	1856	M.D. Columbia P & S, 1881		
Henry Herbert Donaldson	1857	Ph.D. Johns Hopkins, 1885	1885–1887	Munich, Zürich
John Jacob Abel	1857	M.D. Strassburg, 1888	1884–1890	Leipzig, Strassburg, Heidelberg, Würzburg, Vienna, Berlin, Bern
Frederick Warren Ellis	1857	M.D. Harvard, 1881		
Frederic Schiller Lee	1859	Ph.D. Johns Hopkins, 1885	1885–1886	Leipzig
Clifton Fremont Hodge	1859	Ph.D. Johns Hopkins, 1889		
William Henry Howell	1860	Ph.D. Johns Hopkins, 1884		
George Theophilus Kemp	1861	Ph.D. Johns Hopkins, 1886		
		M.D. Long Island, 1891		
Hobart Amory Hare	1862	M.D. Pennsylvania, 1884	ca. 1885–1889	Leipzig, Bern, London
Franklin Paine Mall	1862	M.D. Michigan, 1883	1883–1886	Heidelberg, Leipzig
George Henry Falkiner Nuttall	1862	M.D. California, 1884	1886–1890	Breslau, Göttingen
		Ph.D. Göttingen	1893–1899	Göttingen, Berlin
William Townsend Porter	1862	M.D. St. Louis M.C., 1885	1887–1893	Kiel, Berlin, Breslau (partial years)
Joseph Jastrow	1863	Ph.D. Johns Hopkins, 1886		
John Pendleton Campbell	1863	Ph.D. Johns Hopkins, 1888		

Not identified: S. De Jager, New York City, elected to APS in 1888.

* Individual born in Europe whose uninterrupted training was taken there before coming to North America.

Table 2
Leaders in the Physiological Sciences

	Travelers	Nontravelers
Physiology	25/50*	31/152*
Physiological chemistry	9/16	1/39
Biochemistry	1/9	4/55
Pharmacology	7/30	8/61
Total	42/105	44/307

* Ratios of leaders to nonleaders.

quite small group of physiological scientists entering into their profession went abroad for some kind of advanced training. This generation included many of the founders of American physiology: Bowditch, Ott, Warren, Chittenden, Sewall, Beyer, Lombard, and Abel. For the travelers in this phase, the years abroad frequently constituted not just the most extensive and influential part of the training as researchers but often their only such training.

In the second phase, over the next fifteen years until about 1900, the absolute numbers of those going abroad continued to rise, but the percentage of each cohort fell because many more Americans were finding adequate training here in North America. This group included such men as Lee, Lusk, William Townsend Porter, Phoebus Levene, Edward Mellus, Albert Mathews, Lafayette Mendel, Walter Garrey, Russell Burton-Opitz, Walter James, Yandell Henderson, and Frederick Pratt. Many such sojourners had already had training in physiological subjects before they went abroad for advanced work. Moreover, a sizable portion of travelers in this phase went to Germany specifically for a research apprenticeship in physiological chemistry, which developed rapidly in Germany in the 1880s and 1890s.

The third phase, from about 1900 to the outbreak of World War I, saw the number of travelers drop back to a relatively stable level that constituted about 15–20 percent of each quinquennial age cohort. As a group it contained fewer men and women who were later to carve out distinguished careers for themselves in the physiological sciences, although one could mention Robert Woodworth, Holmes Jackson, John Eyster, Alexander Forbes, Carl Wiggers, and Cattell. Yet the very names also point out a shift in the post-1900 phase; Woodworth, Forbes, and Cattell never even went to Germany, choosing instead physiological laboratories in England. The allure of the German institute had definitely waned.

Just as there was a distinct scientific generation that was trained abroad, overwhelmingly in Germany, from the early 1870s to the mid-1890s and assumed leadership in the American physiological sciences, so was this generation's background geographically delimited. Massachusetts, Connecticut, New York, Pennsylvania, and Maryland were the origins of twenty-one of twenty-eight Society members who studied in Germany: Bowditch, Warren, and Lombard from Boston; Hall, Welch, and Chittenden from southern Connecticut; Beyer, Meltzer, and Donaldson from the New York City area;

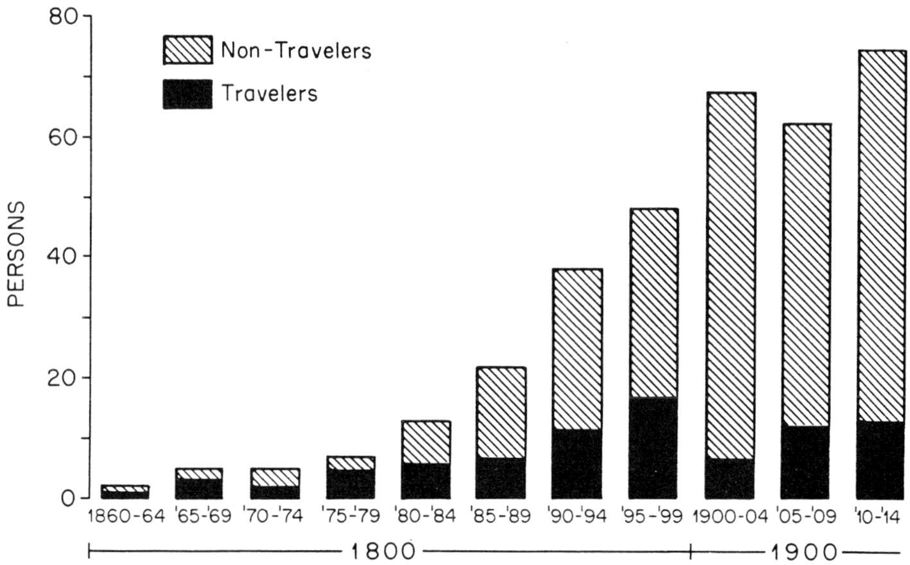

Figure 1: *Number of physiologists beginning their careers (in all physiological sciences) in each quinquennium.*

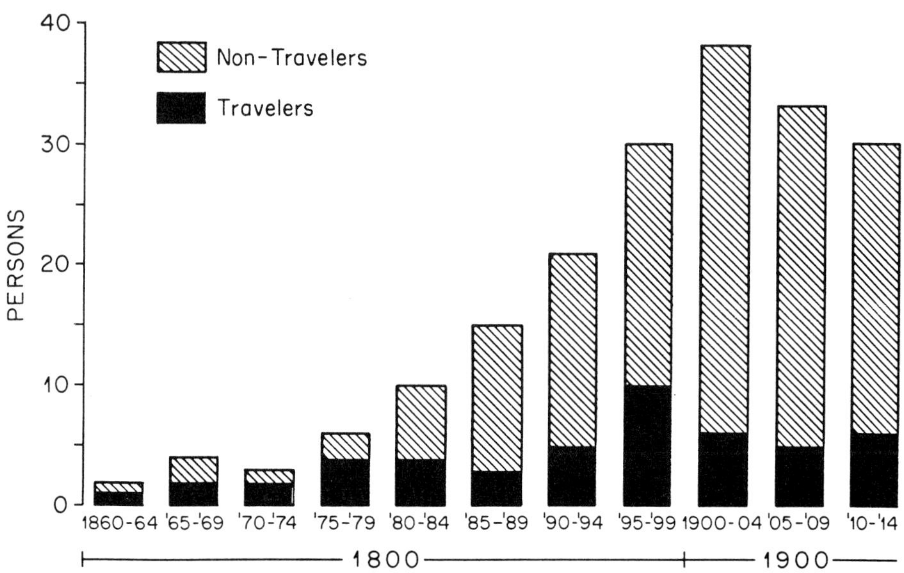

Figure 2: *Number of physiologists beginning their careers (in physiology only) in each quinquennium.*

Chapman, Ott, Smith, Reichert, and Hare from Philadelphia; Councilman and Sewall from Baltimore—all exemplars of the typical origins of travelers to Germany. The larger and more inclusive sample of physiological travelers from *American Men of Science* shows not only the same pattern but that it continued until World War I. In many ways this was to be expected; the same five states dominated the American scientific community generally in the period between the Civil War and World War I. It should serve to remind us that American travelers to Europe, like the "American" Physiological Society, were not so much a national phenomenon, as one expressive of a compact region stretching from Boston to the District of Columbia.

This generation of German travelers in the APS was homogeneous in yet another respect: age. Almost all of them were quite young; their average age was less than twenty-six when they went to Europe.[35] Most were bachelors; the time had not yet come when it was convenient to travel from one university town to another, learning both a new language and a new science, with a wife and/or children in tow. Age did, however, change somewhat with time. By the third phase, after 1900, the average age of the travelers represented in *American Men of Science* had risen almost three years.[36] By that time the relatively few who went abroad were older and more mature.

Insofar as we have any detailed knowledge of the family background and wealth of this German generation, there seems to have been a fair diversity of resources. A few, like Bowditch, James, Lombard, or Hare, came from distinctly well-to-do backgrounds. James could write home wittily from Berlin that travel did indeed bring "moments of keen enjoyment. . . . The contemplation of my letter of credit alone," he said, "makes me chuckle for hours. If I ever have leisure I will write an additional Bridgewater, illustrating the Beneficence and Ingenuity, etc., in providing me with a letter of credit when so many poor devils have none."[37] Bowditch, while buying apparatus and books in Leipzig before his return to Boston, was urged by his father not to stint. A letter of credit for $1,000, about $20,000 in present-day purchasing power, had been forwarded to young Henry just for this purpose and there was more available should the budding physiologist need it.[38] Other travelers, like Dalton and Mitchell before the Civil War and Welch, Smith, and Reichert after it, had physicians as fathers who could underwrite an overseas trip of moderate expense. Naturally some travelers, such as Osler, Hall, or Mall, came from farming or petit bourgeois backgrounds that made a European journey a real sacrifice. Osler later recalled that "there were cobwebs in my pockets in Berlin and Vienna, and only the most necessary textbooks were bought,"[39] although from Osler's later bibliomania one might reasonably suspect his definition of "necessary"!

What could a young man expect to gain from such an outlay of effort and money? Intentions are notoriously hard to discern, but insofar as we have evidence, it seems that motives for such studies were initially quite diffuse and later became more focused—later in both the individual personal development of each traveler, and in time as the 1860s and 1870s wore on into the 1880s and 1890s.

For some, clear intentions to study physiological subjects emerged only after they had arrived in Europe for a course of study that was originally intended to be practical. As Bonner has elegantly sketched out, this was the era in which American doctors, in great numbers, went to the clinical centers of the German-speaking world, such great hospitals as the Allgemeines Krankenhaus in Vienna and the Charité in Berlin, for short, intensive courses in general medicine and in the specialities.[40] They sought there the kind of intense clinical experience, expert teaching, and hands-on opportunities that were difficult to obtain even in the best American medical schools. Vienna was the early favorite; by the late 1870s there were scores of Americans studying there at the "General." As the attraction of Vienna peaked in the early 1890s, it was overtaken by Berlin, which by the turn of the century could not only offer superior clinical facilities, but more rigorous training in the scientific side of medicine.

From the thousands that set their course for Vienna or Berlin, it was natural that a few hundred were seduced into the basic medical sciences and that a few score of these would be diverted into physiology. Bowditch was one whose clinical voyage rather quickly turned into a scientific *Reise.* Mall first went to study ophthalmology and pathology at Heidelberg. After a year there he went on to do another year with the anatomist Wilhelm His in Leipzig. He was about to leave when a researcher's place became available in Ludwig's laboratory, in which he spent yet another year.[41] Lombard went to Europe with the idea of studying obstetrics in Vienna. After learning some German in Weimar, he went to nearby Leipzig to test it out as a guest worker in some clinics. Only by accident did he end up in Ludwig's institute, which he liked so much that he spent almost two years there.[42]

Even if a physician came to Germany intending to focus on the medical sciences, his plans would as often as not be modified or even deflected. When the young Welch set out for Germany in 1876, his teacher at Columbia's College of Physicians and Surgeons, Eduard Seguin, had suggested that he go to Strassburg. Seguin, of French extraction, told Welch that he would find there a good French university only recently taken over by Germany. Welch discovered instead a highly Teutonic institution, deliberately staffed by the Prussian *Kultusministerium* with some of the best researchers and teachers in Germany: Waldeyer in anatomy, von Recklinghausen in pathology, and Hoppe-Seyler in physiological chemistry.[43] Then, wanting a more cellular approach to pathology, Welch traveled on to Leipzig to work in Wagner's institute. To fill in the remaining time, he started some experiments at Ludwig's institute. Soon he found the physiological laboratory more lively and inspiring than its counterpart in pathology, a fact reflected by the praise he lavished on Ludwig in his letters home, as well as by his return to Leipzig for another research stint eight years later.[44]

Chittenden, from Yale's Sheffield Scientific School, came to Strassburg two years later, in 1878, to work with Hoppe-Seyler in physiological chemistry, but he was rather less entranced with the city than Welch. He found the facilities antiquated and overcrowded and Hoppe-Seyler less than accessible. So Chittenden moved north, down the Rhine and up the Neckar, to Heidelberg, where he spent a highly productive year in Kühne's physiological laboratory.[45] Also consider Sewall, one of Martin's first doctoral students at the newly founded Johns Hopkins University. After taking his Ph.D. in Baltimore, Sewall was sent by the expatriate Englishman to his old mentor at Cambridge, Michael Foster. Foster in turn advised Sewall to work first with Ludwig in Leipzig and then with Kühne in Heidelberg, before coming back for a final stint of research at Cambridge.[46]

Such patterns of movement—from one German university to another, and sometimes even elsewhere in Europe—were by no means exceptional. Most of the physiological travelers already had an advanced degree, either an M.D. or Ph.D., so they did not need to feel tied down fulfilling course and degree requirements. They could seek out the person, the circumstances, and the training that they wanted with no goal in mind other than getting the best possible scientific education. Even if, like Warren or Abel, the American was taking his full medical education abroad, that too was no impediment to movement. Universities recognized the work done at their sister centers of learning. Mobility was the order of the day for German medical students as well; Americans simply followed suit. In the initial cluster of APS members, those who studied abroad visited on the average 2.7 universities in the course of their European travels.[47] If anything, such a figure probably understates the case, because in standard biographical notices a scientist would sometimes fail to mention a shorter period of study. A similar, but slightly lower figure, is also characteristic of the broader group of men defined by *American Men of Science.* Interestingly, one can see in this group a marked tendency for the number of places visited to decline, as American institutions for physiological training came to the fore (Table 3). After the turn of the century, Americans who did make a foray abroad seem to have done so with rather more focused goals in mind, and hence spent less time sampling the wares at a number of universities.

Traveling from place to place in Germany, searching out the best laboratories and research training, took time. It is therefore not surprising that American physiologists, once in Europe, stayed there for a relatively long time. The APS group of travelers had spent, on average, almost three years in foreign universities.[48] Unlike many of their clinical confreres bound for Vienna or Berlin, who went there for advanced training after a few years of practice, the nascent physiologists tended to go abroad relatively quickly, about one and a half years after completing an M.D. or a Ph.D.[49]

Moreover, as shown in Table 4, these figures, like those for average age and number of destinations, show some interesting patterns over time, especially for the broader community of physiological scientists encompassed by *American Men of Science.*

The time elapsed between degree and the beginning of foreign travel lengthened slightly, but the time spent abroad decreased markedly. By the beginning of this century it was no longer necessary for American medical

Table 3
Travelers: Average Number of Destinations

Period	Number of Travelers	Number of Destinations
1860–1884	19	2.42
1885–1899	35	2.00
1900–1914	32	1.56

Table 4
Travelers: Time Characteristics

Period	Time Lapse*		Total Time Abroad	
	Number of travelers	Years	Number of travelers	Years
1860–1884	14	2.14	19	4.00
1885–1899	22	3.23	35	2.94
1900–1914	24	2.71	32	1.90

* Average time elapsed between degree and beginning of foreign travel.

scientists to spend such a long time abroad. The kind of solid and general grounding that once could be obtained only in Germany was becoming more and more available in North America.

These four numerical measures—proportional numbers, age, destinations, and length of time abroad—point to a change in the late 1880s that can also be seen qualitatively. Study in Germany began to be undertaken, not as a complete education in physiology, but as a kind of finishing school. Men already trained in the subject sought out European laboratories in a highly selective way. They knew, through their reading and contacts with American mentors, where they should go to acquire what kind of special techniques, experience, or outlook. When Lee went to Leipzig during the year 1885–1886, he had just taken his Ph.D. with Martin at Johns Hopkins. Thus his year with Ludwig and Max von Frey could be spent in specialized studies of the electrophysiology of muscle.[50] Lee's compatriot at Johns Hopkins, Donaldson, had also completed his Ph.D. with Martin and Hall in 1885, on a topic that lay between neurophysiology and sensory physiology. Hall put Donaldson on the staff in psychology and almost immediately suggested that he go abroad to study neurocytology with von Gudden at Munich. After the young American had completed three months of research there, von Gudden was killed by his patient, "mad" King Ludwig II of Bavaria, in a bizarre murder/suicide. So Donaldson continued his training for fifteen months with Forel in Zürich, punctuated by short working visits with Golgi in Pavia, Meynert in Vienna, and Flechsig in Leipzig.[51] Ten years later yet another Hopkins man, Lewellys Barker, was sent to Leipzig, this time by Mall, to polish his physiological acumen with Ludwig. The ambassadorial mission was, however, tinged with tragedy. In April 1895, only a few weeks after Barker's arrival, Ludwig died at the age of seventy-eight. Barker finished his *Arbeit* with Ludwig's assistant, von Frey.[52] By the 1890s the scientific landscape of Europe was well known; Americans had very clear notions of what was to be had and where.

THE ATTRACTION: UNIVERSITY AND INSTITUTE

Stories of young Americans crisscrossing Germany bring us naturally to the questions: What did they find? What was the magnet? The primary attraction was intellectual. They found there an unparalleled richness of research and teaching in the basic medical and clinical sciences. The German world of science was open to all, without regimentation, roll calls, quiz sections, and all the pettifoggery that seemed to be associated with undergraduate colleges and/or medical schools in America. This was the famous *Lernfreiheit* of academic life in central Europe, each student free to select courses and instructors as he wished. Any one university often constituted more than a full meal, and one was free to travel from city to city in search of the best. Foreigners, such as Americans, were in a sense the freest of all. They could shape their courses of study and research unhampered by the specter of the German medical licensing examination, the *Staatsexamen*, which in reality regulated and channeled the lives of their German medical student friends.

Hall, one of the most indefatigable of sojourners, is a good example. On his third German trip of 1878–1880, he first worked on a research project at the physiological laboratory in Berlin with Hugo Kronecker. At the same time he could also follow lectures by Emil DuBois-Reymond in physiology, Helmholtz in physics, and Westphal in psychiatry. When Hall moved on to Leipzig he again started a daily schedule of research, this time at Ludwig's institute, collaborating with the chief assistant, von Kries. To fill out the day he naturally audited the professor's lectures, as well as those in chemistry by Julius Mayer and Jacobus van't Hoff, and those in experimental psychology by Wilhelm Wundt.[53]

However, it is probably Abel who holds the record for gluttony at the intellectual table. From 1884 to 1886 he was at Leipzig studying physiology under Ludwig and von Frey, histology under His, pharmacology with Böhm, along with inorganic and organic chemistry under Johannes Wislicenus. Abel thereby completed the basic work for his Ph.D., which he never bothered to finish. He then spent the winter of 1886–1887 in Strassburg taking courses under Kussmaul in internal medicine, and von Recklinghausen in pathology and infectious diseases. During the summer semester of 1887 he studied at Heidelberg with Erb in medicine and Czerny in surgery. Not one to be idle during the summer, he attended clinics in Würzburg. Then he went back to Strassburg for the academic year 1887–1888 to work with Hoppe-Seyler in physiological chemistry as well as Naunyn and Schmiedeberg in pharmacology. As a hedge against having to practice when he got back to the United States, Abel spent a year of clinics with Nothnagel and others in Vienna; in the interstices he did clinics in Berlin. To top it all off he spent 1889–1890 in Bern, working in von Nencki's biochemical laboratory. Rather en passant, Abel got his M.D. at Strassburg in 1888.[54] Seven universities in six years may be carrying things to extremes, but Abel's peregrinations serve merely to emphasize how great an attraction the principle of freedom and choice was for most visiting Americans.

Germany was rich in another important respect: physical resources. From the middle of the nineteenth century on, the German academic ambiance was the beneficiary of four converging influences: (1) the late, rapid industrialization of Germany, with its concomitant growth of population; (2) the effects of traditional German decentralization; (3) longstanding German esteem for the intellectual life and the focus of that respect upon the university; and (4) the changing nature of chairs at the universities and some fortuitous timing in their disciplinary definition.

In 1815 the pastiche of German states stretching from the Rhine to the steppes of Russia was economically backward in comparison to England or France.[55] The land was overwhelmingly rural; even though Germany had a population almost the equal of France, its twelve largest cities added together did not equal the population of Paris or London. The only nascent industrial areas were those of small-scale textile manufacturing in Upper Silesia and Schwabia and some metallurgical and extractive industries in Upper Saxony and the region around Breslau.

Fifty years later the contrast was striking. The Prussian-led customs union (*Zollverein*) had unfettered internal trade and food distribution. Railroad building had exploded in the 1840s and 1850s, tying together the multitude of German states and creating a domestic demand for steel, coal, and machine works. The overall population had risen dramatically and swarmed into the cities. Older trading and administrative centers had grown like boomtowns. Berlin, the new capital and focal point of the *Reich*, grew from about 150,000 inhabitants at the beginning of the century, to 1.2 million by the 1870s. Leipzig went from 34,000 in 1815 to 150,000 in 1880 and Breslau from 77,000 to 273,000 during the same period. The Ruhr and the Saar regions, sitting atop some of the richest anthracite coal deposits in Europe, rapidly became industrial centers. As early as the 1860s, moreover, it was already clear that Germany's growth in the future would come disproportionately, not just from traditional heavy industries such as coal and steel but also from the new chemical and electrical industries—manufacturing and distribution enterprises that required a greater number of technically educated functionaries.

Unlike England and France, however, the benefits of this wealth and growth were distributed in the geographically dispersed and decentralized patterns that Germany had inherited from the Reformation and early modern period. Even after unification under Prussia in 1871, German regional autonomy and resources remained strong. This decentralization was mirrored in the universities. Each region had its own, almost all of venerable age (Table 5). In addition there were Germanophone universities that lay beyond Germany proper: Vienna, Prague, Graz, and Linz in Austria-Hungary; Zürich, Bern, and Basal in Switzerland; Leiden, Groningen, Utrecht, and Amsterdam in the Netherlands; and even tiny Dorpat in distant Russia.

Until the 1830s all of these local universities had been small, some even miniscule.[56] The division of resources and separation of intellectuals had worked against academic achievement. When the same score of universities became the beneficiaries of new industrial wealth, however, and were pressed into service to educate a growing population, they could erect handsome new buildings, compete with each other to hire the most prestigious professors, and provide a multiplicity of both lower level (*Assistenten* and *Docenten*) and higher level (professors and *Extraordinarii*) academic positions.[57] As can often happen, money and people converted a weakness into a strength.

Such a buildup was made possible because the German principalities had a strong cultural and governmental tradition of support for higher education. In their sleepy pre-Napoleonic condition, the universities had acquired a monopoly on the training of the aristocracy, priests, doctors, jurists, and a host of bureaucrats—the so-called *Bildungsbürgertum*.[58] The professoriate had

Table 5
German Regions and Their Universities

East Prussia	Königsberg (1544)
Silesia	Breslau (1811)
Brandenburg	Berlin (1808)
Pomerania	Greifswald (1456)
Mecklenburg	Rostock (1419)
Schleswig-Holstein	Kiel (1665)
Hannover	Göttingen (1737)
Saxony	Leipzig (1408)
	Halle (1694)
Thuringia	Jena (1548)
Hesse	Würzburg (1582)
	Marburg (1527)
	Giessen (1607)
Rhenish Prussia	Bonn (1818)
Baden	Heidelberg (1386)
	Freiburg (1456)
Alsace	Strassburg (1621/1872)
Württemburg	Tübingen (1477)
Bavaria	Erlangen (1743)
	Munich (1826)

been converted into docile civil servants who had traded financial independence for the intellectual freedom to pursue their own research and govern their own affairs but within the more limited domain of the university.[59] University teachers were seen as the epitomes and carriers of culture.

Finally, physiology was the beneficiary of a set of changes in academic definition made possible by wealth, decentralization, and societal support: the separation of physiology into independent chairs within university medical faculties.[60] Before the 1820s the subject of physiology had invariably been subsumed as part of the duties of professors whose main preoccupation was anatomy, pathology, comparative anatomy, or general medicine. As shown in Table 6, beginning slowly in the 1820s, and continuing more rapidly by the 1860s and 1870s, physiology was separated from its mother disciplines and given independent status.[61]

Perhaps as important is that by the 1860s governmental salaries for professors were sufficient to enable an *Ordinarius* in one of the basic medical sciences to support himself without the need to practice on the side. As Welch noted in a letter home, this common American pattern would have been simply incomprehensible in Germany.[62]

With independent status came independent resources, especially rooms and supplies for experimental physiology. Although the first institutes for physiology were established by Carl Schultze at Freiburg in 1821 and Evangelista Purkinje at Breslau in 1824, physiological laboratories were usually housed in a few rooms in a larger building devoted to another purpose, typically anatomy. Although there were new laboratories at Breslau in 1839 and Prague

Table 6
Independent Chairs in Physiology

Breslau	1811	Göttingen	1860
Königsberg	1826	Würzburg	1865
Marburg	1837	Leipzig	1865
Tübingen	1852	Rostock	1865
Freiburg	1857	Kiel	1868
Munich	1857	Halle	1870
Heidelberg	1858	Greifswald	1872
Berlin	1858	Erlangen	1872
Jena	1858	Strassburg	1872
Bonn	1859	Giessen	1891

in 1851, the real explosion in growth came in the *Gründerjahre* from the 1860s to the 1880s (Table 7).[63]

Americans, at heart great lovers of all that was efficient and modern, were very adept at sniffing out the best facilities. Although the institute at Leipzig was by far the most popular destination for traveling American physiologists in training, a few spread themselves more widely across Germany; almost always, they managed to show up not long after a new laboratory facility had been dedicated. Kühne's new institute at Heidelberg was only a few years old when he welcomed Chittenden in 1878, and Sewall in 1879. James, Henry Cadwalader Chapman, and Joseph Weatherhead Warren attended DuBois-Reymond's laboratory when it still shared quarters with anatomy, and Reichert and Porter did so after the new facilities opened. Porter then went on to the new institute at Kiel. Munich became an attraction for such Americans as Lusk only in the 1880s, as the university, and its laboratory facilities, grew rapidly.

From the very beginning, Americans were appreciative of the facilities and support that German physiological institutes lavished upon their students and guest workers, seemingly without regard to cost. In the winter of 1867–1868 James was in Berlin listening to, among others, the "brilliant" lectures of DuBois-Reymond.[64] He had hoped to do laboratory work at Leipzig with Ludwig, but problems with his back reduced him to an auditor's status in the Prussian capital.[65] He wrote home to Bowditch, then in his final year at Harvard Medical School, that the "opportunities for study here are superb, it seems to me. Whatever they may be in Paris, they can*not* be better. The physiological laboratory, with its endless array of machinery, frogs, dogs, etc., etc., almost 'bursts my gizzard,' when I go by it, with vexation."[66] The Berlin institute was, he wrote to another friend, "a bully physiological laboratory, the sight of which, inaccessible as it is to me in my present condition, gave me a sharp pang."[67] Bowditch, on his part, could see the difference in attitude when he arrived at his first German university. Despite the fact that Bonn was a small and sleepy town, leading him to complain wryly of "the monotony of my existence,"[68] he thought it could match Paris and Boston in one respect:

Table 7
German Laboratories of Physiology: New Independent Facilities

Breslau	1839	Kiel	1879
Prague	1851	Halle	1881
Tübingen	1868	Strassburg	1884
Leipzig	1869	Würzburg	1888
Munich	1874	Greifswald	1889
Heidelberg	1875	Marburg	1889
Berlin	1877	Freiburg	1890
Bonn	1878		

> I went all over the Chemical Laboratory here in Bonn a few days ago. It is wonderful to see the magnificence with which Science is treated here. The building is fully as large as the Technological school & all devoted to Chemistry. In one part of it the Professor & his family have their rooms which are very elegant indeed. The government is very liberal to all such Establishments. I hope that some time or other we shall be able to make *our* legislators take an equally enlightened view of the matter & endow the higher educational establishments with the same liberality with which they have already treated the common schools.[69]

Bowditch found that it was exactly this sort of "liberality," this time by the government of Saxony, that had provided Ludwig with what "is universally acknowledged to be the most complete establishment of the kind in Europe."[70] In March 1870 Bowditch wrote back to *The Boston Medical and Surgical Journal* to describe where he had been working for the last six months. It was very large—an open "U" in shape, 120 feet on a side, with two stories and a basement. Working space was about equally divided among the three subjects under Ludwig's ken: microscopy-cum-histology, experimental or physical physiology, and physiological chemistry. The histological section had a large room for beginning microscopists, as well as a smaller one for advanced workers, which was fitted out with sophisticated injection apparatus. Experimental physiology had a large room just for the storage of apparatus, three rooms for general experiments, and two additional large ones for grand-scale experiments on respiration, acoustics, and metabolism. Rooms were fully

> furnished with operating tables, with bellows attached for keeping up artificial respiration on *curarized* animals, registering apparatus of various sorts for recording the pressure of the blood, water baths where any required temperature may be kept up indefinitely, an injecting apparatus like that in the microscopical department, evaporating closets, glass cases for apparatus, &c.[71]

Physiological chemistry, the third department, occupied two large rooms on the third side of the institute.

This experimenter's heaven above was supported by its own hades below. In the basement there was a gas engine that drove the respiration apparatus and registering instruments, a refrigerated room for performing low-temperature experiments, furnaces for glassblowing, storerooms, animal quarters ("one room being devoted entirely to frogs"), and a workshop. The latter was the domain of the full-time laboratory mechanic, who not only altered and repaired the existing instruments but also built new apparatus for special investigations.[72] As Abel wrote home from Leipzig just after his arrival there in October 1884: "Our laboratory is a perfect physiological paradise, with a fine janitor and a trained mechanic who has been with Ludwig for 30 years."[73]

This penultimate figure, the *Diener*, was, in the minds of the students, perhaps the most important resource of the institute. Abel translated the name as "janitor" but the person was in essence the laboratory technician of today. Chittenden, who had seen the ilk both at Yale's Sheffield Scientific School and in Kühne's Physiologisches Institut at Heidelberg during 1878–1879, wrote that the *Diener*:

> is apt to become an indispensable part of the laboratory equipment, gradually making for himself a position of high importance, recognized by instructors and students alike. He is not to be confounded with the man who simply sweeps and dusts, who merely keeps things clean. Rather he is a mine of wealth, rich in experience with a store of knowledge of many things connected with the experimental work of a laboratory, ready and able to help in the setting up of a complicated apparatus, even to advise in many of the emergencies that arise in experimental work.[74]

Ludwig's *Diener*, Salfamoser, had come with him from Vienna, and had worked with him so long that he knew even the most complicated experimental methods and could teach them to each incoming student. Lombard, who knew the pair in 1883–1885, reminisced that when the professor himself was operating on an animal, Salfamoser would assist—and offer unsolicited advice. "I recall seeing Ludwig draw himself up and say to Salfamoser, 'Who is the professor here, you or I?' 'Oh, you are, Herr Professor; nevertheless, I am right.'"[75] The *Diener* could be as opinionated as he was beloved and useful.

ETHOS OF THE GERMAN RESEARCH LABORATORY

In concluding his detailed description of the Leipzig Institute, Bowditch explicitly linked physical resources with the intellectual values—the ideology—that permeated the stone, mortar, and apparatus:

> It will thus be seen that abundant facilities are here offered, not only for learning the existing state of physiological science, but also for becoming familiar with the manner in which physiology is at present studied in Germany. The patient, methodical and faithful way in which the phenomena of life are investigated by the German physiologists not only inspires great confidence in their results, but encourages one in the hope that the day is not far distant when physiology will take its proper place as the only true foundation of medical science.[76]

What were these habits of thought and action, and the normative ambiance, that were perceived by visiting Americans as accounting for the success of the medical sciences in Germany and physiology in particular?

At the most general level, Americans saw it as a cultural value that prized rigor and completeness, and they admired the ability of Germans to translate this value into institutions. The word "Germanic" applied to scholarship or science has come in the twentieth century to have pejorative overtones—of investigations conducted with too much precision and empty systematization. In the 1860s, 1870s, and 1880s, however, it expressed the very best goals to which medicine, long relegated to the scrap heap of empiricism, should aspire. James perhaps described it best when he wrote to Bowditch from Berlin:

> The general level of thoroughness and exactness in scientific work here is beyond praise; and the abundance of books on every division of every subject something we English have no idea of. It all comes from the thorough mode of educating the people from childhood up. The *Staats Examina*, before passing which no doctor can practice here in Prussia, exact an amount of physiological, and what we at home call 'merely theoretical' knowledge of the candidate which a young doctor at home would claim and receive especial distinction for having made himself master of. But the men here think it but fair; gird about their loins and set about working their way through.[77]

Welch, writing from Leipzig, echoed some of the same themes. Modern physiology was "by far the most exact of any of the branches of medicine, this position of exactness having been obtained more through the efforts of Prof. Ludwig than of any living man." Welch hoped to follow Ludwig's precept and practice, that every man of science ought "not to be satisfied with loose thinking and half proofs, not to speculate and theorize but to observe closely and carefully facts."[78]

This cultural value of precision in medicine was expressed in the characteristic instrument of physiology in the second half of the nineteenth century: the kymograph.[79] It had been invented by Ludwig early in his career, in 1846, when he was still an assistant in the anatomical institute at Marburg. In the 1850s its use spread slowly, but by the 1860s and 1870s it had come to be, especially in Germany, the prime instrumental intermediary between the physiologist and his experiment. Although it was the Frenchman Etienne Jules Marey who popularized the term "graphic method" (*la méthode graphique*), and developed into its philosopher, it was in Germany, perhaps under Ludwig, that Marey had learned kymograph techniques.[80] Graphic records became the new language of physiology, one that required an apprenticeship to learn how to read. In 1878 Hall was working with Kronecker, DuBois-Reymond's assistant and a former student of Ludwig, at the magnificent, newly opened Physiologisches Institut in Berlin. Their project was the graphic recording of muscle reactions to succeeding electrical stimuli delivered to the spinal cord. Kronecker could almost make him believe, Hall wrote back to James at Harvard, that one could write "the whole philosophy of life" in the different forms of muscle curves.[81] Kronecker was simply being Ludwig's student. Eight years earlier, Bowditch had been in the Leipzig laboratory for only about six weeks when he wrote home in a similar vein about the professor's pleasure in a new instrumental development:

> I enjoy the work in the laboratory very much. Prof. Ludwig is a very amiable and agreeable man. He must be between fifty and sixty years old but he retains his youthful enthusiasm and a remarkable faculty of finding pleasure and amusement in trifling matters. I arranged a little apparatus yesterday attached to a metronome for the purpose of marking time on a revolving cylinder covered with smoked paper (an instrument much used in various physiological experiments) and it was real fun to see how delighted the professor was with it.[82]

Ludwig had a better sense than Bowditch of the significance of the innovation; by the end of the century it was used all over Europe and America.

Precise methods meant nothing, however, if one did not have a well-defined problem and a proven set of techniques by which it could be attacked. Ludwig tried to teach his students how much hard thinking had to go into picking a problem that was worth doing. Lombard recalled with approval the estimation made by His that the real "weapons of research" in the Leipzig laboratory were Ludwig's "uncommon sharpness of analysis of living processes under observation, an always clear formulation of the question, and an absolute reliability of the method of attack."[83] Sometimes such clarity of expectations could lead to disappointments. Abel used to recount that when he arrived in Leipzig, Ludwig put him onto testing a pet idea of whether a temperature change would be an effective stimulus to a nerve. When Abel "rigged up an apparatus" that showed the effect, "Ludwig was delighted with the demonstration of the correctness of his theory." When Abel found that the cause of the excitation had been the electricity generated by a copper/iron contact in his preparation, Ludwig was crestfallen for a day or two.[84]

Moreover, as Lombard reminisced, Ludwig was "remarkable for his ability to utilize the work of the young and inexperienced."[85] He was a master at tailoring the proposed investigation to the background, interests, and capabilities of the investigator. In their first interview, for example, Ludwig asked Welch what he had studied. Finding that Welch was particularly interested in microscopical work, Ludwig set him on a study of the ganglion cells of the heart that combined physiology with histology.[86] Because Mall's background and interest had been chiefly in anatomy and embryology, Ludwig put him onto studying the structure of the villus of the intestine, and how it functioned.[87] Ludwig did not, like many laboratory directors, merely give his students the minor and subsidiary parts of his own research program.

Finally, when the work was done, Ludwig was the one who would take the product, in whatever state, and transform it into a publishable (and published) paper. In Hall's case, Ludwig took the data upon which both he and Hall had worked daily, and wrote the entire long article himself.[88] Lombard got to the stage of writing up his work in a fragmentary form, but Ludwig rewrote parts of it and sewed it together.[89] He also no doubt corrected Lombard's German, since the Abels, who overlapped with the Lombards at Leipzig in 1884–1885, noted that the Bostonian spoke the language fluently, but with many blunders.[90] In Mall's case, Ludwig merely had the English text translated into German; but when he discovered after Mall's departure that the article needed more data and more anatomical preparations for illustrations, he provided these, along with the necessary text.[91] In all three cases, as no doubt in hundreds of others, the resultant articles appeared solely under the student's name.

Ludwig was adamant on this point. Work from the institute went out under the student's name and/or that of one of the *Assistenten*, but never with Ludwig's, even though he may have suggested the problem, designed the apparatus, helped execute the crucial experiments, analyzed the results, and written the article. Mall thought this was doing too much, but Ludwig only replied genially that Mall should not "scold" an old man.[92] When Albert Bleile completed, back in Columbus, Ohio, the work that had largely been carried out in Leipzig, Ludwig wrote: "I approve; publish it in your own name; leave mine off."[93] Lombard, for his part, told Ludwig "that I felt I had no right to let the paper appear under my name, for I had been only his hands; that it was really his work, and not mine. 'It is all right,' he [Ludwig] said, 'it has been your work.' Then he added, 'But if you never do anything else, it will be thought that you did not do this.'"[94] It would be hard to imagine a stronger stimulus to a continued career in physiology.

Most of all, Ludwig exemplified how the best investigator gave his life to his investigations and to the research of the students he directed. He was there literally twenty-four hours a day, almost the entire year. Bowditch found that, in contradistinction to Paris, where none of the major professors deigned to return from the country until early November, the institute at Leipzig hummed even in the dog days of summer. He wrote from Saxony in September 1869:

> When I first arrived here I only expected to stay a day or two just to take a look at the place and then to go to Dresden or Berlin for a few weeks while waiting for the term to begin, but I found that Prof. Ludwig was already at work in his laboratory and that I could go to work with him as soon as I pleased. So I concluded to settle down here at once and go to work.[95]

Each morning Ludwig would visit individually the tables of the nine or ten men and women who were working in the physiological section of the institute. He would discuss their results and data and go over the possible next steps with them. It was no mystery how he knew exactly what each worker was doing day-by-day, because each night he would take the records and protocols of the investigations in progress upstairs to his apartment to study. Lombard thought it was remarkable, not just how many people he could oversee, but "how many different forms of research he could supervise at the same time and keep them clearly in mind."[96] Abel contrasted the thoughtful, ever-present Ludwig, to Martin back in Baltimore, who "spent twenty minutes a day going hurriedly from desk to desk."[97]

This kind of atmosphere made the institute a little world unto itself. In Bowditch's time he "joined a sort of club table composed of four or five of the young men who work in the laboratory. We dine together every day at five o'clock. . . . It is," he wrote back home, "quite an international affair. There are two Germans [including Hugo Kronecker], one Russian [Ustimovitch], one Norwegian, one Swiss and one American."[98] Within a month the dinner club had evolved into a little "physiological society":

> We have just formed a sort of society composed of the workers in the laboratory, the object of which is to meet once a week and to give everyone an opportunity to report anything new or interesting which has fallen under his notice. This will be extremely useful for me as almost all my companions are more experienced than myself in physiological work. I mentioned in our last meeting the famous case of the man who had a crowbar driven through his head and lived twelve years after it. I think many of my hearers were inclined to be politely incredulous.[99]

By Welch's time, seven years later, Kronecker was Ludwig's first assistant and he introduced Welch into the still flourishing physiological society, which met once a week to listen to papers. After the discussion of the presentation, Welch explained to his sister, "the converse becomes more lively although not less instructive in an adjoining beer saloon."[100]

The academic year was also punctuated with more formal social/scientific occasions. Bowditch reported that their "physiological society . . . closed its sessions for the Winter with a supper to which Prof. Ludwig and several other professors were invited. We had a very jolly occasion but the result is a decidedly seedy feeling this morning."[101] The festivities had not declined in *Gemütlichkeit* by the time Welch arrived seven years later. At the end of his term there

> the society gave a dinner in honor of four of us who leave here at the close of this semester. We, the departing members, were seated in chairs decorated with flowers and were made the subjects of toasts and presents. Among other things I was presented with a large wooden model of a ganglion cell of the heart, which was of course intended to be humorous. I was also given a very pretty beer mug (decorated in the blue which is so much admired now a days) and a photograph of the physiological institute. Three professors were present but were as jolly as any one towards the small hours of the night. A humorous poem was read describing the events at the institute during the winter. The Germans understand how to mingle a little formality and a great deal of joviality on such occasions so as to make it thoroughly enjoyable.[102]

Is it any wonder that so many of Ludwig's students looked back upon their time in Germany as a *goldene Zeit*?

Any memory, however, no matter how golden, must be assayed. Perception of another culture, especially when mediated through another language, is capable of many kinds of distortion. Kohler has very perceptively analyzed how American biochemists saw Germany with a very selective eye.[103] We also must ask, to what degree were the experiences of visiting American physiologists a true and balanced reflection of what actually existed in the physiological sciences in Germany? In short, the answer is they were reasonably true but decidedly unbalanced.

To a degree that few sojourners fully realized, they saw German science at its best, from its best side, and at its best time. Merely the concentration of American experience in Ludwig's Physiologisches Institut at Leipzig served to unbalance the picture. It was by far the largest, the most well equipped, and the most active physiological laboratory in the world at the time. It was headed by a man who, while not the most original physiological investigator in Germany, was surpassingly unique in the breadth of his knowledge of ideas and techniques in his field, and who gave himself with unparalleled energy to his students and his science. There were physiological laboratories at nineteen other German universities and they were, in one facet or another, defective in comparison to Leipzig. Not that they failed to produce good research; all did that in greater or lesser measure. They seemed deficient in that their facilities were more cramped, or their stock of instruments not so full, or their professors more self-absorbed, or their assistants less helpful, or their atmospheres lacking in élan and excitement.

Moreover, the status of Americans as visitors allowed them to see much that was good in the German physiological sciences and tended to obscure some things that were distasteful. As foreigners they had no need to worry about passing university or state examinations; they could enjoy the student's life, without any of the student's obligations. Because they had an M.D. or Ph.D., they were accorded a relatively high status. Some sensitive people, like Welch, found it embarrassing to be treated more deferentially than a German medical student who might well know more science and more medicine.[104] Most sojourners, however, simply found it pleasing. Many German professors were, in return, flattered to find that Americans had traveled so far to work with them.[105] Quite a number responded by going out of their way to be kind to their apprentices from across the water. Indeed, they were relatively safe in doing so. Americans were outside the German system of social status, and attention extended to them need not compromise one's community standing vis-à-vis German students or one's colleagues. Moreover, making disciples of foreigners brought many benefits and few obligations. A professor did not need to worry about taxing his reputation and connections to find an American a job in Germany; the opportunity to publish his research, and a laudatory letter, were all that was expected in recompense.[106] This, in turn, raised the mentor's reputation in the United States, a land that exerted an almost mesmerizing attraction in Germany from the 1830s through the 1870s; one needs only to recall that German emigration here totaled over 6.1 million in the course of the nineteenth century.

Finally, it is clear in a way verifiable only in retrospect that impressionable Americans were most exposed to German medical science at exactly the time of its exuberant expansion. The peak of the creation of chairs and the building of institutes preceded the peak in American sojourners by only about a decade, or two at most.[107] These travelers arrived and mistook a recently hard-won set of institutions for the established norm. As the growth of German institutions tended to level out toward 1900 and beyond, so did the flow of Americans. There were, indeed, a multiplicity of ways in which an American could get a beguilingly idealized picture of the German physiological sciences.

EPILOGUE: BACK HOME

Sojourners, unless they are to become expatriates, must eventually come home. Those in what I have labeled the "German generation," came back to a United States in some ways similar, but in many ways vastly different from Bismarck's *Reich*. Both nations were rapidly industrializing, growing in population, and becoming predominantly urban cultures, but they had, at least initially, quite different attitudes to culture, scholarship, and medicine. High culture—art, music, poetry—led a precarious existence in so commercial a society as that of the United States. American colleges, with a few nascent exceptions, sought to produce not scholarship, but students.[108] Leading institutions of higher education were not state supported but private. American medicine prided itself on being practical, not theoretical or scientific, and found its natural expression in that most accommodatingly practical of educational institutions, the proprietary medical school. Within this environment, which parts of their collective experience in Germany could and did these sojourners bring back?

Most of all, they brought back a strong feeling for the primacy of research.

Investigation, they believed, was the defining characteristic of modern physiology. Teaching, certainly of the classroom variety, was only a service role. Could it also be transformed, as Merriley Borell has demonstrated in chapter XIII in this book, into laboratory teaching that might snare future devotees of the research career? In any case, the most important activity of the physiologist was to run his laboratory, to recruit workers for it, and to see that it was properly constituted and operated. Bowditch, in negotiating a position at Harvard for himself upon his return, at first rejected a purely lecturing job, holding out for one in which he could develop and furnish his own laboratory.[109] Within five years of his return in 1871 he had a small, but devoted coterie of investigators.[110] This was similar for Lee at Columbia, although during a transitional period he had to work as second in command to John Green Curtis, an enthusiastic supporter of experimental physiology but no experimentalist himself.[111]

Dozens of sojourners had learned that an important part of the experimentalist's armamentarium was instrumentation. In physiological institutes such as Ludwig's, they had become skilled in using many kinds of apparatus: the kymograph, *Stromuhr*, plethysmograph, spirometer, sphygmograph, sphygmomanometer, and dozens more. Moreover, their apprenticeships abroad provided them with detailed knowledge of who on the Continent could supply which instruments at what prices, which pieces of glassware in what quality and quantity. Bowditch, Welch, Curtis, Lee, and Mall all made return trips to Europe in the 1880s and 1890s, partly to buy equipment and supplies. During two such visits, in 1879 and 1880, Bowditch jaunted at more than a sedate pace from one European laboratory to another seeing who was using which instruments for what, and where they could be obtained.[112] Not only did this emphasis on instruments suit the personal proclivities of men such as Bowditch—he had a good mechanical sense and a dexterous touch with apparatus—but it fit in well with American cultural prejudices. The manufacturing and mercantile circles of the Eastern seaboard felt comfortable with machines. Moreover, an array of equipment, especially that which yielded quantitative results, might well have seemed to make physiology, in itself a somewhat abstract and theoretical subject, more concrete and practical.

Having laboratories for research, and instruments to carry it out, helped to perpetuate in these travelers a feeling of international camaraderie, of participation in an enterprise that stretched across borders and oceans. Merely feeling comfortable with the German language opened up a massive body of scientific literature. A research apprenticeship abroad also conferred another advantage. Because German laboratories not only trained scientists from the *Reich*, Austria-Hungary, and Switzerland, but also attracted students from England, Italy, Scandinavia, Russia, Belgium, and the Netherlands (almost every country, in fact, except France), one could become friends and colleagues with physiologists who, upon maturity, would lead the discipline throughout the world. Bowditch's little dining club was merely the beginning of friendships that he renewed on return visits to Europe, especially after 1889, at the triennial Congresses of Physiologists.[113] At the fourth such congress, at Cambridge in 1898, the university took the occasion to confer an honorary degree on America's most well-known, and certainly most well-traveled, physiologist.[114]

Travelers abroad, especially in the 1870s and 1880s, also learned the value of the social accoutrements of science. They saw the role of specialized journals with high standards of quality, such as Pflüger's *Archiv*, or DuBois-Reymond's *Archiv*, or Ludwig's *Arbeiten*. They could also see how formal societies might very usefully grow from informal clubs. The Berlin *Physiologische Gesellschaft*, organized by Kronecker under the patronage of DuBois-Reymond, was seen by Americans as a continuation of that small physiological club in Leipzig of which both Kronecker and Bowditch had been members in the early 1870s.[115] Given the additional strong tradition of voluntary organizations that came to be characteristic of nineteenth-century America, it is little wonder that so many foreign-trained physiologists rallied to the banner when Martin and Bowditch sent their circular letter in November 1887.

Laboratories, equipment, friendships, and institutions all emphasized what travelers had seen so well embodied in German physiological institutes: the autonomy of physiology. The physiology they saw was embedded within a strong university environment, but it was institutionally quite separate from the biological sciences of zoology or botany, which were in the philosophical faculty. Likewise they perceived physiology's relationship to medicine: within it, but not overwhelmed by it. No investigator should try to combine physiology with any other kind of career, especially a private clinical practice. The full-time concept for the basic medical sciences was, they believed, validated by the overwhelming productivity that German physiology had demonstrated.

However, such autonomy costs money in an amount that medical schools could not hope to raise by fees; back home there was no obliging King of Saxony or Prussian *Kultusministerium* to pick up the tab. If American medical scientists believed that such autonomous sciences had a right to support, they had to either lobby federal and state governments for funds, or, more commonly, lead the efforts to find private patronage. It is little wonder that men like Minot held up the German model for American philanthropy to emulate, or that Bowditch organized successful fund-raising efforts at Harvard, or that Lee did the same in New York.[116] It was Welch, a great admirer of the German medical sciences, who ultimately proved to be the most successful raider on the bank accounts of America's industrialists.[117]

There was, however, one highly important respect in which the German generation must be judged less than successful in learning their lessons overseas. Not one of them developed into a physiologist who made strikingly important contributions to the intellectual structure of their discipline. Unlike their older colleagues in physiology, they did do research. They published decent articles, addressing modest problems with creditable technique, but their achievements as researchers did not match their accomplishments as organizers of research. Such a sweeping characterization is obviously a personal one and subject to much qualification, but I think that the general import of the judgment can be maintained.

Why the failure of these physiologists to bring back from Germany, as Diana Long phrased it in a provocatively aphoristic way, everything except the physiology?[118] At the moment I can only put forward two tentative explanations. They suggest themselves from the available evidence, but the historical analysis of the physiological sciences in America has not yet reached the stage where they can be tested with any hope of a convincing outcome.

The first explanation—the relative lack in America of a coherent consensus on important problems and ways to their solutions—is suggested by an odd fact about Bowditch's scientific reputation. The two discoveries for which he is most known in the history of physiology, the "staircase" (*Treppe*) effect in the contraction of the heart and the indefatigability of the mammalian nerve fiber, were both made at Ludwig's institute in Leipzig, the first in 1871 and the second on a return visit there in 1890.[119] Each discovery addressed an important current problem in physiology and resolved it with elegance. Some of Bowditch's other experimental work at Harvard showed equal instrumental virtuosity but not the same insight into that particular nexus of question-and-possible-answers that goes in German under the name of *Problemstellung*. First-class success in science requires not just technical expertise and detailed knowledge, but wide reading in the literature of a subject, as well as rich and varied intellectual communications with one's peers. Lacking it, one can still do good research, but never quite arrive at the ideas and discoveries that dictate the direction and pace of a discipline's movement. This skill can be learned, but almost never taught. The only substitute for it, on an individual basis, may be a monomaniacal idée fixe, idiosyncratically pursued in the manner of a Robert Koch. Isolated individuals, however, even if such exist, cannot replace a scientific community that has a coherence not just socially and institutionally but intellectually as well. Indeed this sense of *Problemstellung* is often the last element to emerge within a discipline in a given cultural theater. It could not be imported, ready-made, like the floorplan of an institute or the hierarchy and goals of a department.

The second explanation—the bending of theoretical questions into applied solutions—is also suggested by a tidbit of Bowditch's biography, and may have rather more to do with the characteristics of the native American scene. Once back in Boston, Bowditch's professional life ramified into many more directions than merely research physiology. He organized anthropometric surveys of Boston schoolchildren, wrote on the physiological effects of alcohol, and reported on the physiques of women in Massachusetts.[120] Whenever some public or medical cause needed a tame physiologist, Bowditch seemingly felt he could not refuse. Similarly, the work Lee did at Leipzig was pure electrophysiology of muscle, but when he came back to Columbia, his interest in problems of muscle physiology gradually transformed itself into the study of industrial fatigue, as Richard Gillespie shows in chapter X in this book. Such examples could be multiplied again and again, especially in the first two generations that were trying to carve out a place for the science of physiology in the American medical wasteland.

Such difficulties of intellectual agenda and social place are, from the proper perspective, easy to understand. The process of cultural transplantation is a tricky one indeed. To my mind, the noteworthy thing is not that some of the most fragile flowers did not survive the journey. In any case, they emerged rapidly enough once the ground was prepared. Rather more remarkable is that so many physiologists were, in so delimited a period and in so profound and detailed a way, exposed to a new ideal of laboratory physiology, and that they were able to bring so much of it back across the Atlantic.

Research was supported by a grant from the Research Committee of the Academic Senate, University of California, Los Angeles. A preliminary version of this chapter was presented to the Section of the History of Physiology, American Physiological Society, at Anaheim, CA, April 1985.

NOTES

1. William H. Howell and Charles W. Greene, *History of the American Physiological Society: Semicentennial, 1887–1937* (Baltimore, MD: American Physiological Society, 1938), p. 197.
2. Ibid., p. 56. The quotation is from Article 4 (of 13) in the Society's constitution, as adopted by the founding meeting at the Physiological Laboratory, College of Physicians and Surgeons, Columbia University, Friday, 30 December 1887.
3. Howell, "The American Physiological Society During its First Twenty-Five Years," in Howell and Greene, *History*, pp. 1–89; quote from p. 3.
4. For general overviews of the history of American physiology in the nineteenth century see: Edward C. Atwater, "'Squeezing Mother Nature': Experimental Physiology in the United States Before 1870," *Bull. Hist. Med.* 52 (1978): 313–335; Karl E. Rothschuh, *History of Physiology*, trans. and ed. Guenter B. Risse (Huntington, NY: Krieger, 1973), pp. 187–190; C. I. Reed, "The Maturation of Physiology in America After 1830," *Physiologist* 5 (1962): 35–41; Walter J. Meek, "The Beginnings of American Physiology," *Ann. Med. Hist.* 10 (1928): 111–125; Henry Sewall, "The Beginnings of Physiological Research in America," *Science* 58 (1923): 187–195.
5. See, for example, the curricula discussed in William Frederick Norwood, *Medical Education in the United States Before the Civil War* (Philadelphia: Univ. of Pennsylvania Press, 1944).
6. Anna Robinson Burr, *Weir Mitchell: His Life and Letters* (New York: Duffield, 1929), p. 97.
7. S. Weir Mitchell, "Biographical Memoir of John Call Dalton," *Biogr. Mem. Natl. Acad. Sci.* 3 (1895): 179–185, esp. p. 181.
8. For an overview see Rothschuh, *History of Physiology*, pp. 150–155, 264–275.
9. Burr, *Weir Mitchell*, pp. 61–68.
10. Given below in the quote from Mitchell's obituary of Dalton.
11. William F. Bynum, "Mitchell, Silas Weir," in *Dictionary of Scientific Biography* (New York: Scribner's, 1974), p. 422.
12. Mitchell, "Dalton," p. 181.
13. Ibid., and Frederic L. Holmes, "Dalton, John Call," in *Dictionary of Scientific Biography* (New York: Scribner's, 1978), pp. 107–110.
14. Burr, *Weir Mitchell*, pp. 68–121; Mitchell's disappointments are well detailed and analyzed in W. Bruce Fye, "S. Weir Mitchell, Philadelphia's 'Lost' Physiologist," *Bull. Hist. Med.* 57 (1983): 188–202.
15. Most of Bowditch's notebooks and correspondence pertaining to his European study are in the Harvard Medical Archives, Francis A. Countway Library of Medicine, Harvard Medical School (hereafter, CL-HMS), Boston, MA. Notebooks are in CL-HMS, GA 9; the relatively few letters to Bowditch are in CL-HMS C 5.2, while in CL-HMS C 5.1 are to be found the multitudinous letters he wrote home to his mother, Lucy Orne [Nichols] Bowditch, his father, Jonathan Ingersoll Bowditch, and his uncle, Henry Ingersoll Bowditch (hereafter, Bowditch Papers). Typed extracts and/or summaries of many of the letters were made in the early twentieth century by Bowditch's son, Harold Bowditch, presumably with an eye toward writing a biography of his father; no such work was ever published. The correspondence was used, although often without specific citation, by Walter B. Cannon, "Henry Pickering Bowditch," *Biogr. Mem. Natl. Acad. Sci.* 17 (1922): 183–196. See also Frederick W. Ellis, "Henry Pickering Bowditch and the Development of the Harvard Laboratory of Physiology," *New Eng. J. Med.* 219 (1938): 819–828; Everett Mendelsohn, "Bowditch, Henry Pickering," in *Dictionary of Scientific Biography* (New York: Scribner's, 1973), pp. 365–368. Bowditch's early training and call to Harvard are well treated in Fye, "Why a Physiologist?—The Case of Henry P. Bowditch," *Bull. Hist. Med.* 56 (1982): 19–29.
16. J. M. D. Olmsted, *Charles Edouard Brown-Séquard: A Nineteenth Century Neurologist and Endocrinologist* (Baltimore, MD: Johns Hopkins Press, 1946), pp. 129–130.

17. H. P. Bowditch to H. I. Bowditch, 29 September 1868, Bowditch Papers.
18. Brown-Séquard to H. P. Bowditch, 1 October 1868; H. P. Bowditch to J. I. Bowditch, 1 October 1868; H. P. Bowditch to H. I. Bowditch, 4 October 1868, Bowditch Papers.
19. H. P. Bowditch to H. I. Bowditch, 4 October 1868, Bowditch Papers.
20. H. P. Bowditch to H. I. Bowditch, 15 October 1868 [portion of letter written 1 November 1868], Bowditch Papers.
21. H. P. Bowditch to L. O. Bowditch, 22 and 30 November 1868; H. P. Bowditch to H. I. Bowditch, 13 December 1868, Bowditch Papers.
22. H. P. Bowditch to J. I. Bowditch, 6 December 1868, Bowditch Papers.
23. H. P. Bowditch to H. I. Bowditch, 13 December 1868, Bowditch Papers.
24. H. P. Bowditch to H. I. Bowditch, 26 January 1869, Bowditch Papers.
25. Cf. his long summaries of Bernard's ideas, H. P. Bowditch to H. I. Bowditch, 26 January and 21 March 1869, Bowditch Papers.
26. H. P. Bowditch to Jeffries Wyman, 14 January 1869; cited in Fye, "Why a Physiologist?" p. 23.
27. For an overview of the place of scientific research in French institutions in the first half of the nineteenth century, see Joseph Ben-David, *The Scientist's Role in Society: A Comparative Study* (Englewood Cliffs, NJ: Prentice-Hall, 1971), pp. 88–107.
28. H. P. Bowditch to H. I. Bowditch, 26 January 1869 [portion dated 8 February 1869], Bowditch Papers.
29. H. P. Bowditch to J. I. Bowditch, 12 February 1869, Bowditch Papers.
30. Ibid.; the book is Paul Joseph Lorain, *De la Réforme des Études Médicales par les Laboratoires* (Paris: 1868).
31. H. P. Bowditch to H. I. Bowditch, 21 March 1869, Bowditch Papers.
32. H. P. Bowditch to L. O. Bowditch, 14 March 1869, Bowditch Papers.
33. H. P. Bowditch to H. I. Bowditch, 12 April 1869, Bowditch Papers.
34. This sample of physiological scientists is drawn from a larger prosopographic project now in progress. Because the focus of the project is medicine and the medical sciences, the larger population includes (1) all scientists with an M.D., regardless of field and (2) all Ph.D.s and equivalents in medical science subjects, as long as they were either trained or functioned within a medical environment. This would exclude, for example, dairy or soil bacteriologists, agricultural biochemists, and "physiologists" who were purely biological both in training and in employment. The grouping defined as physiological scientists includes the subject categories of: physiology; other physiological areas such as applied physiology; physiological chemistry; biochemistry or biological chemistry; other biochemical fields such as nutrition; pharmacology; and toxicology. The word "individual" is used here to mean a person with a given subject identifier. Because a minority of persons would give two or (rarely) three subjects, such a person might be counted twice in subject-related cross tabulations. Abel, for example, listed his subjects as "Pharmacology, Physiological Chemistry"; Chittenden listed his as "Physiological Chemistry, Physiology."
35. Average age, 25.7 years; *N*, 29 (excludes Martin and Meltzer); SD < 3 years.
36. First phase (1860–1884), 25.3 years; *N*, 19. Second phase (1885–1899), 25.8 years; *N*, 35. Third phase (1900–1914), 28.3 years; *N*, 32.
37. William James to Oliver Wendell Holmes, Jr., 17 September 1867, *The Letters of William James*, ed. Henry James (Boston, MA: Atlantic Monthly, 1920), p. 101.
38. J. I. Bowditch to H. P. Bowditch, 22 March 1871, Bowditch Papers.
39. Harvey Cushing, *The Life of Sir William Osler* (New York: Oxford Univ. Press, 1946), p. 114.
40. Thomas Neville Bonner, *American Doctors and German Universities: A Chapter in International Intellectual Relations, 1870–1914* (Lincoln: Univ. of Nebraska Press, 1963); on the attractiveness of the clinical centers of Vienna and Berlin, esp. pp. 69–106.
41. Florence Rena Sabin, *Franklin Paine Mall: The Story of a Mind* (Baltimore, MD: Johns Hopkins Press, 1934), pp. 30–64.
42. Horace W. Davenport, "Physiology, 1850–1923: The View From Michigan," *Physiologist* 25 suppl. 1 (1982): 50; for the best treatment of Lombard's life and work, see 50–76.
43. On the attention paid to the University of Strassburg, see J. E. Craig, "A Mission for German Learning: The University of Strassburg and Alsatian Society, 1870–1918" (Ph.D. diss., Stanford Univ., 1973).

44. Simon Flexner and James Thomas Flexner, *William Henry Welch and the Heroic Age of American Medicine* (New York: Viking, 1941), pp. 78–90, 140–145; the transition is also discussed in Owsei Temkin, "The European Background of the Young Dr. Welch," *Bull. Hist. Med.* 24 (1950): 308–318, esp. 308–312.
45. Hubert B. Vickery, "Russell Henry Chittenden, 1856–1943," *Biogr. Mem. Natl. Acad. Sci.* 24 (1947): 59–104, esp. 65–66. Vickery, who was Chittenden's pupil and friend, cites the story of the brief and unhappy sojourn in Strassburg from Chittenden's *The Development of Physiological Chemistry in the United States* (New York: Chemical Catalogue, 1930), p. 30, although it is not to be found in Chittenden's manuscript autobiography, "Sixty Years of Service in Science: An Autobiography" (1936) in the Chittenden Papers, Yale University, Sterling Library, Department of Manuscripts and Archives, manuscript group 611, box 2, folder 35.
46. Gerald Webb and Desmond Powell, *Henry Sewall, Physiologist and Physician* (Baltimore, MD: Johns Hopkins Press, 1946), pp. 45–50.
47. Average, 2.69 destinations per person; *N*, 29.
48. Average, 2.82 years abroad; *N*, 29.
49. Average, 1.54 years elapsed since doctoral degree; *N*, 22. *N* is smaller because some travelers, for example James, Hall, Warren, and Minot, went abroad before their doctorates, and were excluded to avoid having them be tallied as a negative elapsed time.
50. Frederic Schiller Lee, "Über die elektrischen Erscheinungen, welche die Muskelzuckung begleiten," *Arch. Physiol.* (1887): 204–223; cf. his reflections on his teacher in Lee, "Carl Ludwig," *Science* n.s., 1 (1895): 630–632.
51. Henry Herbert Donaldson, "Memories for My Boys" (1931), MS autobiography in the Library of the American Philosophical Society, Philadelphia, PA, pp. 78–87; the standard published biography is Edwin G. Conklin, "Henry Herbert Donaldson, 1857–1938," *Biogr. Mem. Natl. Acad. Sci.* 20 (1938–1939): 229–243.
52. Lewellys F. Barker, *Time and the Physician* (New York: G. P. Putnam's Sons, 1942), pp. 57–61.
53. Dorothy Ross, *G. Stanley Hall: The Psychologist as Prophet* (Chicago, IL: Univ. of Chicago Press, 1972), pp. 80–89.
54. William D. MacNider, "John Jacob Abel, 1857–1938," *Biogr. Mem. Natl. Acad. Sci.* 24 (1947): 231–257. Abel's notebooks from his studies in Germany are in the Abel Papers, Johns Hopkins Medical Archives, Baltimore, MD, boxes 66–69 (hereafter, Abel Papers).
55. For brief overviews of the economic transformation of Germany see, for example, Theodore S. Hamerow, *Restoration, Revolution, Reaction: Economics and Politics in Germany, 1815–1871* (Princeton, NJ: Princeton Univ. Press, 1958), esp. pp. 3–20, 199–255; or Hajo Holborn, *A History of Modern Germany, 1840–1945* (Princeton, NJ: Princeton Univ. Press, 1969), pp. 3–22, 122–129, and 367–390.
56. See Charles E. McClelland, *State, Society and University in Germany, 1700–1914* (Cambridge, UK: Cambridge Univ. Press, 1980).
57. On the "enrollment explosion" in nineteenth-century German universities, see Konrad H. Jarausch, *Students, Society, and Politics in Imperial Germany: The Rise of Academic Illiberalism* (Princeton, NJ: Princeton Univ. Press, 1982), pp. 23–77.
58. McClelland, *State, Society and University in Germany.*
59. On the origins of a research tradition in German universities see R. Steven Turner, "The Growth of Professorial Research in Prussia, 1818 to 1848—Causes and Context," *Hist. Stud. Phys. Sci.* 3 (1971): 137–182.
60. On the growth of medical disciplines in nineteenth-century Germany see: Hans-Heinz Eulner, *Die Entwicklung der medizinischen Spezialfächer an den Universitäten des Deutschen Sprachgebietes* (Stuttgart, FRG: Ferdinand Enke, 1970); Ben-David, *Scientist's Role in Society*, pp. 108–138; and A. Zloczower, *Career Opportunities and the Growth of Scientific Discovery in 19th Century Germany, With Special Reference to Physiology* (Jerusalem, Israel: Hebrew Univ., ca. 1960).
61. Information derived from Eulner, *Entwicklung der medizinischen Spezialfächer*, pp. 52, 54, 58, and 61.
62. William Henry Welch to Emma Welch Walcott (sister), 18 November 1876, Welch Papers, Johns Hopkins Medical Archives, Baltimore, MD (hereafter, Welch Papers).
63. On the development of facilities for physiology in Germany, see Eulner, *Entwicklung der medizinischen Spezialfächer*, pp. 46–65; also Rothschuh, *History of Physiology*, pp. 152–153.

64. W. James to Thomas W. Ward, ca. November 1867, *Letters*, p. 118; Ibid., W. James to H. P. Bowditch, 12 December 1867, p. 121.
65. Ibid., W. James to Henry James (father), 5 September 1867, p. 98; ibid., W. James to O. W. Holmes, Jr., 17 September 1867, p. 100.
66. Ibid., W. James to H. P. Bowditch, 12 December 1867, p. 121.
67. Ibid., W. James to Ward, ca. November 1867, p. 118.
68. H. P. Bowditch to L. O. Bowditch, 23 May 1869, Bowditch Papers.
69. Ibid., H. P. Bowditch to J. I. Bowditch, 19 July 1869.
70. H. P. Bowditch [letter from Leipzig, 8 March 1870], *Boston Med. Surg. J.* 82 (21 April 1870): 305–307.
71. Ibid., p. 306.
72. Ibid.
73. John Jacob Abel to Frances Hinman, 25 October 1884, Abel Papers, box 62.
74. Chittenden, "Sixty Years," pp. 22–23.
75. Warren P. Lombard, "The Life and Work of Carl Ludwig," *Science* 44 (1916): 363–375, quote on 369.
76. Bowditch, [letter from Leipzig], 307.
77. W. James to H. P. Bowditch, 12 December 1867, *Letters*, p. 121.
78. W. H. Welch to William Wickham Welch (father), 25 February 1877, Welch Papers.
79. On Ludwig see Heinz Schröer, *Carl Ludwig: Begründer der messenden Experimentalphysiologie, 1816–1895* (Stuttgart, FRG: Wissenschaftliche Verlagsgesellschaft, 1967); on the invention and development of the kymograph, esp. pp. 104–114.
80. Biographical information on Marey is neither plentiful nor certain. The trip to Germany before 1860, and Marey's acquaintance with Ludwig, are mentioned in the obituary in *Br. Med. J.* i (28 May 1904): 1289–1290. The anonymous writer mistakenly places Ludwig in Leipzig, although he did not move there from Vienna until 1865. Marey's major statement on the graphic method was *La Méthode Graphique dans les Sciences Expérimentales et Principalement en Physiologie et en Médicine* (Paris: Masson, 1878), although he had been developing many of the same ideas for almost two decades before 1878.
81. Ross, *Hall*, G. Stanley Hall to W. James, ca. December 1878, pp. 81–82.
82. H. P. Bowditch to L. O. Bowditch, 7 November 1869, Bowditch Papers.
83. Lombard, "Ludwig," 366.
84. Philip Bard to E. M. K. Geiling, 28 December 1961, Abel Papers, box 124.
85. Lombard, "Ludwig," 367.
86. W. H. Welch to W. W. Welch, 18 October 1876; W. H. Welch to E. W. Walcott, 18 November 1876, Welch Papers.
87. Sabin, *Mall*, pp. 56–59.
88. Hall, *Life and Confessions of a Psychologist* (New York: Appleton, 1923), p. 205.
89. Lombard, "Ludwig," 369; Davenport, "View from Michigan," pp. 52–54.
90. Mary Hinman Abel to Frances Hinman, 31 October 1884, Abel Papers.
91. Sabin, *Mall*, pp. 60–61.
92. Ibid., pp. 60–64.
93. Cited from quotation of letter from Ludwig to Bleile, n.d., in Anton W. Oelgoetz, "'Those Were Thrilling Days,'" *Ohio State Med. J.* 36 (1940): 873–877, citation on 873.
94. Lombard, "Ludwig," 369.
95. H. P. Bowditch to L. O. Bowditch, 26 September 1869, Bowditch Papers.
96. Lombard, "Ludwig," 368.
97. M. H. Abel to Katharine Daniels, 22 October 1884, Abel Papers, box 62.
98. H. P. Bowditch to L. O. Bowditch, 24 and 31 October 1869, Bowditch Papers.
99. H. P. Bowditch to L. O. Bowditch, 28 November 1869, Bowditch Papers.
100. W. H. Welch to E. W. Walcott, 18 November 1876, Welch Papers.
101. H. P. Bowditch to L. O. Bowditch, 13 March 1870, Bowditch Papers.
102. W. H. Welch to E. W. Walcott, 12 March 1877, Welch Papers.
103. Robert E. Kohler, *From Medical Chemistry to Biochemistry: The Making of a Biomedical Discipline* (Cambridge, UK: Cambridge Univ. Press, 1982), esp. pp. 93–120.
104. W. H. Welch to Emily S. Welch (mother), 20 July 1877, Welch Papers.
105. Cf. Welch's remark that, because an American "is something of a rarity in Breslau," he was receiving much more attention from Julius Cohnheim than might otherwise be the case, W. H. Welch to E. S. Welch, 15 May 1877, Welch Papers.

106. Cf. the testimonial from Ludwig for Bleile in Oelgoetz, "Thrilling Days," 875.
107. See Eulner, *Entwicklung der medizinischen Spezialfächer*, pp. 46–138.
108. A good introduction to the issues involved in American higher education in this period may be found in the essays in Alexandra Oleson and John Voss, eds., *The Organization of Knowledge in Modern America, 1860–1920* (Baltimore, MD: Johns Hopkins Univ. Press, 1979).
109. On Bowditch's negotiations, see Fye, "Why a Physiologist?" pp. 26–29.
110. For a participant's description of that laboratory, see Ellis, "Henry Pickering Bowditch."
111. For the development of laboratories at Harvard and Columbia, see chapter III by Alejandra C. Laszlo in this book.
112. See, for example, Bowditch's notebook kept in the summer of 1879 (CL-HMS GA 9), and the letters, numbering almost fifty, that he wrote back home to his wife in the summer of 1880 from Paris, Heidelberg, Würzburg, Leipzig, Berlin, Dresden, London, and Cambridge (CL-HMS C 5.1).
113. See K. J. Franklin, "A Short History of the International Congresses of Physiologists," *Ann. Sci.* 3 (1938): 241–335.
114. Ibid., 258–261; "The Triennial International Congress of Physiologists: Fourth Meeting," *Nature Lond.* 58 (15 September 1898): 481–486.
115. H. P. Bowditch to Selma Knauth Bowditch, 1 August 1880, Bowditch Papers.
116. Cf. Minot's reports on German biology as a way of stimulating emulation on the part of Americans: "The Study of Zoology in Germany. I. The Laboratories. II. The Methods Used in Histology and Embryology," *Am. Nat.* 11 (1877): 330–336, 392–406; On Bowditch, see Cannon, "Bowditch," 192–193; On Lee see chapter III by Laszlo in this book.
117. On Welch as a fund-raiser, see Flexner and Flexner, *Welch*, pp. 234–296; also Donald Fleming, *William H. Welch and the Rise of Modern Medicine* (Boston, MA: Little, Brown, 1954), pp. 131–202.
118. Dr. Long's phrase occurred in discussion at the conference at which these papers were read, National Library of Medicine, Bethesda, MD, January 1986.
119. H. P. Bowditch, "Über die Eigenthümlichkeiten der Reizbarkeit welch die Muskelfasern des Herzens zeigen," *Arbeiten aus der physiologischen Anstalt zu Leipzig* 6 (1871): 652–689; H. P. Bowditch, "Ueber den Nachweis der Unermüdlichkeit der Säugethiernerven," *Arch. Physiol.* (1890): 505–508.
120. Cannon, "Bowditch," 188–194. The range of Bowditch's activities may be seen very clearly in a reprint that gathers together all of his publications. I. Bernard Cohen, ed., *The Life and Writings of Henry Pickering Bowditch: An Original Anthology*, 2 vols. (New York: Arno, 1980).

II

Growth of American Physiology 1850–1900

W. BRUCE FYE

During the second half of the nineteenth century, American physiology emerged from the amorphous institutes of medicine to become, in Abraham Flexner's words, "the central discipline of the medical school."[1] This scientific discipline developed within the context of several interrelated reform movements in late nineteenth-century America. In this era, physiology was the critical link between the larger educational reform movements in America and the efforts to elevate the standards of medical education and make medicine more scientific. I explore some of the more important individual, institutional, and social factors that contributed to the professionalization of American physiology. Although physiology remained critically dependent on medicine throughout the nineteenth century, its devotees, among the most vocal proponents of medical school reform, became increasingly influential. Physiologists, along with other basic medical scientists, ultimately became the spokesmen for scientific medicine on this continent and helped shape the current structure and philosophy of medical education. To this group we owe the introduction of the full-time faculty concept and what I have chosen to call the research ethic into American medical education. By the research ethic, I refer to the belief that a medical teacher has an obligation to society to attempt to expand medical knowledge through research.[2]

Although Americans made occasional contributions to physiology before the middle of the nineteenth century, their efforts were inspired by individual curiosity, required little in the way of scientific apparatus or assistants, and were usually undertaken in isolation.[3] As with most scientists in antebellum America, the few individuals who pursued biomedical research had no consistent source of funding for their studies. With very few exceptions there was no organized effort to support research. The scientists did the best they could; those without means often developed elaborate and innovative methods to subsidize their investigations. Virtually all American scientists in antebellum America were amateurs in the sense that they did not earn their living from the prosecution of science. To be sure, some held teaching positions in institutions of higher learning, but they were paid to teach, not to do research. Most biomedical research (and there was relatively little published in America during the first half of the century) was undertaken by men who were active medical practitioners. Income from medical practice was used to subsidize their modest experimental studies. Although revisionist historians have demonstrated that antebellum Americans were not totally indifferent to research, that science was being taught in many colleges, and that the so-called warfare of science and religion was more rhetoric than reality, the opportunities were not really all that good for the handful of individuals who wanted to do physiological research.[4]

During the second half of the century the situation gradually improved, and eventually there were opportunities for individuals to devote themselves exclusively to teaching and research in physiology, and to be paid for it. This transformation was gradual, however, and resulted from a complex series of interrelated developments. To illustrate the diverse factors that contributed to the professionalization of American physiology, I first briefly consider the careers of four pioneers of this scientific field: John Call Dalton, Jr., S. Weir Mitchell, Henry P. Bowditch, and H. Newell Martin, each of whom contributed something unique to the development of physiology in this country. After reviewing some of their achievements, I describe several of the factors that led to the widespread acceptance of the scientific methods of teaching and research these men advocated.

American physiology as we now think of it did not develop de novo; like much of American culture it was imported from Europe and then shaped by forces unique to this nation. Throughout the nineteenth century, American physicians went to Europe for postgraduate training.[5] As the century progressed, their goals changed as did the cities and institutions they visited. By midcentury a few had been exposed to Claude Bernard, who was emerging as the leader of French physiology.[6] Pursuing the agenda of his mentor François Magendie, Bernard emphasized the critical role of experimentation in medical research. He argued that observation and speculation were not meaningful ways to elucidate the complex functions of the living organism. Experimentation, with heavy emphasis on vivisection, was the only true method to replace theory with fact. Two young American medical graduates, John Call Dalton, Jr. and S. Weir Mitchell, were exposed to this new philosophy when they studied under Bernard around 1850. Each was inspired by the French physiologist and sought to introduce something of his approach into American physiology.

Dalton, the son of a prominent physician, could have been a successful practitioner in Boston but chose to devote himself to a career in physiology and, in Mitchell's words, became "our first professional physiologist."[7] There was no ready-made opportunity for Dalton to devote himself exclusively to physiology upon his return to America, however. When Dalton graduated from the Harvard Medical School in 1847 there were fewer than 200 medical professors in America to teach the nearly 4,000 pupils enrolled in the nation's thirty-three regular medical schools.[8] Virtually all of the medical professors were practicing physicians who devoted a few hours a week to their lectures. Upon his return from Paris, Dalton took teaching positions in Boston, Buffalo, and in Woodstock, Vermont, before he was elected to the chair of physiology and microscopic anatomy at the College of Physicians and Surgeons of New

York in 1855.[9] Dalton had been impressed by the value of demonstrations and vivisections in Bernard's lectures, and he introduced this method of teaching physiology at each of the institutions where he taught upon his return from Paris. This was the first time vivisection was used in medical instruction in this country and, as had been the case in France, it proved to be quite popular with the students. In addition to being a lecturer, Dalton was an enthusiastic researcher and a prolific author. Like earlier scientifically oriented physicians in America, he was a lone investigator, however. There is no evidence that he encouraged pupils or colleagues to undertake research in his modest private laboratory at the College of Physicians and Surgeons.

Mitchell, another of Bernard's American pupils, returned from Paris at about the same time as Dalton. Mitchell, the son of a prominent Philadelphia physician, medical teacher, and scientist, had been introduced to physiology at Jefferson Medical College by Robley Dunglison, one of America's few full-time medical teachers.[10] Dunglison had performed research and published several successful medical works including a comprehensive textbook of physiology. As Dalton had done in New York, Mitchell introduced vivisection into the teaching of physiology in Philadelphia when he taught at a private medical school known as the Summer Association from 1856 to 1860. The school closed at the onset of the Civil War, however, and Mitchell was left without an institution in which to continue his physiological activities. He had already published several significant papers on the toxicology of venom and related subjects and was recognized worldwide as one of America's most productive and most promising medical investigators.

Mitchell aggressively sought the chair of the institutes of medicine at the University of Pennsylvania and the Jefferson Medical College in Philadelphia in the 1860s. In both cases Mitchell failed to win the post, however. Despite strong letters of support from leading scientists, including Joseph Henry, Jeffries Wyman, Oliver Wendell Holmes, and Louis Agassiz, Mitchell lost the chair at Pennsylvania in 1863 to Francis Gurney Smith, Jr. Smith had some claim to the chair, however; he had taught physiology at several Philadelphia institutions for more than a decade. Both Smith and Mitchell were active medical practitioners. The biggest difference between the two candidates was Mitchell's deep commitment to, and demonstrated success in, original research. Mitchell and Henry discussed the significance of the research ethic in the context of the Pennsylvania canvass. Henry declared Mitchell's defeat unfavorable not only for the young physiologist, "but also [for] the cause of American science."[11] The lay board of trustees had no appreciation of the research ethic in an era when America's professors were expected to transmit knowledge rather than produce it.

Frustrated by his defeat, Mitchell devoted himself to the new demands and opportunities created by the Civil War. When his former teacher Dunglison resigned in 1868, Mitchell saw another chance to regain an institutional base for his physiological activities. This time his opponent had more formidable scientific credentials. James Aitken Meigs, Mitchell's main rival for the Jefferson chair, was a busy practitioner but had made several valuable contributions to ethnology and anthropology.[12] He had been an assistant to Francis Gurney Smith, Jr. and had also taught physiology in various Philadelphia institutions.

Although Mitchell and his supporters tried to point out that Meigs' scientific interests were not focused on experimental physiology as it was now defined by leaders in European science, their arguments were not persuasive. Meigs won the contest, and Mitchell informed Wyman, "Dr. Meigs by whom I have been defeated is not as you know a physiologist so that as you infer the contest turned on other than relative claims as biologists."[13]

Mitchell attributed his defeat to social and political pressure brought to bear upon the Jefferson trustees. He had claimed that his earlier defeat to Smith had resulted from similar factors, and in both cases there is evidence to support Mitchell's interpretation of the events. Physiology was still broadly defined in America, however. In most American medical schools it was still taught as part of the institutes of medicine, an eighteenth-century relic. Occasionally the interests of the professor led to some anomalous unions, and physiology was joined with some unlikely companions such as the diseases of women and children.[14] It is unfair, therefore, to be too harsh on the trustees at Pennsylvania and Jefferson in the 1860s. Although they had been informed of the new direction physiology was taking in Europe by Mitchell and his scientific supporters, this was as yet of little relevance in the United States.[15]

Mitchell viewed the Jefferson chair as his final opportunity to regain an institutional focus for his physiological interests. Most successful American physicians were unwilling to risk relocating their medical practices in this era of provincialism. Moreover, most of the established medical schools appointed well-known local practitioners to their chairs, rather than outsiders. Mitchell, a Philadelphia native, had inherited his father's busy medical practice and had a family to support. Unlike Dalton, a bachelor who had chosen not to practice medicine, Mitchell was not free to relocate simply to improve his prospects for a career in physiology. The Civil War afforded Mitchell an opportunity to devote himself to neurological disorders. Many wounded soldiers suffered from nervous and mental diseases, and with the encouragement of his friend and former colleague William Hammond, the new surgeon general, Mitchell was given a ward for neurological cases at the Turner's Lane military hospital in Philadelphia. This led to a significant career shift that was reinforced by his failure to win a physiology chair. Mitchell retained his deep devotion to physiology, however, and two decades later would propose the formation of the American Physiological Society (APS). Nevertheless, Philadelphia lost an opportunity to take the lead in introducing modern experimental physiology into America through Mitchell.

Experimental physiology in America suffered another setback in the 1860s. Dalton and Mitchell had used vivisection in their teaching and research for more than a decade before the practice was openly attacked as being cruel to animals. In 1866, however, Henry Bergh inaugurated the antivivisection movement in America. A wealthy New Yorker, Bergh had recently returned from Europe outraged over the cruel treatment of animals he had witnessed abroad. He was inspired by the activities of the British Royal Society for the Prevention of Cruelty to Animals to initiate a similar reform movement in America.[16] Animal experimentation was an early and obvious target and Dalton bore the brunt of the attack. In the first of many eloquent and persuasive responses to the antivivisectionists, Dalton sought to demonstrate to a broad audience the benefits to mankind that had resulted from vivisection.[17] Dalton included the discovery of the circulation, the development of techniques for artificial respiration and blood transfusion, and the recognition of the function of the spinal nerves in his list of the benefits to mankind that had resulted from vivisection. This theme was picked up by many medical editors and correspondents and the American medical profession began hearing about the utility of experimental physiology. Dalton could not depend on the handful of individuals who performed vivisection in America to help him defend the practice, however. He sought to form a coalition and looked to organized medicine to lend its support to this practice. The New York physiologist soon managed to secure the support of local medical societies as well as the American Medical Association (AMA) in defense of vivisection in research.

This was the first time the American medical profession was formally asked to lend its support to the research ethic. They paid lip service to it by passing resolutions declaring the importance of animal experimentation in the advance of medical science and practice. The antivivisectionists were persistent, however, and Dalton expended much time and energy throughout his career combating their efforts to limit or halt the practice. It is possible that any plans Dalton might have had to expand his physiology program at the College of Physicians and Surgeons to include laboratory work for students were affected by the antivivisectionist movement. Dalton continued to use the technique in demonstrations and in his research but there was no attempt to introduce practical work using vivisection at the college in this era.

An important event in the history of American physiology and medical education occurred in 1871 when Henry Pickering Bowditch returned from Carl Ludwig's Physiological Institute in Leipzig to found the first physiology laboratory in America open to advanced workers. Holmes had taught anatomy and physiology at the Harvard Medical School since 1848, but the poet-anatomist disliked vivisection and routinely emphasized anatomical teaching at the expense of physiology.[18] The peripatetic French physiologist Charles Edouard Brown-Séquard had been appointed to the newly created chair of physiology and pathology of the nervous system at Harvard in 1866. Brown-Séquard was a popular lecturer and renowned investigator who enthusiastically proclaimed the benefits of experimental physiology wherever he taught. Although Bowditch's interest in anatomy and physiology can be traced in part to his former teacher Wyman, one of America's leading naturalists, Brown-Séquard inspired the young Harvard medical student's enthusiasm for the new European interpretation of experimental physiology.

After graduation from Harvard Medical School, young Bowditch traveled abroad to study with Brown-Séquard who had recently returned to Paris. When this proved impossible, Bowditch took courses from Bernard and worked in his small, damp laboratory. Bowditch had heard of the large and well-equipped German laboratories of physiology from his friend and former classmate William James who had preceded him to Europe for postgraduate study. Increasingly dissatisfied with his opportunities in Paris, Bowditch outlined a course of physiological study in the German universities with the help of Wilhelm Kühne, a Continental physiologist who had worked in several

European laboratories. Both James and Kühne advised young Bowditch to work with Ludwig, widely acknowledged as the leader of German experimental physiology. Before long, he was at work in the Leipzig Physiological Institute under Ludwig's direction.

Bowditch was aware of the difficulty of finding a position in America where he could devote himself to physiology. He received encouragement from several sources, however. His family, former teachers Wyman and Holmes, and his friend James all told Bowditch to persist in his desire to become a full-time physiologist. During the two years Bowditch was abroad, important changes took place at the Harvard Medical School that provided him with the opportunity he sought. In 1869 Charles Eliot, a chemist and educator, was elected president of Harvard to the surprise of many observers.[19] Only thirty-five years old when he became Harvard's president, Eliot had recently spent two years abroad where he had the chance to study European approaches to higher education. Shortly after his election it became apparent to everyone concerned that Eliot planned sweeping reforms at Harvard. His efforts would not be limited to the Cambridge campus; he wanted a voice in the affairs of the Harvard Medical School as well. Located in Boston, the medical school had traditionally been controlled by its faculty, comprised primarily of influential Boston practitioners. This was to be no longer; Eliot insisted that the university take control of the affairs of this school.[20]

Eliot had taught Bowditch when he was a student at the Lawrence Scientific School before the Civil War. Eliot was kept informed of Bowditch's progress in Europe by his supporters who hoped that a position might be found for him at Harvard upon his return. At this time Harvard had more full-time scientists on its staff than any other American institution of higher learning, and several of them had medical degrees but had given up practice to devote themselves to scientific careers. Among Harvard's professional scientists who held medical degrees were Oliver Wendell Holmes, Jeffries Wyman, Louis Agassiz, and Asa Gray. In addition to these role models, Bowditch came from a family with prominent scientists on both his mother's and father's sides. His prospects at Harvard were also aided by the influential positions held by his father and uncle. His father, a wealthy businessman, was a member of the board of overseers of the university, and his uncle was one of Boston's most prominent medical practitioners and a former member of the medical school faculty.

Although Eliot tried to get Bowditch to return to lecture at Harvard, the young physiologist refused to leave Ludwig's laboratory prematurely.[21] When Bowditch finally accepted an offer from Eliot to become assistant professor of physiology in the medical faculty, it was with the understanding that physiology would become an independent department and that he would play a role in reforming the medical curriculum. Bowditch returned to Boston in the summer of 1871 with a collection of apparatus to establish a modest laboratory of experimental physiology in the attic of the medical school. Just as he had subsidized his son's European education, Bowditch's father paid for most of his apparatus and supplemented his Harvard salary.

Before long, a number of scientifically inclined young physicians and graduate students were at work on a wide variety of problems in Bowditch's

Harvard laboratory. Although it was small by German standards, this facility represented a dramatic departure in American physiology. A medical school in this country now had a laboratory where those interested in learning the new European approach to experimental medicine could go not only to see demonstrations of experimental physiology, but to learn these techniques and perform original research. Although Harvard medical students were still taught mainly by illustrated lectures and demonstrations, senior students could, for a fee, undertake research under Bowditch's direction. It would be several years before other American medical schools offered these opportunities. Harvard was first because of a unique set of individual and institutional circumstances that facilitated the introduction of the European interpretation of experimental physiology into an American medical school.

The significance of the establishment of the Harvard Physiology Laboratory and its dependence upon favorable individual and institutional circumstances is emphasized by the contrast between Bowditch's and Mitchell's attempts to find an opportunity to devote themselves to physiology. Just three years before Bowditch returned from Ludwig's laboratory, Mitchell had explained the significance of his loss of the Jefferson chair to Wyman:

> Of course I felt bitterly the defeat of certain intellectual ambitions, but most of all the fact that I must continue to work as hard as ever at practice and so see it year after year encroaching on the little time I can give to science. Had I gotten this chair I meant to organize a physiological laboratory with a paid assistant and to have spent on this scheme at least a thousand a year. I could thus have accomplished a great deal of systematic research which is out of my power so long as only the afternoons of summer can be given to work.[22]

Mitchell's plan reflected his dedication to research but did not suggest a commitment to training others to undertake original investigations. Bowditch, emulating his teacher Ludwig, would encourage advanced workers to enter his laboratory for training and to initiate research projects.

By the early 1870s those interested in reforming higher education in America generally acknowledged the importance of endowments in achieving their goals. Those who were skeptical of the value of endowments were convinced of their merits by the creation and early progress of the Johns Hopkins University.[23] Here, America's second physiology laboratory would be organized under circumstances that were, in several respects, even more favorable than those that facilitated Bowditch's efforts at Harvard. The Johns Hopkins University and the Johns Hopkins Hospital were established by a bequest of $7 million by a Quaker merchant whose interest in endowing an educational institution had been stimulated by his friend George Peabody. Not only was the Johns Hopkins University the wealthiest institution of higher education in America, it had no traditions to overcome. It could adopt the newer principles of higher education that were proving so successful in Europe, where government subsidies permitted innovations too expensive for America's poorly endowed colleges and so-called universities.

Recognizing their opportunity, the Johns Hopkins trustees studied the recent American and European literature on higher education and sought the opinions of the leaders of America's most progressive universities. They consulted three young university presidents: Eliot of Harvard, James B. Angell of the University of Michigan, and Andrew D. White of Cornell. Each of these men advocated a curriculum that emphasized science and encouraged the trustees to select Daniel Coit Gilman as president of their new university. Shortly after graduating from Yale, Gilman had traveled abroad as an attaché of the American legation in St. Petersburg, Russia. While abroad, he had been encouraged to make a study of the universities and technical schools of Europe by Henry Barnard, an 1830 Yale graduate and leading educational reformer.[24] After returning to America, Gilman helped his former teacher James Dana coordinate a fund-raising effort to endow the scientific school that had been established at Yale. After a brief and somewhat frustrating tenure as president of the University of California, Gilman came to Baltimore in 1875 to organize the Johns Hopkins University.

Gilman enthusiastically began his work in Baltimore and, following the precedent established by the trustees, consulted several leaders of education and science in America and abroad. It quickly became apparent that suitable professors for the scientific branches at the university might not be found in the United States. Gilman was seeking progressive teachers who were familiar with the newer methods of instruction used in Europe. Moreover, he wanted specialists who were capable of, and deeply committed to, research in their chosen fields. While abroad Gilman elicited suggestions for potential faculty members. Thomas Henry Huxley, one of England's leading scientists and its most vocal spokesman for the cause of scientific education and research, encouraged Gilman to hire his assistant, Martin.[25] Martin, an Irishman, had studied at University College, London, where his devotion to science was inspired by William Sharpey and Michael Foster. When Foster went to Cambridge to establish a laboratory of experimental physiology he took Martin with him as his assistant.[26] Although his teaching responsibilities had prevented Martin from performing much research, the young physiologist had collaborated with Huxley in the preparation of a popular guide to laboratory work in biology and he was highly recommended by Huxley and Foster. After negotiations, which included a demand for a higher salary, the promise of a suitable laboratory, and a voice in the organization of the curriculum, Martin accepted the professorship of biology at Johns Hopkins. He departed for America in the summer of 1876 accompanied by Huxley, who was going to deliver a series of lectures in America, including the keynote address at the opening of Johns Hopkins.

The special circumstances at Johns Hopkins made it possible for Martin to establish a successful program of advanced training and research in a remarkably short period of time. He was given funds for state-of-the-art apparatus, and the Hopkins' program of fellowships made it possible for him to attract highly qualified graduate students. Gilman viewed the fellowship system as a critical component of the Hopkins philosophy of graduate education and research. By providing economic support the institution could attract bright and energetic scholars who might otherwise have been unable to pursue graduate work. In less than a decade Martin assembled an outstanding group of advanced pupils and junior faculty who were productive investigators. With their help he developed a unique isolated mammalian heart preparation that was widely used and ultimately led to many important discoveries in cardiac physiology. Martin's research program was acclaimed by educators, scientists, and scientifically oriented physicians. As readers of the *Medical Record* were told in 1881, "opportunities for physiological and morphological study and research at the university in question are probably the best to be found in this country. And it is a centre which has already shown evidences of good scientific work. The medical profession should feel gratified that opportunities for such original work now exist in this country."[27]

There were now two institutions in America where individuals interested in careers in biomedical science could study under teachers trained in modern experimental physiology. Bowditch and Martin encouraged their most promising students to complement their Harvard or Hopkins educations with further postgraduate study abroad. When they returned to America, however, their pupils found few opportunities for employment as biomedical scientists. This underscores the unique circumstances that allowed Bowditch and Martin to establish their pioneering programs in physiology in the 1870s. The job market for basic medical scientists grew slowly but steadily in the 1880s. Addressing delegates to the International Medical Congress in London in 1881, John Shaw Billings said

> Seven years ago, Professor Huxley declared that, if a student in his own branch showed power and originality, he dared not advise him to adopt a scientific career, for he could not give him the assurance that any amount of proficiency in the biological sciences would be convertible into the most modest bread and cheese. To-day, I think he might be bolder, for such a fear would hardly be justifiable—at all events, in America—where such a man as is referred to could almost certainly find a place, bearing in mind the professor's remark that it is no impediment to an original investigator to devote a moderate portion of his time to giving instruction either in the laboratory or in the lecture-room.[28]

When the University of Michigan, where emphasis was placed on the scientific curriculum, sought a physiologist in 1881, Martin and Bowditch both proposed candidates for the post. Charles Minot was Bowditch's first pupil to be awarded the doctorate in science at Harvard, and Henry Sewall was Martin's first pupil to receive a doctorate degree from Johns Hopkins. Both candidates had studied abroad and came highly recommended by their teachers. Victor Vaughan of the University of Michigan later recalled that "Professor Martin took a deep personal interest in the matter. He was intent not only on building up his own department in Baltimore, but in seeing the seeds of the new school of physiology then just coming into bloom in England . . . planted and nourished in the universities of America. . . . I then regarded that movement as the first real awakening of scientific medicine, and I have since had no occasion to reverse my opinion on the matter."[29]

Sewall was chosen over Minot and established a physiological laboratory at the University of Michigan based upon the model Martin had developed in Baltimore. Unlike Bowditch, whose family supplemented the modest budget for the Harvard physiology department, or Martin, who received consistent and liberal support from the Hopkins trustees, Sewall had to deal with the economic realities of an institution financed by a state government. Indeed, Sewall's appointment at Michigan was delayed a year because of an unexpected budget deficit at the university. Shortly after his arrival in Ann Arbor, he informed Gilman, "I am making it my business at present to try & show Faculty, Regents & Legislature that the physiology department is a worthy beggar."[30]

In the half century since William Beaumont had published his classic book on digestive function, the methods of experimental physiology had evolved dramatically. Beaumont's pioneering studies of his subject Alexis St. Martin were performed under primitive conditions, without instruments of precision, and in isolation. By 1880 it was no longer possible to undertake significant physiological research in this manner. The requirements for experimental physiology had been redefined in the European university-based physiological institutes: a laboratory, elaborate apparatus, and assistants were now required. Sophisticated research in physiology, however, also required specialized training, time, and money in addition to interest, ingenuity, and manual dexterity. Practicing physicians had neither time, nor training, nor access to laboratories; they could not hope to compete with the full-time European medical scientists. German physiologists were provided with facilities, assistants, and adequate salaries that allowed them to devote their whole time to teaching and research. It was this combination that led to their remarkable productivity.

The defects of the American system of medical education were well known. They had been described, and reforms suggested, for nearly half a century.[31] The nation's centennial stimulated reflection, introspection, and anticipation. In this context, the state of American science, higher education, medical practice, and medical teaching were all scrutinized and their prospects assessed. The American Academy of Medicine and the Association of American Medical Colleges were formed in 1876 in an attempt to improve the quality of medical education. These organizations achieved few of their goals, however. The most critical factor retarding the reforms had become apparent as had the solution. As long as medical schools were dependent solely on student fees, little could be accomplished.

The pupils of Martin and Bowditch entering the job market in the 1880s hoped to find positions where they could teach the new laboratory methods and do research. This was impossible in virtually all of America's medical schools; it was simply too costly. Very few of America's medical schools had any endowment in this era. Short didactic courses were most suitable for busy practitioners who sold tickets to students who wanted to attend their lectures. The larger the enrollment in a professor's course, the more income he derived from his medical teaching. The practitioner-teachers had no incentive to introduce expensive and time-consuming laboratory training. Only when an alternative method of paying for medical education evolved would it be possible to introduce laboratory training and other curricular reforms into the nation's medical schools. The few full-time biomedical scientists, scientifically oriented practitioners, faculty members, and educators agreed—the only answer was endowment. Throughout the nation, reformers pointed to the success of the Johns Hopkins University to support their claim that endowment was the key to elevating American institutions of higher learning to a level comparable with the best European universities. An American model already existed, argued the reformers, now it had to be generalized to other institutions throughout the country. This movement, initiated in the universities, began to extend to the nation's better medical schools as well.

During the closing decades of the nineteenth century there was a gradual but distinct increase in support for the notion that American medicine had to become more scientific. It was in this context that physiology achieved greater significance than ever before in medical education and medical practice. The French physiologist Bernard had declared in 1865 that "experimental physiology is the most scientific part of medicine, and . . . by studying it, young physicians will acquire scientific habits."[32] Anatomy would not serve Bernard's purpose, or that of the reformers who sought to implement his philosophy in America; it was a descriptive, not an experimental science. In Bernard's view, scientific medicine was synonymous with experimental medicine and this could only be established by "spreading the scientific spirit more and more among physicians."[33] Twenty years ago Owsei Temkin claimed, "During the last hundred years, medicine has come to place an almost unprecedented reliance on physiology. This juncture was reached when experimental physiology became the foundation of experimental medicine."[34]

America's physiologists were active participants in the movement to gain endowment for medical education. Dalton was a leader of the effort to establish a $100,000 endowment at the College of Physicians and Surgeons of New York in 1874 to build laboratories at the school. Ironically, Mitchell was elected to the board of trustees at the University of Pennsylvania in 1875, little more than a decade after that body had refused to elect him to its physiology chair. As a university trustee he led the movement to reform the medical curriculum at Pennsylvania and to establish an endowment for laboratory facilities. In each instance these efforts were directed at providing laboratories for limited practical training of medical students.

The development of bacteriology was a major stimulus for laboratory training in medical schools in the 1880s and 1890s. Bacteriology provided the reformers who sought to make medicine more scientific with a persuasive argument for placing greater emphasis on science in the medical curriculum and for encouraging research in American medical schools. The examples Dalton had used to demonstrate the benefits of experimentation to mankind in his rebuttal to the antivivisectionists were, for the most part, rather ephemeral in the utilitarian atmosphere of America. The discovery of the circulation of the blood and the function of the spinal nerves, however significant as advances in fundamental knowledge, had little relevance to the practice of medicine a century ago. The majority of American physicians were poorly trained and more concerned about remaining competitive in an increasingly glutted medical market than they were about abstract scientific discoveries.[35] All of this changed with the advent of bacteriology, however. In the 1880s the medical world was plunged into an era of "bacteriomania," a term chosen by Abraham Jacobi to describe the extraordinary response of the medical profession to the microscopic identification of bacteria and related developments in microbiology in this era.[36] After a brief period of skepticism it was widely acknowledged that the germ theory of disease had been proven by scientific investigation. What made this discovery special, however, was its perceived practical value. Although specific cures of bacterial diseases were not available, enthusiasts claimed these would soon follow; more pragmatic individuals used the knowledge at hand to institute public health and sanitary reforms.[37] It was now easier to convince the profession and the public of the value of medical science. As Russell Maulitz has said, "More than any other laboratory domain, bacteriology spoke to the clinician."[38] The scientific reformers' argument was no longer based merely on idealism; they could demonstrate how research could directly benefit the public and the profession.

Bacteriology was not the only science that promised practical rewards to physicians and their patients a century ago. American physicians had never been reluctant to administer drugs to their patients. Although the trend away from heroic therapy toward a more conservative approach steadily gained in popularity during the middle of the nineteenth century, most agents were administered without any proof of their efficacy. At least the heroic therapies had been dramatic; this was not the case with many of the agents dispensed during the second half of the nineteenth century. One medicine that did produce dramatic effects was nitroglycerin. Its universal side effect of headache suggested that it was a potent drug and its prompt relief of angina pectoris revealed its value as a specific treatment for that distressing symptom. Advocates of scientific medicine could point to the discovery of the efficacy of this agent in the treatment of angina as an example of the benefits of research to practice. Shortly after the British physician and pharmacologist William Murrell published his report describing the use of this agent for angina, an enthusiastic editor declared, "the applications of Nitro-Glycerin to the treatment of disease are directly deducible from physiological study—another proof, if more were needed, of the value of the physiological method."[39]

Bacteriology and therapeutics were merged later in the century when diphtheria and tetanus antitoxins were developed by Emil von Bering. Immunology, according to Frederick P. Gay, was more dependent on animal experimentation than any other medical science. Reviewing the events of the late nineteenth and early twentieth centuries he explained, "In view of the present and potential value of applied immunology in the diagnosis, the prevention, and the cure of human and animal disease, skilled animal experimentation necessitates the encouragement of every humanitarian."[40]

Ordinary American doctors, who were facing increasing competition from sectarians in addition to the hoards of new graduates flowing from the nation's regular medical schools, were told that science could undermine the credibility of the quacks and bring patients into their offices. Alonzo Palmer, dean of the medical faculty at the University of Michigan, declared in 1882, "In view of the present state and drift of science . . . how absurd become the dreams of the exclusive systems; particularly of this pretended universal therapeutical principle of 'like-cure' [i.e., homeopathy] and its dependent doctrines. Such unfounded speculations, originating in the darkness of the past and kept alive by the visionary or interested of the present, must recede more and more into the obscurity of past follies as science advances."[41] In a popular guide on how to be a successful physician, Webster Cathell advised his readers, "If, at your office and elsewhere, you make use of instruments of precision—the stethoscope, ophthalmoscope, laryngoscope, the clinical thermometer, magnifying glass and microscope, make urinary analyses, etc., they will not only assist you in diagnosis, but will also aid you greatly in curing people by heightening their confidence in you and eliciting their co-operation."[42]

American nationalism was also exploited in an attempt to encourage the support of scientific research. American physicians had been criticized by the medical reformers for their lack of interest in research for many years. Mitchell declared in 1858, "considering the adventurous ingenuity of the national mind, as well as the solid character of its achievements in other lines of scientific research, it is not easy to see why a science so eminently experimental as physiology, should be able to boast so few active laborers and so small a number of conspicuous results."[43] In the quarter century after Mitchell made this statement, hundreds, even thousands of American medical graduates made the pilgrimage to Europe where they witnessed the elaborate system of medical training, the emphasis on specialization, and the enthusiasm for research that characterized the German university-based medical schools and clinics.[44] More and more returning Americans grew impatient with the state of medical education and practice in their own country. The New York ophthalmologist Henry Noyes declared in 1885, "During the past year, it has been permitted me to see many medical men and medical institutions in various parts of Europe. One gets a new impulse from such an opportunity, and one may be lifted to a wider horizon in the field of medicine. . . . The impression which, to me, has come uppermost, is that the line of advance must hereafter be through acute, exacting, and deep research."[45] European scientists and clinicians were developing new diagnostic techniques, instruments, and operative procedures at an extraordinary rate, and the call for Americans to participate in the new scientific medicine grew louder.[46]

The reformers, comprised of the few full-time basic medical scientists, several scientifically oriented physicians, medical editors, and educators, argued that the nation's medical schools had to adopt two new concepts if Americans were to become competitive in the increasingly sophisticated world of scientific medicine. First, the teachers of the scientific branches in the medical schools, like their German counterparts, must devote all of their time to their chosen scientific field. Second, they should do more than simply teach. They had an obligation, as did the institutions in which they were employed, to advance medical knowledge through research. Endowments could be used to pay the salaries of the full-time faculty and build the laboratories required for research and the practical training of the students. Young American physiologists who had seen the European system were aware of the financial and philosophical aspects of the problem. Few were as fortunate as Bowditch, whose independent wealth removed economic concerns from his decision to become a physiologist. Writing to his former physiology professor at St. Louis Medical College, Missouri, William Townsend Porter explained in 1890:

> I wish to become a professional physiologist. . . . There can be no question that I can be more useful in St. Louis as a physiologist than as a practitioner. The making of a physiological institute in our community is worth living for. It is not possible to succeed in such an undertaking and to succeed in practice at the same time. To practice medicine and experimental physiology is to be an amateur in two things. I must make a choice. I must be one thing or the other. Physiology means absolute poverty for some years, comparative poverty during life. Practice means giving up the best things in sight for the sake of material comforts. These are the horns of the dilemma. I believe that I have chosen wisely.[47]

The founding of the APS was a signal to the world that a new scientific discipline had emerged in America. As was the case with the recently formed American Society of Naturalists—Martin and Bowditch were members of both societies—the APS required candidates for membership to have a commitment to research.[48] This soon became synonymous with being a professional physiologist. The slow growth of the society reflected both the unwillingness of the officers and members to admit amateurs and the small number of professional biomedical scientists in America in this era. Frederick Ellis, one of the founders, recalled, "In those days all the real physiologists in the country could be counted on the fingers and toes, and a number of the gentlemen present at the first meeting did not claim to be physiologists, but were interested in closely related subjects."[49] Physiologists viewed themselves as scientists and sought to emulate other scientific societies. This was not the case with the majority of America's anatomists in this era. Their society was larger and grew more quickly than the physiologists' because they did not emphasize the research ethic, nor did they attempt to restrict their society to full-time teachers. Three years after its organization in 1888, according to Charles Bardeen, the Association of American Anatomists consisted of "84 members. Of these 44 were professors and instructors in various medical schools, scarcely half a dozen of whom could properly be called scientific investigators."[50] Many of the individuals elected to the American Physiological Society were, or would shortly become, leading advocates for reform in medical education and spokesmen for scientific medicine. In addition to Mitchell, Bowditch, and Martin, Russell Chittenden, William Howell, Warren Lombard, Charles Minot, Victor Vaughan, and William Welch were among the original members.

The annual meetings of the society provided an opportunity for this small group of predominantly full-time biomedical scientists to discuss their research, visit local laboratory facilities, witness demonstrations of new apparatus or techniques, and hear their colleagues present the results of their investigations. The social and political aspects of the meetings were as important as the scientific agenda. Martin, Mitchell, Bowditch, and Dalton's successor John Curtis had all been threatened by Bergh and his disciples. The APS could coordinate the response of the experimentalists to the continuing attempts to restrict or stop animal experimentation. The society also served as a clearinghouse for jobs. Membership in the society was used as a marker signifying a commitment to scientific medicine and research. When Sewall wrote to the University of Michigan Board of Regents to suggest Joseph Warren as his substitute during a leave of absence he explained, "Dr. Warren is well known as a physiologist in this country and abroad. He is one of the original members of the American Physiological Society. . . . [and] is the author of numerous original papers."[51] An informal network of biomedical scientists had existed for several years and had been based primarily upon common institutional experiences in America or in Europe. Now this network was more formalized and it became increasingly effective.

Full-time physiologists often became influential faculty members in their medical schools—in part because they were not busy in private practice, unlike their clinical counterparts. When Eliot encouraged Francis Minot, the Hersey

Professor of Medicine at Harvard, to become dean of the medical faculty in 1883, he declined and suggested that Bowditch be selected. Minot explained to Eliot:

> I think Dr. Bowditch would make an admirable Dean. He is well versed in the affairs of the school, is methodical and industrious, and will be at the college a large part of the day and accessible to students and others. Moreover he teaches the first class, and has thereby a better acquaintance with the whole body of students than one who merely lectures to the third year men. I am most gratified by your expressed preference for me to fill such an important office, but I am very unwilling to take it unless it is absolutely necessary for the welfare of the school. I am sure Dr. Bowditch is much better qualified for it than I am. While the incessant interruptions incident to my practice would seriously interfere with the punctual discharge of the duties of a dean.[52]

America's physiologists were busy as well, but in their laboratories. As their numbers grew and facilities for research improved, American biomedical scientists became more productive and made more frequent contributions to the scientific literature.[53] When it appeared in 1896, the *American Textbook of Physiology*, edited by Howell, contained contributions by America's leading physiologists, all of whom were active members of the APS.[54] Increasing specialization within the biomedical sciences and growing scientific productivity led to the establishment of the *American Journal of Physiology* in 1898. The American physiologists were now being acknowledged by European medical scientists who read their publications and met them at meetings. The Americans were pleased with this recognition and some actively sought to strengthen their ties with their peers in England and Europe. Informing his former physiology teacher Gustav Baumgarten about the upcoming International Physiological Congress in Bern in 1895, Porter, who now was an associate professor of physiology at Harvard, claimed, "Doctor Bowditch is going over and I suppose I must too—it is a bad idea to stay away from those things."[55]

By the early twentieth century more and more physicians joined the ranks of those demanding that American medicine become more scientific. The reformers wanted additional full-time faculty positions for biomedical scientists, more and better equipped laboratories, and larger endowments. The rhetoric in support of the research ethic became stronger and came from a broader constituency. William Keen, a colleague of Mitchell's during the Civil War, proclaimed the accomplishments of scientific medicine to the members of the AMA in his presidential address delivered in 1900. Keen insisted, however, that more could be done and that it was necessary to reform medical education in order to accomplish this goal:

> The function of a medical school, as I have said, should not be limited merely to the imparting of existing knowledge. No school is worthy of the name that does not provide for greater or less research work, by which substantial additions to our knowledge may be made and the facilities and the results of the healing art made more efficient for the welfare of mankind. Twenty-five years ago there were practically few young men who were fitted for research work, especially laboratory work. Now every well-equipped school has attached to it one way or another a score or more of young men who are eager to work, longing for the opportunity for usefulness and distinction if they can only obtain a bare living.[56]

The AMA played a crucial role in promoting scientific medicine in the early years of the twentieth century and physiologists were involved in this effort. Two of the five members of the Council on Medical Education of the AMA formed in 1904 were members of the APS. Although Victor Vaughan was a physiological chemist and William Councilman was a pathologist, they spoke to the needs of all basic medical scientists when they reported to, and spoke on behalf of, the AMA committee. As Abraham Flexner would do four years later, the AMA committee sent representatives to visit America's medical schools in 1906 and "found schools which are absolutely worthless, without any equipment for laboratory teaching, without any dispensaries, without any hospital facilities; some which are no better equipped to teach medicine than is a Turkish-bath establishment or a barber shop."[57] Anticipating the role that the General Education Board of the Rockefeller Foundation would soon play, the committee declared, "The public must be taught the necessities and possibilities of medical education, and philanthropists shown that medicine well deserves the same support that has been given to theology, to colleges of liberal arts, to libraries, etc."[58] They advocated supporting only the better schools that encouraged laboratory work and had full-time basic science faculties.

Eliot put the accomplishments of the reform movement in perspective when he told the Council, "The whole laboratory method . . . has come in during these [last] forty years. The [Harvard] medical school as I first knew it, had but one laboratory and that was a perfectly disgraceful, dirty and abominable dissecting room, but it was infinitely precious since the only laboratory instruction given was in the horrible dissecting room. Now, nineteen-twentieths of the instruction in the medical school is either laboratory work or clinical work which in a sense is also laboratory work."[59] The goals of Bowditch and Martin were finally achieved in the twentieth century when there was widespread acceptance of the research ethic in medical schools and universal adoption of the full-time faculty system for basic medical scientists. Consistent support for these reforms was finally available and, as patriotic medical editors had predicted half a century earlier, Americans became the world leaders in scientific medicine.

America's first professional physiologists, their pupils, and their protégés had played a major role in this dramatic reshaping of the philosophy and reorganization of the structure of medical education in this country. They had also convinced many American physicians and their patients to accept the importance of biomedical research. By the middle of the twentieth century scientific medicine was held in increasingly high esteem, was well funded, and was attracting talented workers. American biomedical scientists have become the most productive in the world by virtue of the development of efficient institutional models for teaching and for research and through consistent financial support for their work.[60] These were the goals of the nineteenth-century pioneers of American physiology.

NOTES

1. Abraham Flexner, *Medical Education in the United States and Canada: A Report to the Carnegie Foundation for the Advancement of Teaching*, Bull. no. 4 (New York: Carnegie Foundation,

1910). Background on the early history of American physiology is provided by Edward C. Atwater, "'Squeezing Mother Nature': Experimental Physiology in the United States Before 1870," *Bull. Hist. Med.* 52 (1978): 313–335; John Harley Warner, "Physiology," in *The Education of American Physicians: Historical Essays*, ed. Ronald L. Numbers (Berkeley: Univ. of California Press, 1980), pp. 48–71; and Walter J. Meek, "The Beginnings of American Physiology," *Ann. Med. Hist.* 10 (1928): 111–125.

2. For background on American medical education see Numbers, ed., *Education of American Physicians*; Martin Kaufman, *American Medical Education: The Formative Years, 1765–1910* (Westport, CT: Greenwood, 1976); and Kenneth M. Ludmerer, *Learning to Heal: The Development of American Medical Education* (New York: Basic Books, 1985). An important but controversial history of American medical research is Richard H. Shryock, *American Medical Research: Past and Present* (New York: Commonwealth Fund, 1947). For an overview of the relationship between science and medicine see Warner, "Science in Medicine," *Osiris*, 2d ser., 1 (1985): 37–58; Samuel E. D. Shortt, "Physicians, Science, and Status: Issues in the Professionalization of Anglo-American Medicine in the Nineteenth Century," *Med. Hist.* 27 (1983): 51–68; and W. Bruce Fye, *The Development of American Physiology: Scientific Medicine in the Nineteenth Century* (Baltimore, MD: Johns Hopkins Univ. Press, 1987).
3. Background on antebellum American science is provided by Alexandra Oleson and Sanborn C. Brown, eds., *The Pursuit of Knowledge in the Early American Republic: American Scientific and Learned Societies From Colonial Times to the Civil War* (Baltimore, MD: Johns Hopkins Univ. Press, 1976); George H. Daniels, "The Process of Professionalization in American Science: The Emergent Period. 1820–1860," *Isis* 58 (1967): 151–166; Sally Gregory Kohlstedt, *The Formation of the American Scientific Community: The American Association for the Advancement of Science 1848–1860* (Urbana: Univ. of Illinois Press, 1976); David D. Van Tassel and Michael G. Hall, eds., *Science and Society in the United States* (Homewood, IL: Dorsey, 1966); and Clark A. Elliott, "The American Scientist, 1800–1863: His Origins, Career and Interests" (Ph.D. diss., Case Western Reserve Univ., 1970). A guide to the growing literature on the social and intellectual factors that led to the progressive professionalization of science in nineteenth-century America is Paul T. Durbin, ed., *A Guide to the Culture of Science, Technology, and Medicine* (New York: Free Press, 1980).
4. For example, see Nathan Reingold, "American Indifference to Basic Research: A Reappraisal," in *Nineteenth-Century American Science: A Reappraisal*, ed. Daniels (Evanston, IL: Northwestern Univ. Press, 1972), pp. 38–62; Stanley M. Guralnick, *Science and the Ante-Bellum American College* (Philadelphia, PA: American Philosophical Society, 1975); and Numbers, "Science and Religion," *Osiris* 2d ser., 1 (1985): 59–80.
5. Thomas Neville Bonner, *American Doctors and German Universities: A Chapter in International Intellectual Relations, 1870–1914* (Lincoln: Univ. of Nebraska Press, 1963).
6. J. M. D. Olmsted and E. Harris Olmsted, *Claude Bernard & the Experimental Method in Medicine* (New York: Henry Schuman, 1952) and John E. Lesch, *Science and Medicine in France: The Emergence of Experimental Physiology, 1790–1855* (Cambridge, MA: Harvard Univ. Press, 1984).
7. S. Weir Mitchell, "Memoir of John Call Dalton. 1825–1889," *Biogr. Mem. Natl. Acad. Sci.* 3 (1890): 177–185.
8. *Proceedings of the National Medical Conventions, New York, May 1846, and Philadelphia, May 1847* (Philadelphia, PA: T. K. & P. G. Collins, 1847), pp. 63–77.
9. John Shrady, ed., *The College of Physicians and Surgeons New York and its Founders, Officers, Instructors, Benefactors and Alumni: A History*, 2 vols. (New York: Lewis, [ca. 1903]).
10. Anna R. Burr, *Weir Mitchell: His Life and Letters* (New York: Duffield, 1929); Fye, "S. Weir Mitchell: Philadelphia's 'Lost' Physiologist," *Bull. Hist. Med.* 57 (1983): 188–202; and Jerome J. Bylebyl, "William Beaumont, Robley Dunglison, and the 'Philadelphia Physiologists,'" *J. Hist. Med.* 25 (1970): 3–21.
11. Joseph Henry to Mitchell, 7 May 1863, Reynolds Historical Library, Univ. of Alabama, Birmingham, AL.
12. Fye, "S. Weir Mitchell." See also Lawrence Turnbull, *Memoir of James Aitken Meigs, A.M., M.D.* (Philadelphia, PA: Pugh Madeira, 1881).
13. Mitchell to Wyman, 10 June [1868], Wyman Papers, Rare Books and Manuscripts Department, Francis A. Countway Library of Medicine, Harvard Medical School, Boston, MA (hereafter, Countway).

14. The spectrum of chairs held jointly with physiology in this era is revealed in "Address to Medical Students," *Am. Med. Times* 1 (1860): 253–269.
15. Some individuals, especially those who had studied in Europe, did appreciate the growing trend toward specialization in science and clinical practice. See, for example, "Specialties and Specialists," *Am. Med. Gaz.* 11 (1860): 168–180.
16. See Richard D. French, *Antivivisection and Medical Science in Victorian Society* (Princeton, NJ: Princeton Univ. Press, 1975); James Turner, *Animals, Pain, and Humanity in the Victorian Mind: Reckoning with the Beast* (Baltimore, MD: Johns Hopkins Univ. Press, 1980); Gerald L. Geison, "Social and Institutional Factors in the Stagnancy of English Physiology, 1840–1870," *Bull. Hist. Med.* 46 (1972): 30–58; and Russell C. McCrea, *The Humane Movement: A Descriptive Survey* (New York: Columbia Univ. Press, 1910).
17. John C. Dalton [Jr.], "Vivisection: What It Is, and What It Has Accomplished," *Bull. NY Acad. Med.* 3 (1866): 159–198.
18. Thomas Francis Harrington, *The Harvard Medical School: A History, Narrative and Documentary*, ed. James Gregory Mumford, 3 vols. (New York: Lewis, 1905). See also Fye, "Why a Physiologist? The Case of Henry P. Bowditch," *Bull. Hist. Med.* 56 (1982): 19–29.
19. Hugh Hawkins, *Between Harvard and America: The Educational Leadership of Charles W. Eliot* (New York: Oxford Univ. Press, 1972).
20. Ludmerer, "Reform at Harvard Medical School, 1869–1909," *Bull. Hist. Med.* 55 (1981): 343–370; Walter B. Cannon, "President Eliot's Relations to Medicine," *N. Engl. J. Med.* 210 (1934): 730–738; and Michael F. Negro, Jr., "Reform at Harvard Medical School in 1871" (Senior thesis, Harvard Univ., 1966).
21. For an overview that emphasizes European physiology in this period see Karl E. Rothschuh, *History of Physiology*, trans. and ed. Guenter B. Risse (Huntington, NY: Krieger, 1973).
22. Mitchell to Jeffries Wyman, 10 June [1868], Wyman Papers, Countway.
23. See Hawkins, *Pioneer: A History of the Johns Hopkins University, 1874–1889* (Ithaca, NY: Cornell Univ. Press, 1960); Shryock, *The Unique Influence of the Johns Hopkins University on American Medicine* (Copenhagen: Munksgaard, 1953); and Lawrence R. Veysey, *The Emergence of the American University* (Chicago, IL: Univ. of Chicago Press, 1965).
24. Richard Emmons Thursfield, *Henry Barnard's 'American Journal of Education'* (Baltimore, MD: Johns Hopkins Press, 1945).
25. Fye, "H. Newell Martin: A Remarkable Career Destroyed by Neurasthenia and Alcoholism," *J. Hist. Med.* 40 (1985): 133–166.
26. Geison, *Michael Foster and the Cambridge School of Physiology* (Princeton, NJ: Princeton Univ. Press, 1978).
27. "Biological Study at the Johns Hopkins University," *Med. Rec.* 19 (1881): 408–409.
28. John S. Billings, "An Address on our Medical Literature," *Br. Med. J.* 2 (1881): 262–268.
29. Victor Vaughan, *A Doctor's Memories* (Indianapolis, IN: Bobbs-Merrill, 1926).
30. Henry Sewall to Daniel Gilman, 8 April 1882, Gilman Papers, Special Collections, Milton S. Eisenhower Library, Johns Hopkins Univ., Baltimore, MD.
31. J[ohn] S[haw] B[illings], "Higher Medical Education," *Am. J. Med. Sci.* 76 (1878): 174–189.
32. Claude Bernard, *An Introduction to the Study of Experimental Medicine*, trans. Henry Copley Greene (New York: Macmillan, 1927).
33. Ibid.
34. Owsei Temkin, "Basic Science, Medicine, and the Romantic Era," *Bull. Hist. Med.* 37 (1963): 97–129.
35. Geison, "Divided We Stand: Physiologists and Clinicians in the American Context," in *The Therapeutic Revolution: Essays in the Social History of American Medicine*, ed. Morris J. Vogel and Charles E. Rosenberg (Philadelphia: Univ. of Pennsylvania Press, 1979), pp. 67–90. See also William G. Rothstein, *American Physicians in the Nineteenth Century: From Sects to Science* (Baltimore, MD: Johns Hopkins Univ. Press, 1979).
36. Abraham Jacobi, "Inaugural Address," *Med. Rec.* 27 (1885): 169–174. See also Joseph Eichberg, "Bacteriology and the Germ Theory of Disease," *Med. Rec.* 29 (1886): 197–202.
37. See Thomas H. Grainger, Jr., *A Guide to the History of Bacteriology* (New York: Ronald, 1958); George Newman, *The Rise of Preventive Medicine* (London: Oxford Univ. Press, 1932); George Rosen, *A History of Public Health* (New York: MD Publications, 1958); and Judith Walzer Leavitt, *The Healthiest City: Milwaukee and the Politics of Reform* (Princeton, NJ: Princeton Univ. Press, 1982).

38. Russell C. Maulitz, "'Physician versus Bacteriologist': The Ideology of Science in Clinical Medicine," in *Therapeutic Revolution*, pp. 91–107.
39. "Nitro-Glycerin," *Med. News* 40 (1882): 408. See also Fye, "Nitroglycerin: A Homeopathic Remedy," *Circulation* 73 (1986): 21–29; and William F. Bynum, "Chemical Structure and Pharmacological Action: A Chapter in the History of 19th Century Molecular Pharmacology," *Bull. Hist. Med.* 44 (1970): 518–538.
40. Frederick P. Gay, *Immunology: A Medical Science Developed Through Animal Experimentation* (Chicago, IL: American Medical Association, 1911).
41. Alonzo B. Palmer, "The Fallacies of Homeopathy," *North Am. Rev.* 134 (1882): 293–314; for similar claims see Mitchell, "The Annual Oration," *Trans. Med. Chirurg. Faculty MD* (1877): 51–58; and Francis J. Shepherd, "Medical Quacks and Quackeries," *Pop. Sci. Mon.* 23 (1883): 145–163.
42. D. Webster Cathell, *The Physician Himself and What He Should Add to His Scientific Acquirements*, 2d ed. (Baltimore, MD: Cushings & Bailey, 1882).
43. Mitchell, "Report on the Progress of Physiology and Anatomy," *North Am. Med. Chic. Rev.* 2 (1858): 105–119. See also "American Subserviency to Foreign Original Research," *Med. Surg. Rep.* 3 (1859): 293–295.
44. Bonner, *American Doctors and German Universities.*
45. Henry D. Noyes, "Pathfinding in Medicine," *Med. Rec.* 28 (1885): 589–594.
46. See, for example, "The Scarcity of Original Workers," *Med. Rec.* 22 (1882): 378–379.
47. William Townsend Porter to Gustav Baumgarten, 12 August 1890, Baumgarten Papers, Washington Univ. Archives, St. Louis, MO. (hereafter, Baumgarten Papers).
48. William H. Howell, "The American Physiological Society During its First Twenty-Five Years," in *History of the American Physiological Society Semicentennial: 1887–1937* (Baltimore, MD: 1938), pp. 1–89.
49. Frederick Ellis to Howell, Archives, American Physiological Society, Bethesda, MD.
50. Charles R. Bardeen, "Anatomy in America," Bulletin of the Univ. of Wisconsin, Bull. no. 115, Science Series, no. 4, 3 (1905): 85–208.
51. Sewall to the Honorable Board of Trustees [of the Univ. of Michigan], 4 July 1888, Michigan Historical Collections, Bentley Historical Library, Univ. of Michigan, Ann Arbor, MI.
52. Francis Minot to Eliot, 16 May 1883, Eliot Papers, Harvard Univ. Archives, Cambridge, MA.
53. See Joseph Ben-David, "Scientific Productivity and Academic Organization in Nineteenth Century Medicine," *Am. Soc. Rev.* 25 (1960): 828–834.
54. Bowditch, John G. Curtis, Henry H. Donaldson, Howell, Frederic S. Lee, Warren P. Lombard, Graham Lusk, Porter, Edward T. Reichert, and Sewall, *An American Text-Book of Physiology*, ed. Howell (Philadelphia, PA: Saunders, 1896).
55. Porter to Baumgarten, 23 December 1894, Baumgarten Papers. See also Kenneth J. Franklin, "A Short History of the International Congresses of Physiologists," *Ann. Sci.* 3 (1938): 241–335.
56. William W. Keen, "The Endowment of Medical Colleges," *Boston Med. Surg. J.* 142 (1900): 583–586.
57. *Third Annual Conference of the Council on Medical Education of the American Medical Association* (Chicago, IL: 1907).
58. Ibid.
59. "Council on Medical Education of the American Medical Association. Fourth Annual Conference," *Am. Med. Assoc. Bull.* 3 (1908): 219–334, quote from pp. 262–263.
60. R. W. Gerard, *Mirror to Physiology: A Self-Survey of Physiological Science* (Washington, DC: American Physiological Society, 1958).

III

Physiology of the Future: Institutional Styles at Columbia and Harvard

ALEJANDRA C. LASZLO

The new relationship between the American university and medical school that evolved as a result of the medical reform movement proved a determining factor in the growth of the preclinical sciences. Physiology in particular profited from the opportunities for expansion and innovation that the movement provided.[1] Curricular changes, such as the advent of "graded" (i.e., sequential) courses and longer terms, permitted the introduction of time-consuming laboratory instruction. Higher admission standards and prerequisite chemistry and biology courses ensured a better-educated student body; freed from the obligation of providing such basic training, physiologists could teach more advanced research courses. In addition, the growing emphasis on high academic standards, no doubt inspired by the university connection, created a demand for qualified specialists in the preclinical departments. It was no longer acceptable for instruction in physiology to be relegated to inexperienced physicians biding their time until a more prestigious clinical position presented itself. After the Civil War, the increasing number of nonmedical and postmedical graduates drawn to the practical training offered in German laboratories supplied the specialized manpower.[2] Closer university affiliations also came to mean larger budgets and the opportunity to pursue the full-time research and teaching of physiology as an experimental science in properly equipped laboratories.

Whereas the basic structure of reform varied little from institution to institution, the styles of the emerging programs in physiology differed greatly. Local institutional contexts and the disciplinary priorities of individual department heads guaranteed development along certain lines. The elevation of physiology to a central position in the curriculum and the introduction of new teaching methods and research aggravated relations with the clinicians. Physiologists were also troubled by the tensions inherent in balancing a medical service role with disciplinary goals. These tensions found expression in two main areas: (1) differences in opinion regarding the proper functions of the medical school and (2) differences in opinion regarding the methods of teaching to be used.

In the area regarding the proper functions of the medical school, the issue was whether the medical school should simply impart knowledge or if it should also be an institution for extending knowledge by original investigations. More specifically, with respect to physiology, would its function be limited to training future practitioners of medicine or would it also include graduate study? Neither physiologists nor clinicians questioned whether physiology did in fact belong in a medical curriculum; in the later decades of the nineteenth century "the philosophy of the German university, the emerging *Zeitgeist* of scientific medicine and the general belief in the value of science for improving medical practice"[3] had combined to convince educators of physiology's important service role. However, all questioned how physiology should be taught. Thus the second area of discord: should physiology be taught for its own sake, irrespective of its immediate bearing on the work of the clinical years, or should physiology and clinical teaching be correlated from the start? Thorny methodological issues were also raised. What, for instance, was the goal of physiological instruction—the imparting of facts, or training in the scientific method, achieved through general principles and illustrated by experiments and practical exercises?

The following case studies of physiology at Columbia College, New York and at Harvard University illustrate the connection between reform goals and disciplinary growth, and show two basic reactions to the increasing power of the medical context. Columbia and Harvard were old, elite schools controlled by powerful, socially prominent groups of clinicians, and both, led by university presidents and preclinical professors, initiated comparable reforms. Yet despite striking similarities, the development of physiology in these institutions proceeded in different directions. At Columbia, physiology was pursued as a broadly biological discipline whereas at Harvard it was regarded primarily as a medical one. These two styles—biological versus medical—epitomize the tensions between competing disciplinary agendas and strategies in constructing an institutional niche for physiology. Proponents of each style addressed different audiences, and drew upon distinct sets of ideological and material resources in advancing their programs. At issue was who would define physiology's disciplinary boundaries—the university or the medical school? These concerns were crystallized in the institutional styles that emerged at Harvard and Columbia during the first decades of the twentieth century.

COLUMBIA UNIVERSITY AND REFORM OF MEDICAL EDUCATION

As was typical of most medical schools, the reform of medical education at Columbia was carried out by the university president, the medical dean, and the biomedical professors. University presidents played an especially critical role in changing the structure of medical teaching. Eager to align their institutions with reputable medical schools, they willingly served as catalysts in the reform process. Laurence Veysey has argued that by the 1880s and 1890s, research and a utilitarian philosophy were beginning to define the character of higher education in America. Affiliation with good medical schools was important to university presidents because medical schools embodied these two roles of higher education: as potential sources of pure science

of prominent physiologists and scientists asking their opinions on this issue and eventually produced a pamphlet supporting their position.[29]

The significance of this debate rapidly spread beyond Columbia and its selection of a site for the medical school. At stake was the direction of medical education, the functions of the medical school, and who should control it—the clinician or the biomedical scientist (or, as Lee would have it, the physiologist). The issue was serious enough that clinicians, represented by the American Medical Association (AMA), soon entered the controversy. In July 1913, Arthur Bevan, head of the AMA Council of Medical Education, wrote to President Butler outlining his objections to the Lee-Pike pamphlet. Bevan's letter neatly lays out the tension that was developing between preclinical scientists interested in furthering an experimental, laboratory style of medicine and clinical scientists interested in producing practitioners:

> As chairman of the Council of Medical Education of the American Medical Association I have read with a good deal of interest a letter marked confidential sent out by two of your men in the department of Physiology. This gives such a biased statement in regard to the reasons that should control the location of the medical department of Columbia University that I should like to put myself on record as believing that the conclusions of the authors are unwarranted, and from the interests of the department of medicine, not only of Columbia, but of American universities in general, should not only not be accepted but the fallacies of their position clearly pointed out.
>
> Modern Scientific medicine must be developed and controlled by medical scientists; in other words, the highest type of clinicians. It must not be controlled by mathematicians, physicists or physiologists. These men lack the necessary medical knowledge and medical point of view. Many of them know so little about modern scientific clinical medicine that they do not even realize that it is a science. There is a group of PhD men and men with very limited medical training who are teaching the fundamental laboratory sciences of medicine, who pose as representing the scientific side of medicine as distinct from the clinical side, who are attempting to control and develop a scheme of medical education without a knowledge of clinical medicine, and without a proper regard for the requirements of clinical work. It is evident that you have to deal with this same group of men at Columbia just as we have in Chicago and in a number of other places in this country. The letters contained in the Lee and Pike pamphlet are largely from this small coterie. With few exceptions I know these men personally and know how thoroughly incompetent they are to decide the important question of the location of the medical department of a great university.[30]

The selection of a site for the Columbia medical school was only the first of a series of controversies that placed the clinicians and the biomedical scientists on opposing sides. Lee's department itself was not immune from these tensions. In 1918 the conflict in the physiology department over the promotion of Russell Burton-Opitz escalated into a bitter battle that took President Butler, the dean, and the medical faculty more than two years to resolve. Ostensibly at issue was the personal antagonism between Lee and his associate Burton-Opitz; in reality what was on trial was the role of physiology within the medical school and the methods by which it would be taught.

ROLE OF PHYSIOLOGY AT COLUMBIA

Burton-Opitz received an M.D. from Rush Medical College (Chicago) in 1895 before getting a B.S. from the University of Chicago in 1898 and a

Ph.D. in 1905. In 1902, at the behest of Lee, Burton-Opitz left Harvard, where he had been an instructor for a year, and went to Columbia as a demonstrator in the Department of Physiology. Almost immediately, Lee assigned all the teaching of medical students to Burton-Opitz and reserved graduate student teaching for himself because it offered more opportunities for research. This division of labor worked well enough until 1918, when it was proposed to promote Burton-Opitz to full professor. Usually promotion to full professor came after ten years as associate professor. Since Burton-Opitz was appointed associate professor in 1910, he was technically two years short of tenure, but citing the length and character of Burton-Opitz's service and the quality of his research, the Committee on Administration of the medical school recommended an early promotion.[31]

Complaining that he had not been consulted in the matter, Lee immediately appealed to President Butler: in his judgment, Burton-Opitz should not be promoted to full professor. Butler proved a sympathetic audience—that same day he wrote to Dean Samuel Lambert, warning him that the Trustees would be unwilling to approve this promotion over Dr. Lee's objections and rather superfluously suggesting that the action of the faculty be reconsidered.[32] A few weeks later Lee was invited to voice his concerns directly to the Columbia Committee on Administration.[33]

Lee's criticisms of Burton-Opitz ran the gamut from the scientific to the personal. First Lee claimed that a wide gulf existed between Burton-Opitz and the other members of the department: Burton-Opitz rarely engaged in scientific discussions, was unfamiliar with the research work of his colleagues and, apart from the contact necessary for the day-to-day operation of the department, had minimal interaction with the rest of the staff. Burton-Opitz's alleged aloofness extended to other American physiologists as well. In 1902, on Lee's nomination, Burton-Opitz had become a member of the American Physiological Society (APS), an honor reserved for recognized researchers in the field, but he rarely attended meetings and had never presented a paper. Among American physiologists Burton-Opitz was a nonentity.[34] Lee also criticized the quality of Burton-Opitz's research work: his apparently respectable bibliography, or so Lee claimed, consisted of collections of detailed and often trivial facts that never led to any generalizations or principles.[35]

Not content to rest there, Lee attacked the narrowness of Burton-Opitz's teaching methods: Burton-Opitz was a strong believer in the lecture as the chief form of instruction and had even pressed for increasing the number of hours devoted to it, a suggestion that found little sympathy from Lee. Burton-Opitz also denounced the recognized textbooks in physiology and actively discouraged students from consulting other authorities; they were to depend entirely on him. The result, Lee claimed, was a widespread feeling among the students that they were not expected to know physiology but rather physiology as taught by Burton-Opitz. Worse yet, Burton-Opitz had failed to keep up with advances in physiology, so that his courses were outmoded and the facts he presented were frequently incorrect. An outraged Lee wrote: "Specific instances of this have been brought to my attention by persons who have heard them directly from him. Some of them are grotesque." Finally, on the personal side, Lee cited Burton-Opitz for his use of objectionable language before the students, his derogatory remarks about other members of the department, and his German sympathies.[36]

There is considerable evidence that these were not manufactured charges—a year earlier a student petition had attacked Burton-Opitz on the same grounds. This ten-page, two-part document first related incidents and statements showing how Burton-Opitz had misused his position in an attempt to discredit an honor student and second, discussed Burton-Opitz's personality and his relations with students and faculty members. The charges, which included favoritism (Burton-Opitz liked bootlicking students and was partial to members of his own undergraduate fraternity), lectures beginning late, and ungentlemanly and anti-Semitic language, may have just been hearsay, but the students swore by it and had no doubt that others would verify them if necessary. More significant, however, the students had also collected evidence of "unscientific and inaccurate statements" and "lackadaisical and mechanical presentations" by Burton-Opitz. The course itself, they claimed, was open to criticism because of the important subjects that were not covered, specifically metabolism, the physiology of ductless glands, new work on the heart such as that by Lewis and Mackenzie, and new work in digestion (Carlson, Walter B. Cannon, Herz), renal physiology and kidney function, and respiratory exchange (Joseph F. Barcroft, John S. Haldane).[37]

For these reasons Lee was convinced that Burton-Opitz was unfit for even his current position in the physiology department. Rather than promote him, Lee concluded, Burton-Opitz should be encouraged to resign. Lee was wholeheartedly supported in this recommendation by other members of the physiology department. Horatio B. Williams, an assistant professor of physiology on war leave at the time, was concerned enough about the evolving situation to write that unless Lee's recommendation regarding Burton-Opitz went through, "I should be strongly inclined on my return to resign and open an office as I should not care to be in this department if Opitz were to receive such recognition as to make him the dominant influence in moulding the policy of the department." Williams went on to say, "I do not know whether I have told you, but I found in 1911 that the British physiologists . . . did not regard his contributions as of very great importance."[38]

Rather than deny Lee's accusations, Burton-Opitz responded with some of his own. First he charged that Professor Lee was a Ph.D. and as such had no right to be a member of a medical school faculty. Burton-Opitz also insisted that he should hold the professorship of physiology because he was the only one who could teach the subject. On the personal side, Burton-Opitz hinted at corruption and impropriety by suggesting that Lee held his position solely because of Mrs. Lee's donations to the college.[39]

Although the personal antagonism between the two men was indisputable, the underlying issue was the role of physiology within the medical school and medical curriculum. Lee placed a premium on the research function of physiologists and disdained teaching, particularly of medical students. Burton-Opitz, on the other hand, felt that teaching physiology to medical students from a medical rather than biological point of view was his primary function, and that physiologists should be rewarded for teaching, not research.

Lambert, the medical dean at Columbia, acknowledged that Lee and Burton-Opitz were incompatible both personally and scientifically, but did not think that removing Burton-Opitz was a viable option. He wrote President Butler that "there is a very general criticism of the work of the department of physiology and that as Dean, [Lambert was] also a vigorous, but not unfriendly critic."[40] Lambert pointed out that the department taught 200 medical students per year and roughly twenty graduate/research students. Obviously the primary function of the physiology faculty was to teach medical students. Lee's criticisms were merely an admission of his own failure in this respect—medical students were not getting what they needed from him in physiology. In Lambert's considered opinion, Burton-Opitz was an excellent and popular teacher working under a severe handicap within the department; the school could not afford to lose his services.[41]

Lee's failure to have Burton-Opitz removed from his post was symptomatic of the new political alignments in the medical school under Lambert. Lee had made progress in the promotion of a research style of physiology so long as there was a sympathetic coalition of reformers to support him. But Dean McLane had retired in 1902. Curtis, Lee's old mentor in physiology, served as acting dean for only two years until 1904, and retired from the department in 1908. Although President Butler continued to uphold Lee's reform ideals, his support was not sufficient to prevent tensions from arising with the clinicians. When Lambert became dean of Columbia P&S in 1905, the political base of the medical school shifted in favor of the clinicians.

Lambert, as one of his colleagues noted, "had been saturated with medical tradition." The son of a physician, he was educated at a private day school in New York, studied at Yale's Sheffield Scientific School and in 1885 was graduated from Columbia P&S. As a product of P&S and a local clinician popular with practitioners, Lambert had been brought onto the faculty to reorganize clinical teaching. In this respect he was quite effective: he inaugurated a system of clinical clerkships and was instrumental in supplying adequate hospital facilities through an affiliation with Presbyterian Hospital. In other areas of reform, however, he tended to be conservative, jealously guarding the rights and privileges of the clinicians. His frequently nostalgic view of medicine and its practitioners was expressed in his lack of support for the modern emphasis on science in medicine.[42]

Lambert was thus a thorn in Lee's side and Lee did not hesitate to place the blame for problems in the medical school largely at Lambert's door. No doubt frustrated by his battle with Burton-Opitz and Lambert, early in 1918 Lee wrote President Butler a rather undiplomatic letter outlining what he felt were the chief problems at P&S:

> The conversation after the Council meeting on Tuesday indicated to me that you quite fail to grasp the real situation in the Medical School. I would give a good deal if I could make you see it as clearly as I, and others, see it and induce you to put an end to it. You can do it if you only will. Don't you yet realize that the real reason why we can't induce anyone to give money to develop the School and why we don't go ahead and do something really worth while, is because the men of wealth that we are after have a lack of confidence in the medical faculty to devise and carry out a proper plan. This lack of confidence can be traced right back to two glaring defects of the medical faculty: mediocrity and provincialism.[43]

central goal of the young Harvard faculty, several of whom were German trained. Although their commitment to research was strengthened by Bowditch's appointment, the ideals of the scientific investigator were little appreciated in Boston medical circles of the time. Even Eliot, otherwise a leader in medical reforms, supported laboratory science only for its role in providing high quality teaching; he did not as yet consider research an important function of the medical faculty.[66] There was only one "laboratory" in the medical school when Bowditch returned and that was a dissecting room. Thus Bowditch not only had to create his own facilities but also an appropriate atmosphere for the pursuit of research.[67] Partially through a bequest from a deceased classmate of Bowditch and partially through funds provided by the Harvard Corporation, two attic rooms in the overcrowded medical school building were converted into a very modest laboratory. Simply furnished, the main room held a table for experiments, a sink, a workbench, a foot lathe, and a tub for frogs. Another much smaller room that Bowditch shared with an assistant in histology contained cases for instruments and a few books.[68] Equipped with apparatus purchased in Germany (and paid for by Bowditch's father), this became the first teaching laboratory for physiology in the United States.

Despite meager appropriations,[69] Bowditch was able to attract both students and advanced workers to his laboratory. As early as 1871, the Harvard catalog noted that advanced students would be accepted into the lab and by 1872–1873 undergraduates were, for the first time, given the opportunity to conduct original investigations.[70] Bowditch's hospitality and enthusiasm stimulated research not only in physiology but in most areas of experimental medicine. Among the many talented investigators trained by Bowditch were Charles S. Minot in general biology, Isaac Ott and Robert W. Lovett in pharmacology, John C. Warren in pathology, O. K. Newall in experimental surgery, G. Stanley Hall and William F. Southard in psychology, and Warren P. Lombard, James J. Putnam, William James, G. M. Garland, Frederick W. Ellis, and C. H. Williams in physiology.[71] By 1879 the laboratory had become so productive that a volume of collected papers, reprinted from various scientific journals, was issued. In 1886 it was followed by a second volume of twenty-two papers. Bowditch hoped that by drawing attention to the laboratory facilities in this way, he would inspire a larger number of workers to pursue physiological investigations.[72]

As assistant professor, Bowditch was also responsible for physiological instruction in the medical school, a service he performed for the next thirty-five years. In 1876 he was promoted to full professor, and from 1903 until his resignation in 1906, he held the newly established George Higginson Professorship of Physiology. Bowditch was generally regarded as an enthusiastic and stimulating teacher.[73] Polished lectures, a clear expository style, and the use of convincing demonstrations to illustrate fundamental facts contributed to his reputation.[74] Ironically though, given the premium President Eliot placed on laboratory teaching, Bowditch did little to introduce this method in the physiology department, at least as a means of training medical practitioners.[75] Bowditch did, however, make a notable contribution to educational procedures at Harvard. Early in his teaching career he initiated informal weekly conferences in which students presented physiological theses based on a reading of original sources. These discussions were highly valued by the students both for the exposure it gave them to the primary literature (a unique departure in physiological teaching) and for its reinforcement of the lectures.[76]

Bowditch had been instrumental in advancing the position of physiology within Harvard, gaining for it a research role as well as increasing its teaching function. He also did much to professionalize American physiology more generally, serving as a founding member and second president of the APS and supporting the establishment of the *American Journal of Physiology* (1898).[77] Yet for all his ambitious development of disciplinary goals, he never lost sight of physiology's medical context. Over the course of his long career, Bowditch repeatedly demonstrated that his first allegiance was not so much to physiology, but to medical science, and particularly to the Harvard Medical School.

Bowditch was keenly aware that Harvard's emergence as a modern medical school was dependent on more than a change in educational philosophy; the new program initiated by President Eliot required increased expenditures for laboratories, equipment, and faculty salaries. Because tuition fees alone could not supply the needed capital, reformers actively sought outside support in the form of endowment.[78] The medical faculty, led by Bowditch and Harvard surgeon Warren, raised most of the school's endowment in two fund-raising drives in 1874 and 1901. By appealing to the public's self-interest in having competent, well-trained physicians for the community, the faculty was able to justify its pleas for assistance. Harvard, they argued, had embarked on a rigorous but costly program to provide quality medical instruction. Impressed by this new departure in education, the professional and business classes of Boston responded with the necessary funds. Bowditch's and Warren's contacts among the established families of Boston, combined with their entrepreneurial skill in convincing the community of the need for a modern scientific school, led to the construction of new medical buildings in 1883.[79]

Physiology was a major beneficiary of this move. The new Boylston Street laboratory, designed to Bowditch's specifications, was considered palatial and probably the best available in America at the time. Bowditch could now realize his ambition to be a full-time experimental physiologist in facilities specifically designed and equipped for research. At just that point, however, he was persuaded by the Harvard Corporation that he would have "opportunities of increased usefulness to medicine" as dean of the medical school. Bowditch accepted the appointment in 1883, further demonstrating his primary allegiance to the medical school at the expense of physiology.[80] The press of administrative duties eventually caused him to limit his physiological activities. By 1889 he had largely removed himself from teaching responsibilities and by 1890 had ceased doing original research (although not, his many biographers note, before inspiring a new generation of researchers with the ideology and spirit he had imbibed from Ludwig and the Leipzig group).

During his tenure as dean, Bowditch initiated many changes that contributed significantly to Harvard's emerging national reputation as a leader in medical education. Bacteriology was introduced as part of the medical curriculum; a mandatory four-year curriculum was finally adopted; elective courses were offered; and for the first time Harvard went outside the confines of Boston in filling faculty positions in the preclinical sciences.[81] Bowditch also recognized the importance of bringing the medical school into closer relations with area hospitals. A new style of clinical teaching characterized by small-group instruction at the bedside was introduced as a supplement to formal class lectures and clinical professorships were established. These innovations were especially useful in securing valuable hospital facilities for the students.[82] Writing in 1911, William Osler summed up Bowditch's contributions to the development of medicine at Harvard: "It is not too much to say that the present splendid position of the Harvard Medical School is due to him, as dean, more than to any other man. President Eliot started the reforms, but Bowditch was the active promoter." [83]

With Bowditch involved in running the affairs of the medical school, it fell to William T. Porter to maintain and develop instruction in the physiology department. Porter, who came from St. Louis Medical College, Missouri, in 1893 to take over the duties of assistant professor, fit well into that community of scholars characteristic of the Eliot era.[84] As part of his dedicated effort to raise educational standards at Harvard, Porter introduced a "new method" of teaching physiology that relied almost entirely on daily laboratory exercises to disclose, step by step, the fundamental facts of physiology. Porter's method inverted the traditional relationship between lectures and experimentation, thus effecting the shift from reliance on authority to reliance on observation so central to Eliot's 1871 reform agenda.[85]

CANNON, PHYSIOLOGY, AND THE CLINICAL CONNECTION

Porter's considerable achievements were overshadowed, however, by the accomplishments of his young colleague in physiology, Cannon.[86] After receiving an M.D. from Harvard in 1900, Cannon became an instructor in the Department of Physiology. He was promoted to assistant professor in 1902 and in 1906, only six years after his initial appointment, succeeded Bowditch as George Higginson Professor and chairman of the department.[87] As a pupil of Bowditch, and thus an intellectual "grandson" of Ludwig, Cannon's predominant interest was in physiological research. Over the course of his forty-five year career he published eleven books and more than 340 articles in areas ranging from the physiology of digestion and the role of endocrine secretions to the bodily effects of emotion and traumatic shock, and studies of neural functions.[88]

Cannon also had a reputation as an excellent teacher and was concerned with training men "for the high duties of medical practice." [89] He insisted that physiology, as the primary foundation of medical science, had a significant role in teaching prospective physicians to think scientifically. Cannon took this responsibility seriously; besides acquainting students with the fundamental facts of physiology and training them to integrate this knowledge into an understanding of the organism as a whole, the object of the basic course in physiology was "to show [the student] the utility of applying physiological concepts to the study of medicine." [90] In 1921, a reorganization of the medical curriculum provided more time for electives and thus allowed the department to offer several advanced courses in special topics. Much of the advanced work was conducted through a tutorial system introduced in 1923 to give capable students the opportunity and stimulus to pursue concentrated research. These students were permitted to elect physiology for periods of no less than four months and, under the guidance of a faculty tutor, were assigned a specific research problem to investigate. Cannon and his colleagues did not see the tutorial system as a means of weaning men away from medicine and into careers in the preclinical sciences. Instead, they emphatically argued, its purpose was "to aid in what must become more and more a general object in medical education, the sending into medicine of men with scientifically trained minds." [91]

Cannon's commitment to training physiologists and physicians proved to be the determining factor in keeping him at Harvard. Despite many pressing invitations over the years, including in 1921 the very tempting offer of a research position at the Mayo Clinic in Minnesota, Cannon's loyalties remained with physiology—a physiology he specifically linked to medicine. "Where," he asked himself when considering the Mayo position, "could I as a physiologist be of greatest service to medicine?" [92] His answer: "I came to the conclusion that my best services to medicine would be realized in the development of investigators in physiology and in clinical research related to physiology." [93]

Cannon was dedicated to expanding the role of physiology and had long recognized a clinical connection as a means of achieving this end. His interest was already evident as early as 1899 when he was completing his medical degree and President Eliot and Minot, head of the Department of Histology and Embryology, proposed the establishment of the new Faculty of Comparative Medicine. Conceived primarily as a means of consolidating the various medical departments, The Faculty of Comparative Medicine was to serve as a de facto graduate department of the medical school and to incorporate the veterinary school as well. As part of this scheme, Eliot insisted that a new professorship in general physiology be established so as to strengthen ties between the medical school and the university.[94] For financial reasons the plan was never realized,[95] but the general idea was revived a few years later in 1903 when the Committee on University Education in Medicine suggested setting up divisions to coordinate the work of the medical departments.

Cannon agreed in principle with the division plan, but objected to the specifics of the proposal and in a letter to Eliot submitted his own ideas. Under the committee's scheme the medical departments would be grouped into four divisions according to normal or "abnormal" interests. Cannon argued that the proposed grouping was both artificial and arbitrary; he suggested instead an alignment by structure and function. This "natural" arrangement would result in seven divisions—four scientific and three clinical. In Cannon's scheme, physiology occupied a central role; rather than serving as one of several departments within a "normal" division, it was promoted to a division of its own, and the Department of Physiological Chemistry was placed under its administrative wing as well.[96]

Cannon, however, was not willing to settle solely for increased visibility for the physiology department. Already an excellent propagandist for his discipline, he pressed for expansion. Although he could easily have sought a closer connection to the university by pursuing expansion in the area of general physiology, an idea that Eliot was already receptive to, Cannon preferred innovation in a new direction—clinical physiology. Cogently arguing the need for a division to link work in the medical sciences with training in routine tests such as urine, blood, and tissue analysis, he proposed that a Division of Clinical Methods and Clinical Physiology be included in the reorganization. This division would consist of the existing Departments of Pathological Chemistry and Clinical Microscopy and a new clinical physiology department yet to be created. Although Cannon's scheme was not realized, it is significant for two reasons: first, it confirms his interest in disciplinary entrepreneurship and second, it shows his commitment to a clinical role for physiology.[97]

Under Cannon's leadership the role of physiology within the medical school was greatly increased. In 1906 a research Department of Comparative Physiology was set up with Porter as its head, and in 1918 an industrial physiology department was established with Cecil K. Drinker, a Cannon protégé, as professor. This latter development was the result of an intense lobbying effort primarily by Drinker, an assistant in physiology, and Cannon, at that time in charge of the Laboratory for Surgical Research at Dijon, France. Early in 1918 Frederick Cheever Shattuck, Emeritus Jackson Professor of Clinical Medicine, proposed that the school for health officers be enlarged to include a Division of Industrial Hygiene.[98] When he pledged to collect the $25,000 a year necessary to begin the enterprise, Shattuck received the wholehearted support of Harvard President A. Lawrence Lowell and the new dean of the medical school, David Edsall. Confident "that Dr. Shattuck's whole idea was to get something worthwhile underway," Drinker was determined "that physiology should certainly be represented." His own thorough background in medicine, he felt, made him an especially suitable candidate for the job. Thus he wrote Cannon:

> I need not tell you how much the whole thing appeals to me. I could retain my association with you since it would be necessary to use the laboratories for this teaching. Physiology would simply expand greatly and I take it that is a matter very near to the hearts of both of us. If they start this new department in the school and I am lucky enough to get it I shall be finally settled and we may simply look forward to more money and more scope for physiology than we have had in the past.[99]

Cannon of course welcomed any project that advanced the interests of physiology, particularly in a clinical direction, and with a receptive audience on all fronts Drinker was soon successful.[100] In 1923 industrial physiology was transferred from the medical school to the recently founded School of Public Health and became the Department of Applied Physiology. Although it was an autonomous department offering instruction to public health students, applied physiology nonetheless maintained close ties with the medical Department of Physiology in terms of staff, research, and administration. The fact that Edsall functioned as dean of both the Schools of Medicine and Public Health ensured that public health was never really cut off from medicine.[101]

Thus by 1920 physiology at Harvard was represented by four departments bound together in a loose organization known as the "Laboratories of Physiology." Occupying an entire wing of a medical school building (Fig. 1), it included the Departments of Physiology, Comparative Physiology, Applied Physiology, and Physical Chemistry, which was established in 1920 and headed

LABORATORIES OF PHYSIOLOGY, HARVARD MEDICAL SCHOOL

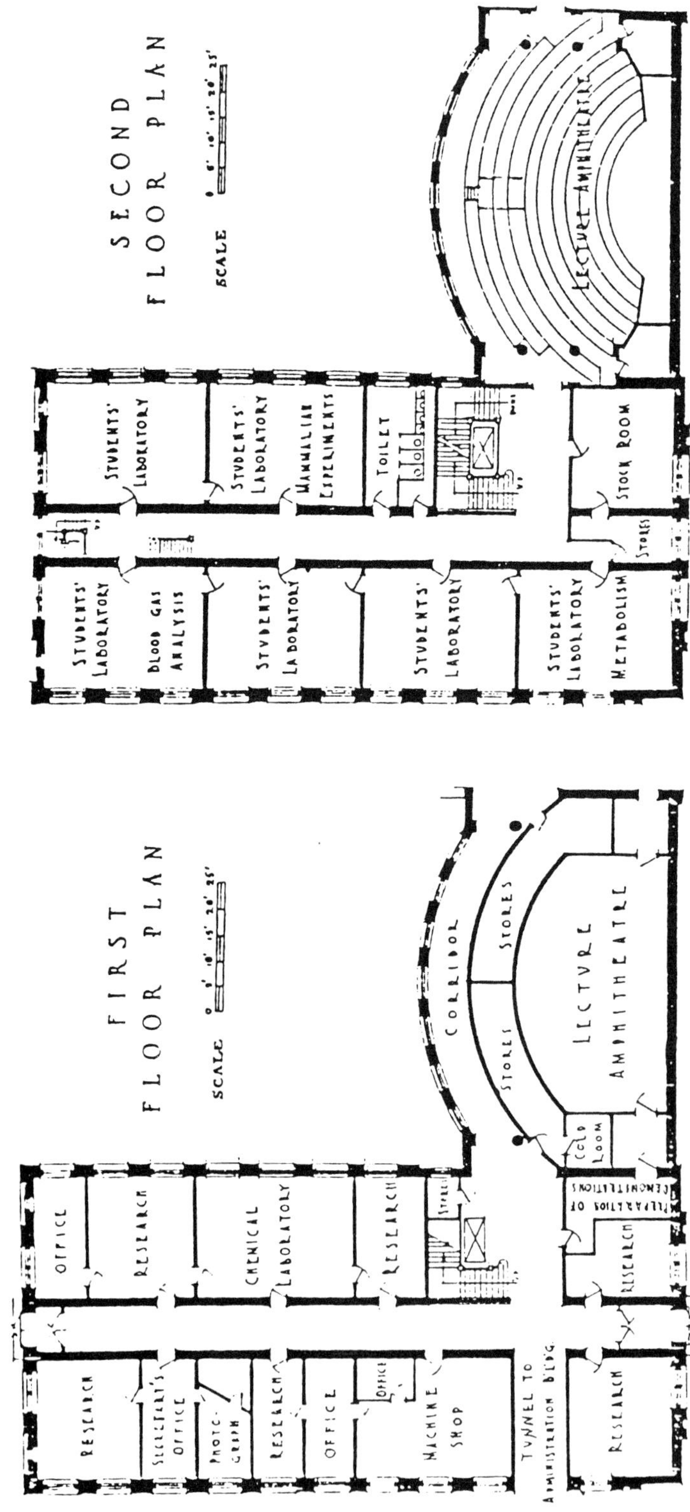

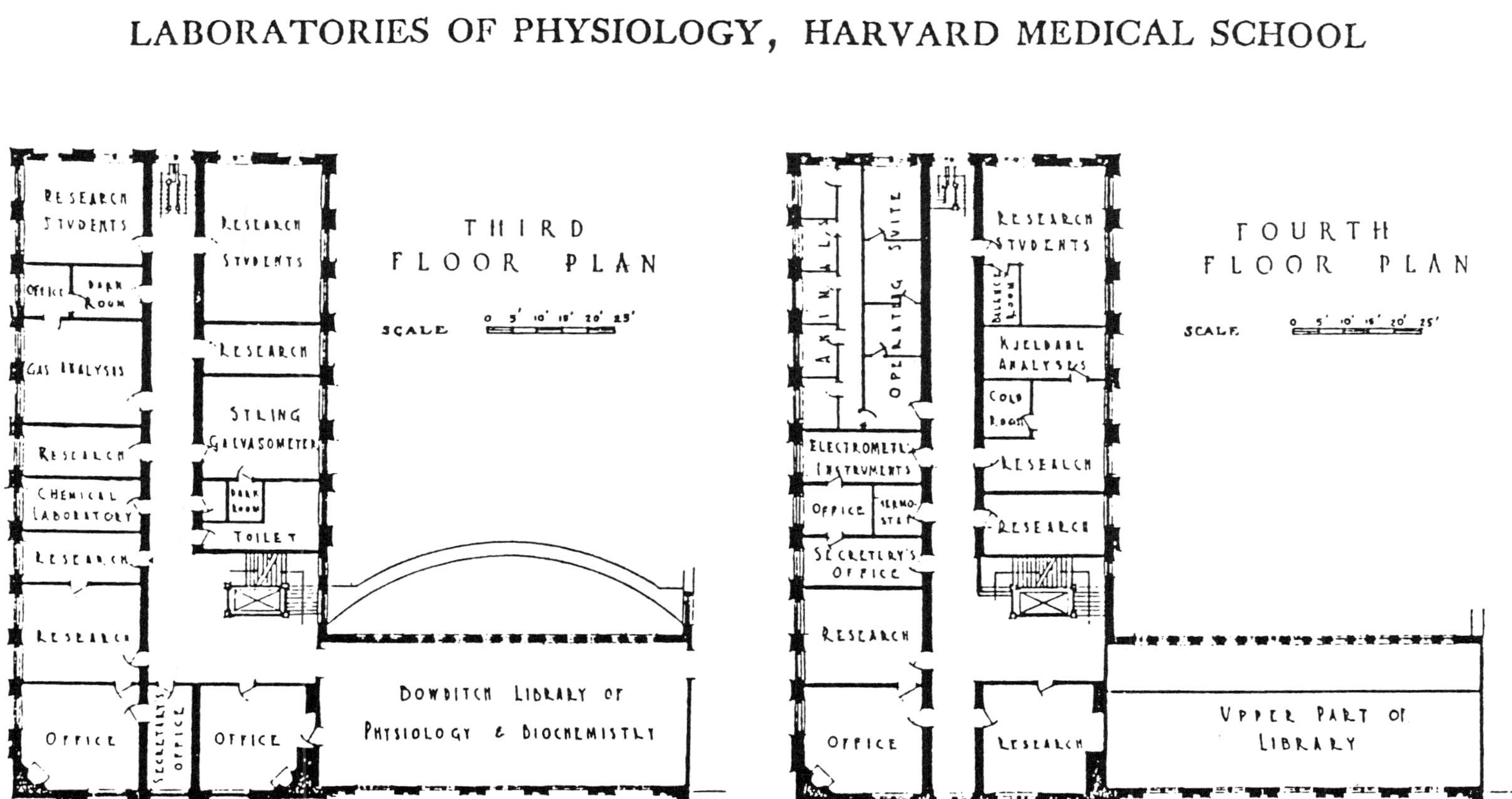

Figure 1: *Floor plans, laboratories of physiology, Harvard Medical School. [From Alfred C. Redfield, "Laboratories of Physiology, Harvard Medical School," in Methods and Problems of Medical Education, 3rd Series (New York: Rockefeller Foundation, Division of Medical Education, 1925).]*

by Lawrence J. Henderson. While the departments had separate budgets and personnel, this organization enabled them to function as a unit with respect to teaching, research, and the exchange of technical talent and equipment. In a 1924 report on the facilities, Alfred C. Redfield, assistant professor of physiology, claimed that "the association of these four laboratories under one roof brings together a larger group of physiologists representing a greater variety of special interests within this field than is to be found elsewhere in this country."[102] Indeed, he argued, the laboratories "really constitute an institute of physiology superior in its scope to any individual department."[103]

This opinion was shared by some outside of Harvard as well. Recognizing Harvard as "one of the outstanding centers in physiology" and Cannon as "a leader with ability to attract and train men," the General Education Board of the Rockefeller Foundation awarded him a three-year grant of $22,500 in 1927.[104] In 1930 the foundation's medical sciences division, headed by Alan Gregg, continued this "aid to a leader" with a $175,000 appropriation to be used by Cannon until his retirement in 1937.[105] Gregg considered this one of the most successful grants made throughout the decade; in a 1938 appraisal of the Harvard project he wrote:

> I can think of no considerable grant made in the MS [Medical Sciences] whose quality and quantity of return surpasses that of the aid to Cannon in physiology. It is undoubtedly true that it is due to this aid that Cannon's laboratory has accepted, trained, and turned out so large a number of competent physiologists. Dr. Cannon's scientific productivity has been a double one. As an individual his continued addition to the knowledge of the physiology of the nervous system is the admiration of all competent physiologists. His influence on juniors is in the aggregate as great if not greater.[106]

Gregg had good reason to be pleased. Since 1930 Cannon had trained a total of thirty-two teaching and research fellows—ten fellows from the National Research Council, five from the Rockefeller Foundation, two Guggenheim fellows and fifteen from various other organizations. Altogether Gregg counted twenty full professorships in physiology held by persons formerly associated with the department. Not surprisingly, he concluded that Harvard had "the leading department of physiology in the United States, and [was] certainly among the first five in the world."[107]

CONCLUSION

If the success of a department is judged by its size, facilities, number of graduates, budget, and the ability to attract outside support, the Department of Physiology at Harvard was eminently successful. Cannon's careful cultivation of a physiological program linked to medicine and the clinic had resulted in a veritable "institute of physiology" at Harvard. By the same standards, the Department of Physiology at Columbia experienced only minor success: Lee's untiring campaign to establish physiology as a biological, university discipline, had resulted in a sorely divided department at loggerheads with the medical hierarchy. The contrast between the two programs by 1920 could not have been greater.

At Columbia, the bitter conflict pitting Lee and his vision of a biological physiology against the priorities of the clinicians was only one aspect of the deteriorating state of affairs in the medical school. Issues of faculty appointments, hospital affiliations, the location of the new school, and a full-time clinical plan had also contributed to the serious crisis in medical education. Lee's desperate plea of 1918 to President Butler for intervention only marked a continuing escalation of hostilities in a situation that had no simple or speedy resolution. In the end neither Lee nor his clinical opponents had won any decisive victories at Columbia, although both scored minor triumphs.

As the Columbia case study suggests, disciplinary goals frequently came into conflict with the utilitarian values of medical education. While the service role of training physicians pulled physiology in a clinical direction, the academic values of biomedical professors such as Lee demanded a broader development of physiology within a university setting. When Lee became executive officer of the Department of Physiology, he adopted a self-conscious policy for the development and expansion of his discipline based on the Johns Hopkins model. Physiology at Hopkins had always been part of the university biology system: the physiological and zoological laboratories shared the same building; physiological research was encouraged on the same basis as research in zoology, chemistry, and physics; and organization of the medical school in no way altered the relationship between the physiology department and the university.[108] Lee compared the Hopkins program with the older physiology department at Harvard that functioned as part of the medical school and concluded that Hopkins was preeminent. The cause of this preeminence, he claimed, "lies in the fact that the environment of the one has been scientific, that of the other medical." [109] Emulating the biological style of Hopkins, Lee established courses in the School of Pure Science and Columbia College and sought to control work in physiology offered by Barnard and Teachers Colleges. Ultimately his goal was the creation of a university Department of Physiology divorced from medicine.

Lee's single-minded campaign to link physiology to the biological work of the university did little to endear him to the clinicians but it was his total lack of accommodation to medical priorities that, more than anything else, alienated his colleagues in the medical school. Lee was very explicit that among departmental priorities, medical teaching ranked no higher than any other function. His inflexible position irritated the clinicians, particularly Dean Lambert, who, years after the initial controversy over Burton-Opitz and the work of the physiology department, renewed his criticisms.[110] Physiology at Columbia, like physiology almost everywhere, was firmly entrenched in the medical school. While it was possible for reform-minded individuals like Lee to establish connections with the natural sciences, physiology's institutional context demanded that proper attention be paid to the medical service role. Lee's strategic mistake was to consistently undervalue and underestimate the clinical connection. His failure to adapt to the medical environment led to perpetual conflict and ultimately to the defeat of his biologically oriented program.

By contrast, the case study of physiology at Harvard is one of successful adaptation to the medical context. Harvard had always had a strong clinical orientation, dominated by local medical dynasties.[111] Rather than challenge the hegemony of the clinician, both Bowditch and Cannon chose to operate within the rich traditions of the medical school. The fact that Bowditch, a scientist-physician rather than a typical clinician, was elected dean and could function so effectively suggests that a rapport developed between the preclinical and clinical sciences. Tensions between disciplinary goals and medical values, laboratory worker and clinician, university and medical school—so evident at Columbia—were unproblematic at Harvard. The strategy of accommodation and adaptation to the medical context, employed by both Bowditch and Cannon, resulted in a medical style of physiology and a greatly expanded role for the department.

Adaptation meant reconciling a research role with training practitioners. Cannon, for one, had always viewed these two roles as complementary; laboratory investigation was never independent of medicine. Indeed, he belonged to a research tradition whose aim was the creation of a physiology of man that would serve as the primary foundation of medical science. Medical students were both the audience and the participants in this enterprise. Lee, however, had been unable or unwilling to balance his scientific research goals with a medical service role. He found teaching medical students an annoying distraction from his first priority, which was physiological research. In the end, his pure science ideals prevented him from cultivating a rich resource for expanding the scope of physiology.

The final lesson to be learned from these case studies is that the style of physiology departments was ultimately shaped by their relation to medical schools.[112] While some flexibility existed for the expression of individual styles, the institutional context demanded alliance with the clinic. Thus the medical reform movement that provided so many opportunities for the growth of physiology as a discipline also placed constraints on its development. Physiologists had been extremely effective in selling physiology as the scientific basis of medicine. With this service role clearly established, attempts to wean physiology from its medical home met with little success. Lee's bold prediction of the 1890s had fallen wide of the mark: medical physiology, not biological physiology, was the physiology of the future.

NOTES

1. This analysis of emerging styles of physiology builds on a similar argument made by Robert Kohler for the development of biochemistry. Kohler's recent book, *From Medical Chemistry to Biochemistry: The Making of a Biomedical Discipline* (Cambridge, UK: Cambridge Univ. Press, 1982), provides a useful model for describing the growth of biomedical disciplines within the context of the medical reform movement.
2. Thomas Nelville Bonner, *American Doctors and German Universities: A Chapter in International Intellectual Relations, 1870–1914* (Lincoln, NE: Univ. of Nebraska Press, 1963).
3. John Harley Warner, "Physiology," in *The Education of American Physicians: Historical Essays*, ed. Ronald L. Numbers (Berkeley, CA: Univ. of California Press, 1979), pp. 48–71, esp. 65.
4. Laurence R. Veysey, *The Emergence of the American University* (Chicago, IL: Univ. of Chicago Press, 1965).
5. The standard histories of Columbia College and the College of Physicians and Surgeons include: Horace Coon, *Columbia: Colossus on the Hudson* (New York: Dutton, 1947); James Brander Matthews et al., eds., *A History of Columbia University, 1754–1904* (New York:

Columbia Univ. Press, 1904); Frederick Paul Keppel, *Columbia* (New York: Oxford Univ. Press, 1914); J. H. Van Amringe, *An Historical Sketch of Columbia College of the City of New York* (New York: Columbia College, 1876); and Edward C. Elliot, ed., *The Rise of a University* (New York: Columbia Univ. Press, 1937).

6. Coon, *Columbia*, p. 266.
7. Details of the negotiations leading to the merger can be found in documents located at the College of Physicians and Surgeons, Health Sciences Library, Special Collections (hereafter, P&S, Spec. MS Coll.) and in the Central Administration Files, Low Library [hereafter, CAF followed by folder (f) name and number] New York, NY. See esp. "Confidential Report for the Use of the Trustees of Columbia College and of the College of P&S," 1890, CAF, College of P&S 1886–1909, f. 1. In addition to materials at the two locations mentioned above, much of this article is based on correspondence and archival sources housed in the Rare Book and Manuscript Library, Butler Library, Columbia University, New York, NY.
8. The medical course was lengthened to four years in 1893, although it took effect with the entering class of 1894. Matthews, *History of Columbia University*, p. 243.
9. President Seth Low to J. G. Curtis, 28 February 1893, CAF, Curtis, f. 1.
10. Kohler convincingly argues that these innovations were not radical changes signifying a break from traditional educational philosophy, but rather were incremental improvements brought about by internal circumstances and the local competition for good students. See his analysis of reform at Columbia in Kohler, *From Medical Chemistry*, pp. 134–137.
11. In September, 1901, President Low was nominated for mayor of New York by an organization called "The Citizens Union"; he accepted the nomination and resigned the presidency. Butler, a forty-year-old professor of philosophy at Columbia, was inaugurated president on 19 April 1902, and held this position for the next forty-four years. Matthews, *History of Columbia University*, p. 170.
12. N. M. Butler to Curtis, 3 March 1904; Butler to Curtis, 1 October 1904, CAF, Curtis, f. 2.
13. Kohler, *From Medical Chemistry*, p. 127. See also Philip J. Pauly, "The Appearance of Academic Biology in Late Nineteenth-Century America," *J. Hist. Biol.* 17 (1984): 369–397. Using Kohler's analysis of biochemistry as a point of departure, Pauly argues that the development of biology in America also had its roots in the medical reform movement. In particular, see his discussion of the Columbia biology program and its relation to physiology, 387–389.
14. Curtis to James Woods McLane, "Report on the Scope and Needs of the Department of Physiology," 19 May 1890, P&S, Spec. MS Coll.
15. Ibid., p. 3.
16. Curtis to Samuel Lambert, "Final Report on the Department of Physiology," 10 June 1909, CAF, Curtis, f. 2.
17. Curtis to McLane, "Report on the Scope," p. 4. Despite these lobbying efforts it was not until 1903 that an elective course of laboratory instruction for medical students was offered; in 1905 this course became mandatory. When Frederic Lee arrived at P&S in 1891 he organized a theoretical and practical course in physiology that was available to M.A. and Ph.D. candidates. Curtis to Lambert, "Final Report," p. 7.
18. Curtis to McLane, "Report on the Scope," p. 6.
19. Curtis was always looking for more space to annex to the Department of Physiology. Citing reasons of increased teaching load and growth of the student body and faculty, he untiringly tried to persuade Low to give him a number of rooms vacated by the anatomy department. Curtis was only partially successful in this goal because President Low was concerned about lessening the common space in the medical building. Curtis to Low, 8 Dec. 1894, 20 Dec. 1894, 23 Jan. 1895, 27 March 1895; and Low to Curtis, 19 Dec. 1894, 17 Jan. 1895, 25 March 1895; CAF, Curtis, f. 1.
20. In 1883 Mr. James T. Swift donated $10,000 for the purchase of the "more expensive, delicate, and complicated instruments and appliances, mainly instruments of precision, requisite for the pursuit of physiological science." The gift was in memory of his brother, Dr. Foster Swift, an alumnus of the class of 1857. Lee, "The School of Medicine," in *History of Columbia University*, p. 323.
21. Curtis outlined a plan, based on the proven German model, for a "staff consisting of the Professor; of at least one expert assistant, often himself a physiologist of standing; of a skilled mechanic; and of a laboratory servant, competent to care for the animals, to set up

batteries, and to do much also beside mere cleaning." Curtis to McLane, "Report on the Scope," pp. 9–11.

22. Ibid., pp. 14–15.
23. Charles M. DaCosta, a trustee of the college and an alumnus of the class of 1855, left a legacy of $100,000 to be used at the discretion of the trustees for the establishment of a professorship. The trustees chose to set up a Department of Biology under the direction of Henry F. Osborn. Osborn was also appointed curator of mammalian paleontology in the American Museum of Natural History, the first of a series of affiliations with important New York institutions that Columbia was to make. Matthews, *History of Columbia University*, pp. 155–156.
24. Curtis to Low, 22 April 1891, CAF, Curtis, f. 1. Osborn also supported the establishment of a physiological branch at the university and encouraged Curtis to discuss this matter with President Low. Osborn to Curtis, 2 May 1895 as quoted in Curtis to Low, 6 May 1895, CAF, Curtis, f. 1.
25. Low to Curtis, 4 November 1892; Curtis to Low, 5 November 1892, CAF, Curtis, f. 1. The Faculty of Pure Science, established in 1892, included the Departments of Geology, Mineralogy, Botany, Zoology, Chemistry, Mathematics, and the Physical Sciences.
26. Lee to Low, 15 November 1894, CAF, Lee, f. 1.
27. Lee to Curtis, "Report on Undergraduate Work in Physiology," 23 April 1902, CAF, Curtis, f. 2.
28. Ibid., pp. 2–3; Butler to Curtis, 29 April 1902, CAF, Curtis, f. 2.
29. See Lee Papers, box 2, f. Location of medical school. For the standard letter sent out to prominent physiologists, see Lee to Charles S. Sherrington, 21 October 1912, Lee Papers, box 2, f. Sherrington.
30. A. D. Bevan to Butler, 25 July 1913, Lee Papers, box 1, f. Columbia University, President Butler. Some of the men Bevan refers to include Sherrington, David Edsall, Henry Donaldson, E. B. Wilson, T. H. Morgan, J. McKeen Cattell, and Franz Boas.
31. "Recommendation from Committee on Administration to the Faculty," 18 February 1918, Lee Papers, box 1, f. Burton-Opitz. Burton-Opitz had served as assistant demonstrator in physiology from 1902–1903, as instructor from 1903–1904, and as adjunct professor of physiology from 1904–1910. In 1910 he was transferred to the equivalent grade of associate professor under a new method of classification. The committee also claimed that Burton-Opitz "is doing most of the undergraduate teaching and some of the graduate teaching" and "is an important if not chief contributor to physiological literature from the university."
32. Butler to Lambert, 5 March 1918, CAF, Lambert, f. 8.
33. Lee, "Statement Made Before the Committee on Administration," 10 April 1918, Lee Papers, box 1, f. Burton-Opitz.
34. Lee claimed that because of this inactivity his expulsion from membership in the American Physiological Society (APS) had been suggested. Lee to Butler, 23 March 1919, Lee Papers, box 1, f. Burton-Opitz.
35. Lee's grievances against Burton-Opitz were formally outlined in two contexts—once before the Committee on Administration and again a year later in a letter to President Butler. Lee, "Statement Made Before the Committee"; Lee to Butler, 25 March 1919.
36. Lee to Butler, 25 March 1919.
37. F. D. Zeman and H. T. Hyman to Medical Faculty, "Petition by Frederic D. Zeman and Harold T. Hyman, 4th yr. Medical Students," 17 May 1917, Lee Papers, box 1, f. Burton-Opitz.
38. H. B. Williams to Lee, 4 May 1919, Lee Papers, box 1, f. Burton-Opitz.
39. Burton-Opitz's comments about Lee can be found in Zeman and Hyman to Medical Faculty, "Petition." Mrs. Lee, the former Laura Billings, was independently wealthy; at the time of the conflict the Lees were paying the salaries of four members of the physiology department, either directly or through the income from the Lee Fund, established in 1917. See Lambert to Butler, "Budget Report," 12 December 1917, CAF, Lambert, f. 7, pp. 14–15.
40. Lambert to Butler, 21 March 1919, Lee Papers, box 1, f. Burton-Opitz.
41. Lambert to Lee, 5 April 1919, Lee Papers, box 1, f. Burton-Opitz.
42. For example, when asked by Butler to compare medical students with two and four years of college preparatory work, Lambert responded that success in medical school depended more on a personal equation than on college training, reminding Butler that raising the

entrance standards had not notably reduced the attrition rate. He then discussed the student of the 1880s who had a less specialized basis for his education than students of today. Emphasis in the past was placed on a broad classical education, a condition that he claimed "placed in a man's soul a better potential for the educated physician to be a Doctor in fact as also in name." For Lambert a physician was not so much a scientist as a cultured gentleman. Lambert to Butler, 24 September 1917, CAF, Lambert, f. 7.

43. Lee to Butler, ca. March 1918, Lee Papers, box 1, f. Burton-Opitz.
44. Ibid.
45. Ibid.
46. Both of these philanthropic organizations played central roles in providing a blueprint and capital for changing the organization of medical education. See in particular Raymond B. Fosdick, *The Story of the Rockefeller Foundation* (New York: Harper, 1952) and Henry Pritchett's "Introduction" in Abraham Flexner, *Medical Education in the United States and Canada*, Bulletin no. 4 (New York: Carnegie Foundation for the Advancement of Teaching, 1910). For a more critical appraisal of the role of foundations in reforming medical education, see E. Richard Brown, *Rockefeller Medicine Men: Medicine and Capitalism in America* (Berkeley: Univ. of California Press, 1979), esp. chapt. 4.
47. The tempest produced by Adrian Lambert's appointment and the various reactions to it are described in: Pritchett to Lambert, 4 March 1918, 12 March 1918; Lambert to Pritchett, 8 March 1918, 20 March 1918; Lambert to Butler, 25 March 1918, 28 May 1918; CAF, Lambert, f. 8. For Lambert's resignation see Lambert to Butler, 20 January 1919, CAF, Lambert, f. 9. See also Lambert, "Report of the Dean of the Medical School," *Columbia Univ. Annu. Rep.* 1919, pp. 107–129.
48. William Darrach, "Report of the Dean of the Medical School," *Columbia Univ. Annu. Rep.* 1920, p. 81. Lee served as Research Professor of Physiology from 1920 to 1928.
49. Pike to Lee, 29 November 1920, Lee Papers, box 2, f. F. H. Pike. Lee at first declined to join this battle against Burton-Opitz, but on his return to the States, a closer examination of the text caused him to change his mind. Lee produced evidence that significant portions of the book were plagiarized, and brought this information to the attention of the Council of the APS for appropriate action. While acknowledging the gravity of the situation, the APS Council was still hesitant to take any action that could only hurt the interests of the discipline. Lee agreed to withdraw his request for a formal investigation. Lee to Pike, 19 December 1920, Lee Papers, box 2, f. F. H. Pike; Lee to Council of the APS, 3 November 1921, Lee Papers, box 1, f. Burton-Opitz; APS Council Minutes, vol. 2, 28 December 1921; APS Council Minutes, vol. 2, C. W. Greene (Secretary) to Lee, 3 March 1922; APS Council Minutes, vol. 2, Lee to Greene, 10 March 1922. Minutes of the APS Council are located in the archives of the APS. Bethesda, MD. I gratefully acknowledge the help of Dr. Toby Appel, APS archivist, in locating this material.
50. "Physiology Report," 16 January 1922, Faculty Discussions, P&S, Spec. MS Coll.
51. Ibid.; see also "Report of the Committee on Physiology," 9 January 1922, Faculty Discussions, P&S, Spec. MS Coll.
52. "Committee on Administration Discussions," vol. 1, 28 November 1921, P&S, Spec. MS Coll.
53. "Statement of Purposes and Functions of P&S," ca. 1920, P&S, Spec. MS Coll.
54. The standard histories of the Harvard Medical School include Henry Beecher and Mark Altschule, *Medicine at Harvard* (Hanover, NH: Univ. Press of New England, 1977); Thomas Harrington, *The Harvard Medical School*, 3 vols. (New York: Lewis, 1905); and Samuel Eliot Morison, *The Development of Harvard University Since the Inauguration of President Eliot, 1869–1929* (Cambridge, MA: Harvard Univ. Press, 1930). For a detailed account of the reforms during Eliot's administration, see Kenneth M. Ludmerer, "Reform at Harvard Medical School, 1869–1909," *Bull. Hist. Med.* 55(1981): 343–370.
55. On Eliot, see Henry James, *Charles W. Eliot, President of Harvard University, 1869–1909*, 2 vols. (Boston, MA: Houghton Mifflin, 1930); Edward Howe Colton, *The Life of Charles W. Eliot* (Boston, MA: Small and Maynan, 1926); and Hugh Hawkins, *Between Harvard and America: The Educational Leadership of Charles W. Eliot* (New York: Oxford Univ. Press, 1972).
56. Although the powerful older members of the medical faculty led by Henry Jacob Bigelow and Oliver Wendell Holmes opposed Eliot's reforms, he did find support among a few

younger members, particularly James Clarke White. See Ludmerer, "Reform at Harvard," 346–347 and Beecher and Altschule, *Medicine at Harvard*, pp. 87–93.

57. Forty-sixth *Annual Report* of the president of Harvard College, 1870–1871, p. 33; Beecher and Altschule, *Medicine at Harvard*, pp. 90, 93.
58. Beecher and Altschule, *Medicine at Harvard*, pp. 91, 94; Ludmerer, "Reform at Harvard," 344, 352. A detailed account of curricular changes is found in Harrington, *Harvard Medical School*, chapts. 33–35.
59. Ludmerer, "Reform at Harvard," 348.
60. Ibid.
61. On H. P. Bowditch's career, see Walter B. Cannon, "Biographical Memoir Henry Pickering Bowditch, 1840–1911," *Natl. Acad. Sci.* 8 (1922): 181–196; Idem, "Henry Pickering Bowditch," *Proc. Am. Acad. Arts Sci.* 46(1912): 739–745; Frederick W. Ellis, "Henry Pickering Bowditch and the Development of the Harvard Laboratory of Physiology," *N. Engl. J. Med.* 219(1938): 819–828; Everett Mendelsohn, "Henry Pickering Bowditch," *Dictionary of Scientific Biography*, vol. 2 (New York: Scribner, 1973), pp. 365–368. See also W. Bruce Fye, "Why a Physiologist?: The Case of Henry P. Bowditch," *Bull. Hist. Med.* 56(1982): 19–29.
62. For details regarding Bowditch's European studies, see chapter I by Robert Frank and chapter II by Fye in this book.
63. Charles E. Brown-Séquard, *Advice to Students: An Address Delivered at the Opening of the Medical Lectures of Harvard University, Nov. 7, 1866* (Cambridge, MA: Cambridge Univ. Press 1867); H. P. Bowditch, "Memoir of Charles Edouard Brown-Séquard, 1817–1894," *Biogr. Mem. Natl. Acad. Sci.* 4(1902): 93–97.
64. Despite his inclinations and the encouragement of his former mentors, Jeffries Wyman and Holmes, H. P. Bowditch was hesitant to pursue a scientific career because of the few opportunities for research in the United States. A pledge of support and financial assistance from his father resolved this dilemma, however, and Bowditch continued his physiological education. H. P. Bowditch to Jonathan Ingersoll Bowditch, 12 February 1869, 18 March 1869; and J. I. Bowditch to H. P. Bowditch, 1 March 1869, in Bowditch Papers, Harvard Medical Archives, Francis A. Countway Library of Medicine, Harvard Medical School, Boston, MA (hereafter, CL-HMS). This exchange of letters has been quoted in nearly every major biographical account of H. P. Bowditch's life; see esp. Cannon, "Biographical Memoir," 184–185 and Fye, "Why a Physiologist," 23–24.
65. In a letter to the editor, *Boston Med. Surg. J.* 82 (1870): 305–307, H. P. Bowditch enthusiastically described the Leipzig laboratory and Professor Carl Ludwig. This essay later appeared in "The Physiological Laboratory at Leipzig," *Nature Lond.* 3 (1870): 142–143. Under Ludwig's direction, H. P. Bowditch not only carried out research that resulted in two published papers, classics in the physiological literature but he also designed several pieces of apparatus. See Cannon, "Biographical Memoir," 185, and Cannon, "The History of the Physiology Department of the Harvard Medical School," *Bull. Harv. Med. Sch. Alumni Assoc.* (March 1927): 15.
66. Ludmerer argues that the Eliot of 1870–1871 valued teaching more than the pursuit of knowledge, although by the 1890s and 1900s he was actively promoting original research. Ludmerer, "Reform at Harvard," 351.
67. Cannon, "Henry Pickering Bowditch," 741.
68. Ellis, "Henry Pickering Bowditch," 821.
69. An illustrative anecdote recounted by a former student, Ellis, suggests what conditions were like in the physiology laboratory in 1878: "On one occasion in my undergraduate days I experienced a desire for a frog for some scientific purpose, and I ventured to approach Bowditch for a donation from his store. He very gravely and courteously informed me that the funds at his disposal did not warrant a reckless extravagance of that kind, and then the natural goodness of his heart overcame him and he went to his tub and seized the most diminutive batrachian in it, with which I departed with mingled feelings of exultation and remorse." Ellis, "Henry Pickering Bowditch," 821.
70. The undergraduate experiments were for the most part chemical and in 1896 their supervision was transferred to the chemistry department. Beecher and Altschule, *Medicine at Harvard*, p. 100.

71. Cannon, "Biographical Memoir," 186.
72. *The Harvard Medical School 1782–1906* [by the faculty? 1906], p. 92. H. P. Bowditch's own investigations were as varied as those of his students. For a complete bibliography of H. P. Bowditch's published works, see Cannon, "Biographical Memoir," 195–196.
73. Cannon, "History of the Physiology Department," 16.
74. Ellis, "Henry Pickering Bowditch," 822.
75. Throughout the 1870s and 1880s students were permitted access to the laboratory twice during their first year for a small amount of practical instruction. When routine laboratory instruction was finally offered in the 1890s, H. P. Bowditch felt regret at the passing of the old methods and warned against too great a reaction from a didactic style "lest useful as well as useless things be swept away." However, he also maintained an open mind toward laboratory teaching and later embraced it "as an important forward step in the improvement of educational methods in medicine." Cannon, "Biographical Memoir," 191; H. P. Bowditch, "Reform in Medical Education," *Science* 8 (1898): 926.
76. See Cannon, "Henry Pickering Bowditch," 742; and Ellis, "Henry Pickering Bowditch," 822.
77. On the founding of the APS, see chapter VI by Toby Appel in this book.
78. On the importance of endowment, see Ludmerer, "Reform at Harvard," 358–359.
79. Ibid., 358–362. H. P. Bowditch and John C. Warren were particularly effective fund raisers because of their dual credentials as scientific and social leaders. Having gained postgraduate experience abroad, they were recognized as representatives of scientific medicine. Yet as blue-blooded Boston Brahmins dedicated to community service they were able to use their social prominence to more practical advantage on behalf of the medical school.
80. Cannon writes that "the conflict between scientific study and administrative activities seems to have disturbed [H. P. Bowditch], for it is recorded that occasionally in the late years of his life he would ask a friend whether his life would not have been of greater service if he had devoted himself exclusively to experimental physiology." Cannon, "Biographical Memoir," 194.
81. Morison, *Development of Harvard*, pp. 563–565. H. P. Bowditch was a strong advocate of the elective course system in medical study. Two of his later addresses deal with this theme. H. P. Bowditch, "Reform in Medical Education," *Science* 8 (1898): 921–927; H. P. Bowditch, "The Medical School of the Future," *Phila. Med. J.* 5 (1900): 1011–1018.
82. Morison, *Development of Harvard*, p. 563.
83. William Osler, "Henry Pickering Bowditch," *Lancet* (1911): 975.
84. A good account of Porter's research and teaching career at Harvard may be found in A. Clifford Barger, "The Meteoric Rise and Fall of William Townsend Porter, One of Carl J. Wiggers' 'Old Guard,'" *Physiologist* 25 (1982): 407–413.
85. Porter, "The Teaching of Physiology in Medical Schools," *Boston Med. Surg. J.* 139 (1898): 647–648; Idem, "The Teaching of Physiology," *Phila. Med. J.* 4 (September 1, 1900); Idem, *Physiology at Harvard*, 2d ed. (Cambridge, MA: Harvard Univ. Press, 1903), pp. 1–7. See chapter XIII by Merriley Borell in this book.
86. A comprehensive account of Cannon's career can be found in Chandler McC. Brooks, Kiyomi Koizumi, and James O. Pinkston, eds., *The Life and Contributions of Walter Bradford Cannon, 1871–1945* (New York: State Univ. of New York, 1975).
87. Porter had been instrumental, first in getting Cannon onto the faculty and second in securing his promotion in 1902. However, Cannon's appointment as Higginson Professor, a position Porter had lobbied to receive himself, caused a rift between the two that continued for thirty years. See Saul Benison and Barger, "Walter Bradford Cannon," in *Dictionary of Scientific Biography*, vol. 3 (New York: Scribner, 1971), pp. 71–77; Benison, Barger, and Elin L. Wolfe, "Changing of the Guard," *Harv. Med. Alumni Bull.* 56 (1982); Barger, "Meteoric Rise and Fall," 412–413.
88. For a bibliography of Cannon's work, see Brooks et al., *Life and Contributions*, pp. 239–263; H. H. Dale, "Walter Bradford Cannon," *Obituary Notices of Fellows of the Royal Society of London* 5(1947): 407–423.
89. Dale, "Walter Bradford Cannon," 412.
90. Cannon, *A Laboratory Course in Physiology*, 4th ed., (Cambridge, MA: Harvard Univ. Press, 1923).

91. A discussion of the tutorial system is found in Alfred C. Redfield, "The Laboratories of Physiology of the Harvard Medical School," in *Methods and Problems of Medical Education*, 3d series (New York: Rockefeller Foundation, 1925), pp. 19–34. Cannon may have been assuming a defensive position against critics who felt that he was trying to entice medical students into physiology. At various other times he argued that the way to attract men to the preclinical sciences was to expose them to research and thus grasp their interest.
92. Quoted in Beecher and Altschule, *Medicine at Harvard*, p. 234.
93. Cannon to Richard M. Pearce, 9 December 1929, Record Group 1.1, 200A, box 89, f. 1069, located in the Rockefeller Archive Center (RF), Pocantico Hills, North Tarreytown, New York 10591 (hereafter, RF series number/box number/folder number). Cannon also justifies his decision not to accept the Mayo Clinic offer in a letter to his colleague, Joseph Erlanger. Cannon to Erlanger, 21 February 1921, Erlanger Papers, box 2, f. 12, located at the Washington University Medical School Archives, St. Louis, MO.
94. Eliot to Charles S. Minot, 13 March 1899; Eliot to Minot, 29 April 1899, Minot Papers, C 21.2, CL-HMS.
95. Several professorships in comparative subjects were created, however, in the intervening years. The Fabyan Professorship of Comparative Pathology had already been established and in 1902 the James Stillman Chair of Comparative Anatomy was founded for Minot. In 1906 when Cannon became Higgison Professor of Physiology, Porter was given the newly established Chair of Comparative Physiology. Beecher and Altschule, *Medicine at Harvard*, p. 167.
96. Apart from the Division of Physiology, the medical sciences would be represented by the Division of Anatomy (Department of Anatomy, Department of Histology and Embryology), the Division of Pathology and Bacteriology (Departments of Pathological Anatomy, Comparative Pathology, and Bacteriology), and by the Division of Hygiene and Therapeutics (Departments of Hygiene, and Material Medical and Therapeutics). Cannon to Eliot, 5 January 1904, Cannon Papers, box 1, f. 3.
97. Ibid.
98. Beecher and Altschule, *Medicine at Harvard*, p. 109.
99. Cecil K. Drinker to Cannon, 28 April 1918, Cannon Papers, box 1, f. 8.
100. Cannon to Drinker, 21 May 1918; Drinker to Cannon, 7 June 1918, Cannon Papers, box 1, f. 8.
101. Morrison, *History of Harvard University*, pp. 577–578; Beecher and Altschule, *Medicine at Harvard*, pp. 196–197.
102. With a total staff of 58 including faculty, graduate students, and fellows, and a combined budget of $77,876 for 1924–1925, it would be hard to dispute his claim. Redfield, "Laboratories of Physiology," pp. 22, 34.
103. Ibid., p. 19.
104. Record of grant, 16 April 1930, RF200A/89/1069.
105. When Cannon was permitted to continue as head of the department despite having reached retirement age, the foundation appropriated an additional $75,000 for the next five years (1938–1942). Record of grant, 20 November 1936, RF200A/89/1069.
106. Gregg, "Harvard University—Appraisal," December 1938, RF200A/89/1075.
107. Ibid.
108. See also Pauly, "Appearance of Academic Biology," pp. 378–382, for a discussion of physiology at Hopkins.
109. Lee to Low, 15 November 1894, CAF, Lee, f. 1.
110. Lambert to Butler, 21 March 1921, Lee Papers, box 1, f. Burton-Opitz.
111. Morison provides a detailed account of families associated with the medical school. For instance he notes that "four of the five generations of Warrens in direct line, four members of five generations of the Jacksons, and four of the Shattuck family from three generations, have had seats on the faculty. Four successive generations of the Reynolds family have been connected with the School. Three families have been represented in three generations: White, Bowditch, and Bigelow." Morison, *Development of Harvard University*, p. 592.
112. Cf. Kohler, *From Medical Chemistry* and Pauly, "Appearance of Academic Biology."

IV

A. B. Macallum and Physiology at the University of Toronto

SANDRA F. McRAE

There is a very old joke about Canada, the country that could have had the best of all possible worlds with British government, French culture, and American know-how, but ended up instead with French government, American culture, and British know-how. Physiology in Canada was also based on foreign models, with varying degrees of success, and the joke can thus serve as a point of departure for this preliminary discussion of the institutional development of physiology in Toronto in the late nineteenth and early twentieth centuries, when Canadian medical education and research were being influenced by American, British, and European examples. Toronto, while geographically adjacent to the United States of America, was at that time still linked ideologically more closely to Great Britain. Canadian physiologists and biologists were able to develop a distinctive blend of scientific research in which the dominant British influence was integrated with strong ties to the United States.

That general conclusion emerges only gradually in the course of this chapter. My more immediate aim is to show that the University of Toronto had a well-established research tradition in experimental physiology from the 1880s onward, with Archibald Byron Macallum playing a leading role in the rise of both physiology and biochemistry in Canada from that institutional base. After a brief summary of Macallum's career, I provide a fairly extensive discussion of the transformation of medical education in Toronto in the late nineteenth century. Macallum's success cannot be understood in isolation from this sequence of events, which opened up a new and larger institutional niche for physiology and the other life sciences at the University of Toronto. I then review the careers of some of Macallum's students, and of other Toronto graduates in the biomedical sciences, in order to demonstrate the vigor and quality of the training offered there by the 1890s. Finally, I show that physiology at the University of Toronto during Macallum's tenure there, from 1887 to 1918, was characterized by a pattern in which the best research students were sent off to the United States, while senior faculty posts at Toronto were filled mainly by recruits from other parts of the British Empire.

One important reason for telling this neglected story is that it helps us to understand how Toronto became the site of perhaps the most celebrated example of a discovery made in a physiological laboratory that had dramatic therapeutic implications: the discovery of insulin, isolated and purified for clinical use in 1921–1922 at the University of Toronto by the research team of Frederick G. Banting, Charles H. Best, John James R. Macleod, and James B. Collip. It was the discovery of insulin that made Toronto famous as a center for physiological research—indeed, it led in astonishingly swift order (in 1923) to the first Nobel Prize in physiology or medicine to be awarded to a North American. Unfortunately, conventional accounts of the famous insulin story and disputes over the awarding of the Nobel Prize to Macleod and Banting jointly, have tended to distort the historical image of the state of physiological research in Canada in the early 1920s. By emphasizing the role of two virtually untrained Canadian "heroes" working in isolation (Banting and Best), while minimizing the crucial work of a thoroughly trained biochemist (Collip) and portraying Macleod as an autocrat whose contributions were negligible, these accounts fostered the illusion of an almost miraculous discovery occurring in the midst of a scientific wilderness.[1] This was hardly the case.

The recent books of the Canadian historian Michael Bliss[2] have provided a more balanced assessment of the insulin story and its proper place in the history of biomedical research in Canada:

> The discovery of insulin is sometimes thought to have caused the first wave of enthusiasm for medical research in Canada. Actually the discovery was in part a consequence of the first wave, which came to Canada earlier than is often realized (and the misunderstanding about its relationship to the discovery of insulin is a nice example of how in history we can sometimes get cause and effect mixed up).[3]

This chapter confirms and develops that point by showing that the University of Toronto had a well-established and even distinguished research tradition in physiology long before anyone had even heard of insulin. Macallum had been appointed to a full-time teaching and research position in physiology at Toronto as early as 1887. By 1909, partly as a result of a reform of the system by which the Province of Ontario funded the University of Toronto, the Faculty of Medicine had no less than three professorships occupied by men trained in physiology at laboratories in England or the United States (namely, Macallum, John Beresford Leathes, and Thomas Gregor Brodie). By the end of the First World War, they had, with their associates and successors, succeeded in establishing Toronto as a creditable center for physiological and biochemical research. The laboratory and intellectual resources that made possible the discovery of insulin at Toronto in 1921–1922 were directly dependent on these earlier developments in physiological research at the University of Toronto and can even be considered the most tangible legacy of the leadership exercised by Macallum, who had left the university only four years before.

WORK AND CAREER OF A. B. MACALLUM

Archibald Byron Macallum (1858–1934) was one of the first professional physiologists and biochemists in Canada, and the story of his education and career in many ways reflects the development of a new scientific discipline in a relatively young country. Macallum was born in 1858 on a farm near Belmont, Western Ontario, nine years before the British North America Act of 1867 gave Canada the status of a Dominion. At the age of fifteen he became a teacher in a rural school in order to earn money to attend University College in Toronto, where he graduated with an Arts degree from the University of Toronto in 1880, winning the silver medal in natural sciences. For three years he taught high school in Cornwall, Ontario, but he continued to carry out research on the anatomy and physiology of the catfish *Amiurus* and other freshwater vertebrates, in collaboration with Robert Ramsay Wright (1852–1933), the professor of natural history at University College, Toronto.[4]

Ramsay Wright, a graduate of the University of Edinburgh, reportedly relinquished the offer of a position aboard the *Challenger* deep-sea expedition in order to come to Toronto in 1874 to teach natural history. He was to remain at the University of Toronto for thirty-eight years, after 1887, as professor of biology, during which time he built up the Department of Biology with the aid of his students, including Macallum, and was a powerful force in advocating medical interests within the university. Ramsay Wright is best remembered for his brilliant teaching and for the group of distinguished students that he trained, but he remained active in research over the years in aquatic biology and parasitology.[5] He was able to build up a strong Department of Biology within the University of Toronto before it reestablished its teaching Faculty of Medicine in 1887, in a situation partially analogous to that at the Johns Hopkins University in the 1880s and at other American institutions where biology flourished briefly in the absence of competing medical interests.[6]

In 1884 Macallum left teaching in Cornwall to return to the University of Toronto as a fellow in biology in Ramsay Wright's department, where he did some undergraduate teaching and carried out research on the digestive and nervous systems of freshwater vertebrates. At about the same time, Macallum also began to work toward his Ph.D. with the physiologist Henry Newell Martin at Johns Hopkins. (It is not entirely clear how Macallum divided his time between Toronto and Baltimore during this period.) Martin had been a student of Michael Foster at University College, London, and later at Cambridge, until Foster sent him to Baltimore to start an American school of physiology.[7] Martin's research at Johns Hopkins focused on the heartbeat and was closely related to the research of Foster, Langley, Gaskell, and others of the Cambridge School, but his teaching covered a much broader range of topics, for he had been brought to America specifically to teach courses that would serve the needs of both medical students and research biologists.[8] Macallum's doctoral thesis under Martin on "Studies on the Blood of Amphibia," was completed in 1888.[9]

During the 1890s Macallum continued research related to his Ph.D. thesis at Johns Hopkins on amphibian blood and, more generally, on iron-bearing compounds in vertebrate fluids and tissues. In Toronto he was aided by Robert Russell Bensley (1867–1956), an assistant and then a lecturer in the biology department who taught biology and histology there from 1892 to 1901.[10] Together, Macallum and Bensley devised a microchemical technique for demonstrating the presence of iron within plant and animal cells and they later extended this work to phosphorus compounds as well.[11] Even this early, Macallum's research was more biochemical than physiological and he consciously related it to an evolutionary framework.[12]

Around 1900, Macallum's research program was radically altered by the construction of a Canadian Marine Biological Station in the Maritimes, owned by the Biological Board of Canada, which had been founded in 1898 with Macallum and Ramsay Wright among its first members.[13] With access to a marine field station, Macallum began to spend his summers analyzing the inorganic ionic constituents of jellyfish, inspired by some of the work of the German-American physiologist Jacques Loeb.[14] This study inaugurated nearly two decades of work on the ionic composition of the circulatory and intracellular fluids of marine invertebrates and, later, on the paleochemistry of vertebrates, that was to dominate the rest of Macallum's research career. He proposed that the evolution of various forms of animal life in the seas could be traced through their development of closed circulatory systems, whose fluids were similar to the oceanic environment in which the animals evolved.[15]

Macallum clearly exerted a profound influence on the growth of scientific medicine at Toronto, not least through his support for and exemplification of the research ideal. However, he would never have had an opportunity to exert that influence in the absence of major reforms in medical education that occurred at Toronto in the 1880s. Indeed, even as Macallum pursued his graduate studies in physiology in the United States, the University of Toronto underwent organizational changes that were to determine the future course of his career.

MEDICAL REFORM AND PHYSIOLOGY AT THE UNIVERSITY OF TORONTO

Education in Canada is under provincial rather than federal control and in 1887 the government of the Province of Ontario passed the University Federation Act, a piece of legislation intended to unite the separate denominational colleges of Ontario under the administration of a single provincial university, similar to the system at the University of London.[16] The Federation Act also allowed the reestablishment of the Faculties of Medicine and Law as teaching departments of the University of Toronto for the first time in thirty-five years, a development that helped to create an institutional niche for physiology hitherto absent in Toronto.

The history of medical education in Toronto in the nineteenth century is a long and convoluted story that is beyond the scope of this paper, but a few salient points deserve attention here because of their bearing on the development of physiology at Toronto as both a biological and a medical science. The original King's College that preceded University College had a small Faculty of Medicine from the 1840s until 1853, when for political reasons, the Faculty of Medicine and the Faculty of Law were abolished on the grounds that professional education ought to be self-supporting and not carried out at public expense. Thus after 1853 the University of Toronto acted only as an examining body granting medical degrees, while the instruction was carried out by private medical colleges. At various times these medical schools included the medical department of the Methodist Victoria College, the Anglican Trinity Medical College, and the independent Toronto School of Medicine but by the 1880s only the latter two rival schools survived.[17]

In 1882 the University of Toronto had tried to elevate the standards of its medical examinations, requiring more familiarity with the basic sciences, but even then medical teaching remained outside of university control. When the 1887 Federation Act allowed the Faculty of Medicine to be reactivated, the University of Toronto arranged to assimilate the existing Toronto School of Medicine, renting its buildings and paying professor's salaries from the medical students' fees, on the understanding that the new medical school would not be a financial drain on the slender resources of the university.[18]

Long before 1887 the old Toronto School of Medicine had taught physiology to its medical students as part of their preparation for the examinations set by the university. However, in the spring of that year Dr. Michael Barratt, who had been teaching physiology at the Toronto School of Medicine, fell ill and died. Professor Ramsay Wright taught the remaining six physiology lectures to the medical students.[19] By October of 1887 the Toronto School of Medicine had been transformed into the Medical Faculty of the University of Toronto and Macallum had been appointed to the new position of lecturer in physiology within Ramsay Wright's Department of Biology, with his salary paid by the Faculty of Arts.

For the first few years of operation of the reactivated Faculty of Medicine, medical students divided their time between the university buildings, in which the basic science courses including biology, chemistry, and physiology were taught, and the old Toronto School of Medicine building located several miles away (near the old Toronto General Hospital), where anatomy and the clinical subjects were taught. A special streetcar service was provided to carry students between the two sites but the arrangement was not very satisfactory.[20] In 1889, however, the new Biological Building was built at the university and at the opening ceremonies, Ramsay Wright, Sir Daniel Wilson (President of the University of Toronto), the Premier of Ontario, and the Minister of Education were joined on the platform by four guest speakers whose presence and speeches made explicit the already close identification of medical and biological interests at Toronto. Those four guests were William Osler and William H. Welch (both at the Johns Hopkins Hospital before the opening of the medical school there in 1893), Charles S. Minot, from the Harvard Medical School, and Victor C. Vaughan, from the University of Michigan.[21]

In 1891 an extension known as the Museum Wing was added to the Biological Building of the University of Toronto. The attic of the Museum Wing contained a dissecting room for teaching human anatomy to the medical students, who were now able to take their first two preclinical years of courses in the university buildings. The new rooms for anatomical instruction were too much for the enemies of the university medical school, particularly Dr. Walter Geikie, dean of the rival Trinity Medical College, who complained that the provincial government was subsidizing one of Ontario's medical schools while refusing to aid the others.[22] A fierce battle ensued in 1892, with Geikie claiming that the Biological Building was in all but name a medical school building, and that the government of the province had been deliberately deceived as to its real function by Ramsay Wright and Sir William Mulock, the vice-chancellor of the University of Toronto and a prominent supporter of medical interests there.

The "Biological Building debate," as I have called this dispute elsewhere,[23] was eventually resolved when the University of Toronto's Faculty of Medicine agreed to pay rent to the university for the use of the dissection room. The anatomical laboratory was allowed to stay in the building and the allegiances that were forged at this time were to have consequences for all of the sciences related to medicine. The new president of the University of Toronto, James Loudon, professor of physics and a longtime ally of Ramsay Wright's in the struggle to introduce laboratory teaching in science courses at University College, had arranged for an investigation into the circumstances surrounding the Biological Building addition in October of 1892. The transcripts of this inquiry, conducted almost in the form of a trial, reveal that the decision-making power with respect to the new facilities rested not with the professors who actually taught anatomy but with Ramsay Wright and Sir William Mulock.[24] Ramsay Wright told the committee that he considered anatomy to be a part of biology, that a human anatomy course would be useful for arts students to take, and that the new anatomical laboratory had seemed to him to be "a very proper addition to the Biological Building, because at the time in Cambridge [England] a similar building and a similar combination was being made."[25]

For a decade or so after 1892, medical and biological interests at the University of Toronto were perceived as one and the same in the eyes of the government and in the competition for funds with other science departments within the university. This alliance between biology and medicine was to prove particularly important for Macallum, whose initial appointment was in the Faculty of Arts but whose physiology classes were directed largely toward the medical students. Macallum even took an M.B. at Toronto in 1889, "feeling it was wise for a teacher of medical students to take the degree in medicine,"[26] although he was never to practice. By 1890 he had been appointed associate professor (later promoted to full professor) of physiology and physiological chemistry, positions that he retained until 1908 when the chair was divided into two positions. Macallum then became the professor of biochemistry at Toronto from 1908 until 1916, worked for the Canadian government during the First World War, and was later the professor of biochemistry at McGill University in Montreal from 1920 to 1928.

During the 1890s the Biological Building housed the medical students for most of their preclinical courses as well as the students in the natural science honors and pass courses. This was precisely the period during which many successful research scientists were trained at Toronto. At the turn of the century a new combined course known as the "B. and P." (the initials first stood for "biology" and "physics", but the "P" later stood for "physiology") was developed to produce an elite stream of medical students trained in scientific research methods.[27] The B. and P. was a six-year program leading to the M.B. degree, in which students earned an Honors B.Sc. after four years, the last two serving as the first two years of the four in the medical course.

Even in the mid-1890s it had been possible to enroll in the last two years of arts and the first two years of medicine simultaneously, as student records indicate.[28] However, the introduction of the B. and P. course gave official recognition to the importance of teaching scientific medicine.

In 1903 a new medical building was opened at the University of Toronto, designed and built according to the unit system of Minot.[29] In that same year Trinity College finally joined the University of Toronto federation. As a result, Trinity Medical College and the Faculty of Medicine of the University of Toronto were united and moved into the new facilities together. As part of the agreement all of the professors from both medical schools were hired by the newly amalgamated Faculty of Medicine and the two groups of students were also united into one enormous body. Among the guests invited to the opening were Osler, Welch, and Minot once again, Professors Keen from Philadelphia, Henry P. Bowditch from Harvard, and Charles Sherrington from Liverpool.[30] By then Toronto had clearly established a reputation for biomedical research and teaching and it could point with pride to an ever-growing body of distinguished alumni mainly, but not exclusively, former students of Macallum.

UNIVERSITY OF TORONTO GRADUATES

Although Macallum had a physiological laboratory in the Biological Building in Toronto where he and his assistants carried out research, he never really developed a recognizable "Toronto school of physiology," despite his attempts to model his department on Michael Foster's school at Cambridge.[31] In 1932 Macallum wrote as follows about the early years of physiology at Toronto:

> In Canada the advancement of the science began in 1885 when the writer, a student of Henry Newell Martin, introduced in the teaching of medical students the modernized type of physiology, all under the influence of R. Ramsay Wright, an outstanding exponent of Biology in the University of Toronto from 1874 to 1912. In 1887 the course in physiology was patterned after that in Cambridge under Sir Michael Foster, under whose auspices the school of physiology there rivalled that anywhere else on the continent. As a result in the succeeding twenty years a number of physiologists were trained, of whom three occupy chairs of the subject in universities in the United States, one in England and one in Canada, all of whom have made, by their experimental investigations, noteworthy contributions to the knowledge of this subject.[32]

Unfortunately Macallum named only one of these five students, the one in Canada, Professor F. R. Miller, who taught physiology at the University of Western Ontario in London, Ontario. Of the three former students occupying chairs in the United States, one can easily be identified as Frederick Hughes Scott; the other two might be Maud L. Menten and/or Ralph Lillie, though that would suggest that Macallum's definition of the term physiologist was a remarkably broad one. The holder of the English chair remains a mystery. In any case, five Toronto-produced professors might seem to reflect a rather low rate of production of new physiologists there before the First World War, but it should be remembered that graduate opportunities were then very limited and the Ph.D. program was slow to evolve, having started only in 1897 with strong support from Macallum.[33]

As a matter of fact, the first Ph.D. awarded by a Canadian university was earned by Macallum's research student and assistant, Scott (1876–1951), who graduated from the University of Toronto in 1900 with a thesis "On the Structure, Micro-chemistry and Development of Nerve Cells, with Special Reference to Their Nuclein Compounds."[34] For the next few years, Scott worked as a demonstrator in physiology at the University of Toronto, until he completed his medical degree there in 1904 and went to England on a MacKinnon Studentship of the Royal Society of London. In 1906 and 1907 Scott published two promising papers that have recently been rediscovered because of their clear enunciation of two basic concepts in the neurosciences: the chemical transmission of nervous impulses and axonal transport.[35] For several reasons Scott did not return from London to the University of Toronto but went on instead to become professor of physiology at the University of Minnesota from 1908–1944. It is certainly Scott to whom Robert Kohler alludes (but does not name) as "a young Ph.D. physiologist from Starling's school" who was hired in 1908 to build up a research school in physiology at Minnesota.[36]

Another Ph.D. student of Macallum's was James Bertram Collip (1892–1965), who received his doctorate in biochemistry at the University of Toronto in 1916 and taught at the University of Alberta from 1915 to 1921. He then returned to Toronto on a research fellowship and achieved fame for his contributions to the purification of insulin there in association with Macleod, Banting, and Best. J. J. R. MacLeod (1876–1935), who succeeded Macallum and T. G. Brodie as professor of physiology at the University of Toronto from 1918 to 1928, divided his share of the 1923 Nobel Prize in medicine or physiology with J. B. Collip, while Frederick G. Banting divided his share with Charles H. Best.[37] Collip left Toronto in 1922 to return to the University of Alberta and went on to become one of Canada's leading medical researchers of the midtwentieth century, at the Universities of Alberta, McGill, and Western Ontario.[38]

Although the postgraduate degree programs at Toronto were slow getting started and fellowship opportunities were extremely limited, the university did excel as early as the 1890s at producing biomedical research scientists trained at the undergraduate level, thanks to the combined efforts of Ramsay Wright and Macallum. Because many of these Toronto graduates went on to successful university or hospital careers elsewhere, particularly in the United States, their early training in Toronto has often been overlooked. The Toronto B.A., B.Sc., and M.B. graduates can be loosely categorized into two main streams: (1) those whose research interests were primarily medical, often in pathology, physiology, or biochemistry and (2) those who were biologists, working mainly in morphology or fisheries research.

The first group of Toronto graduates with medical careers in the United States were described in 1918 as follows by John George Adami (1862–1926), professor of pathology at McGill University from 1892 to 1914[39] and a graduate of Foster's Cambridge physiology laboratory:

> Although belonging to the rival university I willingly admit that in the late eighties and early nineties, for those capable of taking advantage thereof and electing to pursue further studies elsewhere after graduation, the University of Toronto afforded the finest medical education in North America. The clinical opportunities at that time were not of the very first order, but the preliminary training in science was singularly full and thorough. The combination of Ramsay Wright's stimulating and thought-compelling lectures in general biology with A. B. Macallum's practical course in physiology, developed a set of men who today are leaders in American medicine: Llewellys (sic) Barker of Baltimore, Thomas McCrae of Philadelphia . . . , W. G. MacCallum of Columbia and now again of Johns Hopkins, T. Futcher of Johns Hopkins . . . , Oskar Klotz of McGill and Pittsburgh, and John McCrae.
>
> No school of this generation in North America can claim quite so brilliant a band of physicians and pathologists as original alumni.[40]

Adami's list could well have included other names. John Bruce MacCallum (1876–1906) graduated with a B.A. from Toronto in 1896, studied medicine at Johns Hopkins, and became associate professor of physiology at the University of California under Loeb until his life was cut short by tuberculosis.[41] J. B. MacCallum's older brother Will (1874–1944) left Toronto with his B.A. in 1894 to join the first class of the Johns Hopkins Medical School, taught pathological physiology at Johns Hopkins and pathology at Columbia, and returned to Baltimore in 1917 to succeed his mentor Welch as professor of pathology.[42] Thomas S. Cullen (1868–1953)[43] was a classmate of Lewellys F. Barker (1867–1943)[44] in medicine at Toronto, graduating in 1890, and they interned together at the Toronto General Hospital before going to Baltimore to work with the men they later succeeded—Cullen with Howard Kelly and Barker with Osler. John McCrae (1872–1918), who was better known as a war poet than a pathologist, took his B.A. in 1894 and M.B. in 1898 at Toronto and spent a brief period at Johns Hopkins before joining Adami at McGill University in 1900.[45] His brother Thomas McCrae (1870–1935) took his B.A. in 1891 and M.B. in 1895 at Toronto, worked with A. B. Macallum as a fellow in biology before moving to Baltimore in 1899, where he helped Osler with his famous textbooks, and then went finally in 1912 to Jefferson Medical College in Philadelphia as professor of medicine.[46] Robert Russell Bensley taught at Toronto from 1892 to 1901, when he joined the Department of Anatomy at the University of Chicago, at that time headed by Lewellys Barker.[47] Ralph Stayner Lillie (1875–1952), who received his B.A. from Toronto in 1896 and his Ph.D. in 1901 at Chicago under Jacques Loeb, went on to teach at Nebraska, Harvard, Johns Hopkins, Pennsylvania, and Clark Universities before returning to Chicago as the professor of physiology in 1924.[48] Finally, Maud L. Menten (1879–1960), one of the first women to graduate from the integrated medical course at Toronto in 1907, who worked in A. B. Macallum's laboratory, at the Rockefeller Institute, and with Leonor Michaelis in Germany, earned her Ph.D. in biochemistry at Chicago in 1916 and moved to Pittsburgh to teach pathology from 1916–1950.[49]

While perhaps none of these successful graduates of Toronto would have defined themselves exclusively as physiologists by profession, most of them were clearly influenced by the research ideal they had been exposed to as students; by what Macallum called the "Scientific Spirit" in medicine[50] as it was expressed in Toronto. Even those Toronto graduates of the 1890s whose careers were more biological than medical in their orientation acknowledged Macallum's influence in directing them toward research. They included Benjamin Arthur Bensley (1875–1934), who succeeded Ramsay Wright as professor of biology at the University of Toronto in 1914,[51] and Joseph Stafford (B.A., Toronto 1890), who taught biology at McGill.[52] Both worked in fisheries research at the field stations operated by the Biological Board of Canada. Other Toronto biologists migrated to the United States, including Edward Charles Jeffrey (1866–1952), who taught biology and botany courses at Toronto from 1892–1901 and then moved to the botany department at Harvard University.[52] Like his younger brother Ralph, Frank Rattray Lillie (1870–1947) also got his start in the life sciences at Toronto (B.A., 1891), where he had originally intended to study for the Anglican ministry but changed his mind when he became intrigued by natural science. Macallum showed F. R. Lillie how to "devote his life to science" and interested him in physiology, while Ramsay Wright introduced him to embryology and recommended him to C. O. Whitman at the Marine Biological Laboratory at Woods Hole, Massachusetts as a graduate student.[54]

This pattern of migration from Toronto to the United States or Europe had already been apparent in the 1880s among Ramsay Wright's most promising students, including Macallum himself, who had also been sent elsewhere to complete his postgraduate training. Other examples include John Joseph Mackenzie (1865–1922) and James Playfair McMurrich (1859–1939), who received his B.A. in 1879 and M.A. in 1881 from Toronto and earned his Ph.D. in morphology with Walter Keith Brooks at Johns Hopkins University in 1885. He taught briefly at the Ontario Agricultural College in Guelph, then at Johns Hopkins, Haverford College, Clark, Cincinnati, and the University of Michigan, before returning to the University of Toronto to become the professor of anatomy from 1907 until 1930.[55] John Joseph Mackenzie graduated from Toronto in 1886 and went to Europe immediately—first to Leipzig to study anatomy and physiology with von Frey and Ludwig "of whom he was a favorite pupil" and then to Berlin to study "physiological methods and research" with Dr. Gad in Emil DuBois-Reymond's Laboratory and bacteriology with Globig and Frankel at Robert Koch's Institute, where he was also associated with Virchow. Returning to Toronto to work as a fellow in biology under Ramsay Wright, Mackenzie held a number of appointments there during the 1890s, including bacteriologist to the Provincial Board of Health, before joining the University of Toronto as professor of pathology and bacteriology in 1900.[56]

AMERICAN AND BRITISH INFLUENCES

There were relatively few scientists in Canada in the late nineteenth century, but during the 1880s several professional gatherings helped to increase communication among them. The first of these meetings was that of the Royal Society of Canada, founded in 1882, which contained (in addition to its literary sections) two scientific academies: one for members working in the physical sciences, the other for those in the biological and geological sciences. The second major event was the 1884 meeting of the British Association for the Advancement of Science (BAAS) held in Montreal, due largely to the efforts of Sir William Dawson of McGill University.[57] In 1897 the BAAS met once again in Canada, this time in Toronto, with Macallum heading the local arrangements committee. Suggestions about the need for a marine field station in Canada were made on this occasion and led to the founding of the Biological Board of Canada (later renamed the Fisheries Research Board of Canada) in 1898.[58]

Macallum, like most physiologists trained at American universities, became a member of the American Physiological Society (APS). He was also elected to membership in the Association of American Physicians (although he never practiced medicine), was president of the Society of Biological Chemistry, and was a foreign member of the American Philosophical Society.[59] In Canada he served as president of the Royal Canadian Institute (a Toronto-based learned society in whose journal he published numerous papers), as president of the Royal Society of Canada, and from 1917–1920 as the first chairman of the wartime Honourary Advisory Council on Scientific and Industrial Research, which later became the National Research Council of Canada.

Unlike most scientists in the United States, however, Macallum also had close associations with another set of physiological colleagues, namely those spread throughout the British Empire. In 1891 a summary of Macallum's microchemical techniques for the localization of iron within cells was presented to the Royal Society of London by Henry Newell Martin,[60] and by the mid-1890s Macallum was publishing much of his work in English scientific journals. After the 1897 meeting of the BAAS in Toronto, attended by Lord Lister, Sir Michael Foster, and Charles Sherrington, to name but a few, Sherrington presented more of Macallum's work to the Royal Society of London.[61] By 1905 Macallum was traveling to South Africa to attend the BAAS meetings in Cape Town as one of three Canadian representatives. In 1906, the year that the British Medical Association held its annual meeting in Toronto, Macallum was made a fellow of the Royal Society of London.[62] Macallum played more than a token role in these British scientific societies and in 1910 he was named president of the physiological section of the BAAS meeting at Sheffield.[63]

The English influence on physiology in Canada surely helps to explain the strong biological and evolutionary slant of Macallum's own research program and its tenuous links with clinical applications. His work at the Canadian marine field station was not unlike the physiological and biochemical research carried out at similar marine laboratories in Plymouth, England and Naples, Italy. The parallels with British institutions were intentional; as already noted, Ramsay Wright in 1892 had justified the teaching of biology, physiology, and anatomy in the same building at the University of Toronto on the grounds that Cambridge University had made a similar combination of subjects. Whereas biology and medicine were often competing interests in American universities in the 1890s, with strong biology programs only evolving where the building of the medical school was delayed,[64] this was not the case at Toronto during that decade. After 1892, with Ramsay Wright and Macallum strongly influencing the preclinical medical teaching and controlling the biology department, biology and medicine at the University of Toronto shared common interests and both benefitted from increased university support.

After 1903, however, the identification of medical and biological interests at the University of Toronto became less intimate. Macallum's physiology department moved from the Biological Building into the new medical building and the enormous increase in the size of the Faculty of Medicine allowed it greater autonomy from biology. The rapid expansion of the enlarged Faculty of Medicine created new faculty positions at the University of Toronto, most of which were filled by graduates of American or British universities. Many of these scientists were British subjects. In 1909 Leathes (1864–1956) came from London, where he had worked with E. H. Starling, to Toronto as the first professor of pathological chemistry.[65] When Leathes returned to England in 1914 to take the chair in physiology at Sheffield University, he was succeeded in 1915 by Andrew Hunter (1876–1969). A Scotsman educated at Edinburgh and in Germany, Hunter came to Toronto by way of Cornell University and the United States Public Health Service. At Toronto he held the chair of pathological chemistry from 1915 to 1920, of biochemistry from 1919 to 1929 and, after a sojourn at Glasgow, returned to Toronto as head of pathological chemistry from 1935 to 1945.[66]

In the Department of Biochemistry the "Imperial invasion" began in 1918, when Macallum went to work for the Canadian government during the First World War. He was very briefly succeeded by T. Brailsford Robertson (1884–1930), an Australian who had studied with and succeeded Jacques Loeb at the University of California. Robertson returned to Australia in 1919.[67] Andrew Hunter took over the chair of biochemistry at this time and an Englishman, Victor J. Harding (1885–1934), moved from McGill to Toronto to teach pathological chemistry from 1920 to 1934.[68] Hardolph Wasteneys (1881–1965) followed Hunter as professor of biochemistry from 1929 until 1951. Born in England and raised in Australia, Wasteneys worked at the Rockefeller Institute in New York from 1910 to 1916 in association with Loeb, earned his Ph.D. from Columbia in 1916, and came to Toronto in 1918 by way of the University of California, as assistant professor of biochemistry under T. Brailsford Robertson.[69]

In the Department of Physiology, when Macallum's chair was divided in two in 1908 and Macallum was appointed to the new chair in biochemistry, Thomas Gregor Brodie (1866–1916) was called to Toronto from England, where he had held several research and teaching positions, to become professor of physiology.[70] Renowned for his experimental skill, Brodie remained at Toronto only briefly, joining the Canadian Army Medical Corps in 1915 and dying suddenly in 1916 on war service in England. Macallum later wrote of Brodie's few years at Toronto:

> Very soon after he entered upon the duties of his new post he began to develop the Department of Physiology along Biophysical lines specially, with the result that at his death the Laboratory was ranked amongst the three leading Laboratories of Physiology on either side of the Atlantic. It was his ambition to make Toronto a great and renowned School of Physiology, and, had he lived, he would have succeeded in achieving this according to his desire.[71]

After Brodie's death, John James Rickard Macleod (1876–1935) was called from Case Western Reserve to become professor of physiology at Toronto.[72] In spite of his important textbooks and contributions to the understanding of carbohydrate metabolism, Macleod's years at Toronto are now remembered mainly for his association with the discovery of insulin, a story in which he has often been portrayed in an unfavorable light.[73] Although Bliss's work does help us to regain a sense of Macleod's contributions to the insulin discovery, Macleod's other scientific accomplishments and his international reputation are largely forgotten today. Ironically, Macleod was president of the APS at the very time of the work on insulin at Toronto.[74]

CONCLUSION

Returning to the old joke with which this essay began, it should be clear by now that it hardly applies to Canadian physiology, which had only indirect inspiration from French (or German) models, but in other ways did indeed achieve the best of all possible worlds by adapting American and especially British research traditions to the Canadian context. Ramsay Wright and Macallum were able to exploit the intellectual resources in the life sciences of both the United States and Great Britain, through their close geographical and professional associations with the United States and their strong political and educational links with the British Empire at the close of the nineteenth century. Canadian scientists were able to take advantage of specialized scientific societies in Britain and the United States long before their numbers in Canada were sufficiently large to form their own associations. The Canadian Physiological Society was not established until 1934, the year of Macallum's death, and met for the first time in 1935.[75]

There were, however, some disadvantages to this dependence on connections with scientific institutions in other countries. The overall pattern that emerges from the history of physiology during Macallum's tenure at the University of Toronto is one of sending the best research students off to the United States and of recruiting new senior faculty members from the United Kingdom. The absence of a strong graduate program at the University of Toronto until the 1920s seriously impaired its ability to produce graduates that it could hire, while the students sent to the United States for postgraduate degrees seldom returned to Canada. Through such students of A. B. Macallum's and Ramsay Wright's as F. R. Lillie, Lewellys Barker, and Will MacCallum, the University of Toronto exerted an important influence on the shape of North American biological and medical research but that influence was less obviously in evidence in physiology at the University of Toronto itself. It was not until 1929, when Macleod had returned to Scotland, that the next "Canadian," Charles H. Best, was appointed head of the Department of Physiology at Toronto.[76] Best was then only twenty-nine and was perhaps chosen partly for nationalistic reasons because of his role in the discovery of insulin, but he had also received postgraduate training in England with Sir Henry Dale.

For all of its dependence on English and American models, Canadian physiology did develop a somewhat distinctive blend of the two, if one not so strong as to justify the term "national style." However, the ultimate cost of defining oneself primarily in terms of the standards and styles set elsewhere may be found in measuring success only in those terms. Few Canadians are aware that biomedical research was carried out in Canada nearly a century ago; fewer still have heard of A. B. Macallum. The fact that the true story of the discovery of insulin as a collaborative effort at a well-established research institution has only recently begun to emerge, that four of the five physiologists referred to by Macallum in1932 held chairs outside of Canada, and that most of the highly regarded biomedical scientists trained at Toronto in the 1890s are no longer even known to have been Canadians, much less recognized as the products of a distinctive "Toronto school," is telling evidence of this lack of self-consciousness.

In the final analysis, a single picture may be worth a thousand words. The photograph included here (Fig. 1) was taken on the lawn of the Biological Building at the University of Toronto on the occasion of the meetings of the BAAS held in Toronto in August of 1897.[77] It represents a gathering of physiologists attending the meetings, centered around two of the grand old men of the discipline, Foster from Cambridge and Bowditch from Harvard. They are surrounded by an assortment of bright young men from England and the United States, Wesley Mills from McGill, and a couple of European scientists. The Toronto meeting was an important point of intersection for the British and American physiologists but the University of Toronto served mainly as a background for the visitors who dominate the photograph:

Figure 1: *Second meeting of Canadian British Association for the Advancement of Science, August 8–15, 1897, Toronto, Ontario, Canada. Top row (left to right): J. J. Mackenzie; Warren Plimpton Lombard; William Newton Barnhart; Karle Hürthle. Middle row: William Dobinson Halliburton; Paul Carus; Frederic Schiller Lee; Thomas Wesley Mills; Lightner Witmer. Bottom row: Rupert William Boyce; Augustus Desire Waller; Charles Richet; Sir Michael Foster; Henry P. Bowditch; Lewis Erle Shore; Sir Charles Scott Sherrington. [Courtesy of University of Toronto Archives, B71-0006/0001(06).]*

Macallum and Ramsay Wright do not appear and Toronto is represented only by one medical graduate and by J. J. Mackenzie, whose time-consuming minor appointments were to prevent him from attaining the research reputation he once sought. The photograph is in focus but the central Canadian character is missing. I hope that this chapter has put A. B. Macallum and the University of Toronto back in the picture of North American physiology.

Scholarship support from the Hannah Institute for the History of Medicine for the research on which this chapter is based is gratefully acknowledged. The staff of the University of Toronto Archives also provided invaluable assistance, and T. H. Levere and G. L. Geison were kind enough to read and comment on early versions of the manuscript.

NOTES

1. Michael Bliss, "Introduction: What Happened at Toronto?" in *The Discovery of Insulin* (Toronto: McClelland and Stewart, 1982), pp. 11–19.
2. Bliss, *Banting, A Biography* (Toronto: McClelland and Stewart, 1984).
3. Bliss, "The Aetiology of the Discovery of Insulin," pp. 333–346 in *Health, Disease and Medicine: Essays in Canadian History*, ed. Charles G. Roland (Proceedings of the First Hannah Conference on the History of Medicine, McMaster University, June 3–5, 1982). The Hannah Institute for the History of Medicine (Toronto, 1984), p. 338.
4. Biographical information on Archibald Byron Macallum (1858–1934) from J. B. L[eathes], *Obituary Notices of Fellows of the Royal Society of London* 1 (1934): 287–291; J. B. Collip, *R. Soc. Can. Proc. Trans.*, ser. 3, 28 (1934): xix–xxi; *Can. Med. Assoc. J.* 30 (1934): 576–577; and in Thomas McCrae, *Univ. Toronto Monthly* 34 (1933–1934): 217. See also the entry by R. A. Richardson in the *Dictionary of Scientific Biography*, vol. 8 (New York: Scribner, 1973), pp. 583–584.
5. A. G. Huntsman, Obituaries of "Robert Ramsay Wright (1852–1933)" in *R. Soc. Can. Proc. Trans.* ser 3, 28 (1934): iv–vi; *Univ. Toronto Monthly* 34 (1933–1934): 22; see also E. Horne Craigie, *A History of the Department of Zoology at the University of Toronto* (Toronto: Univ. of Toronto Press, 1965), pp. 13–35.
6. See P. J. Pauly, "The Appearance of Academic Biology in Late Nineteenth-Century America," *J. Hist. Biol.* 17, no. 3 (Fall 1984): 369–397.
7. C. S. Breathnach, "Henry Newell Martin (1848–1893), A Pioneer Physiologist," *Med. Hist.* 13 (1969): 271–279; Gerald L. Geison, *Michael Foster and the Cambridge School of Physiology; The Scientific Enterprise in Late Victorian Society* (Princeton, 1978), pp. 140–143; K. Rothschuh, *A History of Physiology* (New York, 1973), pp. 318–319; and the memorial volume dedicated to H. Newell Martin of the *Johns Hopkins Biol. Lab. Mem.* 3 (1895), which includes reprints of many of his American research papers and public addresses.
8. Geison, *Michael Foster and the Cambridge School of Physiology*, p. 141.
9. Published in part as Macallum, "Studies on the Blood of Amphibia," *R. Can. Inst. Trans.* 2 (1890–1891): 221–260.
10. Normand L. Hoerr, "Robert Russell Bensley (1867–1956)," *Anat. Rec.* 128, no. 1 (May 1957): 1–18; Sylvia H. Bensley, "Notes on R. R. Bensley and the Department of Biology, 1892–1901," in *History of the Department of Zoology* by Craigie, appendix C, pp. 81–84.
11. For a review of this research and other microchemical work see Macallum, "Die Methoden und Ergebnisse der Mikrochemie in der Biologischen Forschung," *Ergeb. Physiol. Biol. Chem. Exp. Pharmakol.* 7 (1908): 552–652.
12. Sandra F. McRae, "A. B. Macallum and the Origin of Life: The Case of an Evolutionary Physiologist" (unpublished paper, Institute for the History and Philosophy of Science and Technology, Univ. of Toronto, 1980).
13. Kenneth Johnstone, *The Aquatic Explorers: A History of the Fisheries Research Board of Canada* (Toronto, 1977), esp. chapt. 3, "The First Board and the Moveable Station, 1898–1907," pp. 29–45.
14. J. Loeb, "On the Ion-proteid Compounds and Their Role in the Mechanics of Life Phenom-

ena.-1.The Poisonous Character of a Pure NaCl Solution," *Am. J. Physiol.* 3 (1900): 327–338; and Loeb, "On the Different Effects of Ions Upon Myogenic and Neurogenic Rhythmical Contractions and Upon Embryonic and Muscular Tissue," *Am. J. Physiol.* 3 (1900): 383–396.

15. Macallum, "The Ancient Factors in the Relations Between the Blood Plasma and the Kidneys," *Am. J. Med. Sci.* n.s. 156 (1918): 1–11; "The Paleochemistry of the Body Fluids and Tissues," *Physiol. Rev.* 6 (1926): 316–357.
16. N. Burwash, "The Development of the University 1853–1887," in *The University of Toronto and Its Colleges, 1827–1906* (Toronto, Univ. Library, 1906), chapt. 2, pp. 39–56.
17. George W. Spragge, "The Trinity Medical College," *Ont. Hist.* 58 (1966): 63–98; see also the summary written by one of the participants in these rivalries, Walter B. Geikie's "An Historical Sketch of Canadian Medical Education" (Part 1), *Can. Lancet* 34 no. 5 (January 1901): 225–236.
18. See A. Primrose, "The Faculty of Medicine," in *The University of Toronto and Its Colleges, 1827–1906*, chapt. 10, pp. 168–179.
19. Toronto School of Medicine Minute Book, meetings from 16 March to 16 April 1887, Univ. of Toronto Archives (hereafter, Toronto Archives), B74-007/1, pp. 184–188.
20. "Faculty of Medicine 1888–1892," Toronto Archives, Edward Blake Papers, B72-013/3, f. 3/2.
21. "Opening of the New Building of the Biological Department, University of Toronto, December 19, 1889" *Can. Practitioner* 15 (16 January 1890): 29–46.
22. Geikie to Oliver Mowat (attorney general and premier of the Province of Ontario), 3 November 1891, Toronto Archives, A72-030, President's Papers, James Loudon, box 6, f. M-37, 4 p.
23. McRae, "Is Anatomy a Science? The University of Toronto Biological Building Debate of 1892" (Paper presented to the Canadian Society for the History of Medicine, Learned Societies Meetings, Montreal, Quebec, 1 June 1985).
24. Loudon Papers, Univ. of Toronto Archives, A72-030, box 6, M-65, M-66.
25. Ibid., "The Bursar," f. M-66, p. 11.
26. Thomas McCrae, obituary of Macallum, p. 217.
27. "The New B. and P. Course," *Torontonensis* (1904): 62.
28. University of Toronto Archives, A73-0026/251(69) "John Bruce MacCallum"; A73-0026/251(87) "William George MacCallum."
29. J. J. Mackenzie, "The New Medical Building," *Univ. of Toronto Monthly* 2 (1902): 122–124, 216–217, 221; Macallum, "The New Medical Buildings," *Univ. Toronto Monthly* 3 (1903): 133–134.
30. I. H. Cameron and J. T. Fotheringham, "The Inauguration of New Medical Laboratories," *Univ. Toronto Monthly* 4 (1903): 11–25; "Toronto University," *Montreal Med. J.* 32 (1903): 828–830; "Opening of New University Medical Buildings, Toronto," *Can. Lancet* 36 (1903): 935.
31. F. R. Miller, "Archibald Byron Macallum," *West. Ont. Hist. Notes* 8 (1950): 50–52. Miller became T. G. Brodie's graduate student after 1908 but he had studied with Macallum and worked as his demonstrator.
32. Macallum, "The Development of Physiology and Biochemistry in Canada," *R. Soc. Can. Fifty Year Retrospect; Anniv. Vol. 1882–1932* (1932): 163–165, esp. 163.
33. Macallum, "Professor MacMechan on University Reform; A Reply by Professor Macallum," *Westminster* (September 23, 1899): 332; Macallum, "The University Question; Prof. Macallum Replies," *Westminster* (14 October 1899): 431–432.
34. Published as Frederick Hughes Scott, "The Structure, Micro-Chemistry and Development of Nerve Cells, With Special Reference to Their Nuclein Compounds," *Univ. Toronto Studies Physiol. Series No. 1* (1900), and in the *Trans. Can. Inst.* 6, pts 1 and 2 (1898–1899): 405–438.
35. H. Blaschko, "Frederick Hughes Scott and His Contribution to the Early History of the Transmitter Concept," *Notes Records R. Soc. Lond.* 37, no. 2 (March 1983): 235–247; and F. C. MacIntosh, "Canada's First Ph.D. in Physiology," *Physiol. Can.* 16 (1985): 94–105.
36. Robert E. Kohler, *From Medical Chemistry to Biochemistry: The Making of a Biomedical Discipline* (Cambridge, UK: Cambridge Univ. Press, 1982) pp. 114–115.
37. Bliss, *The Discovery of Insulin*, pp. 12–17, 225–238.

38. R. J. Rossiter, obituaries of "James Bertram Collip, 1892–1965," in *R. Soc. Can. Proc. Trans.* ser. 4, 4 (1966): 73–82; M. L. Barr and Rossiter, *Biogr. Mem. Fellows R. Soc. Lond.* 19 (1973): 234–267; R. L. Noble, "James Bertram Collip," *Dictionary of Scientific Biography*, vol. 3 (New York: Scribner, 1971), pp. 351–354; Barr, "James Bertram Collip (1892–1965): A Canadian Pioneer in Endocrinology," *Publ. Hannah Found. Med.* 1, no. 1 (1977): 6–15.
39. Marie Adami, *J. George Adami, A Memoir* (London, 1930).
40. J. George Adami, "John McCrae," *Brit. Med. J.* (9 February 1918): 190–192, p. 191.
41. Archibald Malloch, *Short Years: The Life and Letters of John Bruce MacCallum, M.D. 1876–1906* (Chicago, 1938).
42. W. L. Longcope, "William George MacCallum, 1874–1944," *Biogr. Mem. Nat. Acad. Sci.* 23 (1945): 339–364; see also *Trans. Assoc. Am. Physicians* 58 (1944): 28–30.
43. Judith Robinson, *Tom Cullen of Baltimore* (Oxford Univ. Press, 1949).
44. Rufus Cole, "Lewellys Franklin Barker, 1867–1943," *Trans. Assoc. Am. Physicians* 58 (1944): 13–17; and Barker, *Time and the Physician* (New York: Putnam, 1942).
45. "John McCrae," Toronto Archives, A73-0026/256(81).
46. "D.Sc.—Thomas McCrae, B.A., M.D.," *Rec. Proc. Celeb. Centenary of Univ. Toronto, 1927* (Toronto, 1929): 33–35; R. T. Woodyatt, "President's Address," *Trans. Assoc. Am. Physicians* 51 (1936): 1–3.
47. Hoerr, "Bensley."
48. R. W. Gerard, "Ralph Stayner Lillie: 1875–1952," *Science* 116 (1952): 496–497.
49. "Maud Lenora Menten," Toronto Archives, A73-0026/318(11a).
50. Macallum, "The Scientific Spirit in Medicine" (Address delivered at the opening of the 76th session of the Medical Faculty at McGill University), *Montr. Med. J.* 37 (1908): 1–21.
51. Robert Falconer, obituaries of "Benjamin Arthur Bensley," *Univ. Toronto Monthly* 34 (1933–1934): 138, 140–141; C. C. Little, *J. Mammal.* 17 (1936): 312–313; and *R. Soc. Can. Proc. Trans.* ser. 3, 28 (1934): xxi–xxiii.
52. Johnstone, *The Aquatic Explorers*, p. 45; "Joseph Stafford," Toronto Archives, A73-0026/439(77).
53. "Edward Charles Jeffrey," Toronto Archives, A73-0026/177(24).
54. B. H. Willier, "Frank Rattray Lillie, 1870–1947," *Biogr. Mem. Nat. Acad. Sci.* 30 (1957): 179–236; R. L. Watterson, "Frank Rattray Lillie," *Dictionary of Scientific Biography*, vol. 8 (New York: Scribner, 1973), pp. 354–360; and Watterson, "The Striking Influence of the Leadership, Research, and Teaching of Frank R. Lillie (1870–1947) in Zoology, Embryology and Other Biological Sciences," *Am. Zool.* 19 (1979): 1275–1287.
55. Huntsman, "James Playfair McMurrich, 1859–1939," *R. Soc. Can. Proc. Trans.* ser. 3, 33 (1939): 143–146; McMurrich scrapbooks, Toronto Archives, B73-013.
56. Hoerst Oertel, "John Joseph Mackenzie, An Appreciation," *Can. Med. Assoc. J.* n.s. 12 (1922): 836; "The Late Professor John Joseph Mackenzie, B.A., M.B.," *Can. Med. Assoc. J.* n.s. 13 (1923): 62–64.
57. For details of this meeting see "A Royal Society of Canada Symposium to Mark the 150th Anniversary of the Foundation of the British Association for the Advancement of Science (1831–1981), Held at York, England, 3 September 1981," *R. Soc. Can. Proc. Trans.* ser. 4, 20 (1982): 471–547. Of particular interest are T. H. Levere, "The British Association Goes West: Montreal 1884," pp. 489–497; Susan Sheets Pyenson, "Better Than a Travelling Circus: Museums and Meetings in Montreal During the Early 1880's," pp. 499–518; Vittorio de Vecchi, "The Pilgrim's Progress, the BAAS, and Research in Canada: From Montreal to Toronto," pp. 519–532; and Richard A. Jarrell, "British Scientific Institutions and Canada: The Rhetoric and the Reality," pp. 533–547.
58. Johnstone, *The Aquatic Explorers*, p. 27.
59. McCrae, obituary of Macallum, p. 217.
60. Macallum, "On the Demonstration of the Presence of Iron in Chromatin by Micro-chemical Methods," *R. Soc. Lond. Proc.* 50 (1891–1892): 277–286. Communicated by Martin.
61. Macallum, "On the Detection and Localization of Phosphorus in Animal and Vegetable Tissues," *R. Soc. Lond. Proc.* 63 (1898): 467–479. Communicated by Professor Sherrington.
62. Macallum was the first University of Toronto graduate to be elected to the Royal Society of London, according to the obituary in the *Can. Med. Assoc. J.* p. 576.
63. Macallum, Presidential Address to the Physiological Section, Report of the BAAS Sheffield

Meeting, 1910. Printed as "Surface Tension in Relation to Cellular Processes," in *Science* 32 (1910): 449–458, 492–502.

64. Pauly, "The Appearance of Academic Biology in Late Nineteenth-Century America," p. 373.
65. Rudolph Peters, "John B. Leathes," *Biogr. Mem. Fellows R. Soc. Lond.* 4 (1958): 185–191; "J. B. Leathes," Toronto Archives, A73-0026/225(14); See also E. Gordon Young, *The Development of Biochemistry in Canada* (Toronto, 1976), pp. 20–25, 44–46 for an excellent summary of biochemical appointments held at the University of Toronto.
66. James A. Dauphinee, "Andrew Hunter, 1876–1969," *R. Soc. Can. Proc. Trans.* ser. 4, 8 (1970): 83–89.
67. Hedley R. Marston, "Thorburn Brailsford Robertson," *Aust. J. Exp. Biol. Med. Sci.* 9 (1932): 1–21.
68. Hendry and Van Wyck, "Victor John Harding," *R. Soc. Can. Proc. Trans.* ser. 3, 29 (1935): iv–vi.
69. A. M. Wynne, "Hardolph Wasteneys, 1881–1965," *R. Soc. Can. Proc. Trans.* ser. 4, 5 (1967): 125–129.
70. "Thomas Gregor Brodie," Toronto Archives, A73-0026/038(93).
71. Macallum, "Professor Thomas Gregor Brodie," *Univ. Toronto Monthly* 17 (1916): 25–26.
72. Noble, "John James Rickard Macleod," *R Soc. Can. Proc. Trans.* ser. 3, 29 (1935): viii–xi.
73. This unfavorable portrayal came particularly from partisan friends of Banting, whose dislike of Macleod was legendary. Bliss (1984) wrote that: "Banting's only success at the University [of Toronto] in these post-insulin years was in continuing to blacken the reputation of J. J. R. Macleod. To the end of his life Banting despised and vilified Macleod." In *Banting, A Biography*, p. 158.
74. Lloyd G. Stevenson, "John James Rickard Macleod," *Dictionary of Scientific Biography*, vol. 8 (New York: Scribner, 1973), pp. 614–615.
75. *Canadian Physiological Society 1937–'38*, 23 p., p. 3. According to the 1956 version of this pamphlet, the society "was founded in 1936."
76. Reginald E. Haist, "Charles Herbert Best, 1899–1978," *R. Soc. Can. Proc. Trans.* ser. 4, 16 (1978): 43–47. Best was born in the United States, but his parents were Canadian.
77. Irving Heward Cameron, photograph collection, Toronto Archives, B71-0006/0001(06).

V

International Relations and Domestic Elites in American Physiology, 1900–1940

GERALD L. GEISON

In studying the history of any scientific discipline in the United States, it is hard to escape some version of the following questions: At what point did American scientists achieve full parity with their disciplinary colleagues across the Atlantic? When did American representatives of this or that discipline no longer feel the need to go abroad in search of proper facilities, training, and intellectual guidance? When did they begin to produce a substantial body of research that attracted international attention and esteem? When did they come to be recognized as equal participants and even leaders at the research front? When, in short, did they cease to be "scientific colonials?"

In the case of physiology, the answer is surprisingly precise: the Great War marked a great divide in the history of physiology as in so much else. By the close of World War I, American physiologists had become a leading force on the world scene, and their international standing has remained high ever since. In the first part of this chapter, I explore the international ascendancy of American physiology between the two World Wars, using quantitative data when possible.

In the second part, I turn to the domestic situation in American physiology from 1900 to 1940. Specifically, I present a body of mostly quantitative data in pursuit of three intertwined goals: (1) to identify the leading individuals and research centers in American physiology up to 1940; (2) to construct from that and other material a set of "social maps" of the disciplinary leaders over half a century, using a sort of time-lapse photography; and (3) to emphasize how few research centers and indeed how few men it took to lead American physiology to a prominent place on the world stage. Hence the heading "Domestic Elites" for the second part, although I end with a plea to recognize the role of social and institutional factors in the production of these elites.

INTERNATIONAL RELATIONS

The rise of American physiology to full international parity was to no small extent the result of a tragedy: World War I. There were, of course, very important developments in American physiology long before the Great War, and they laid some of the essential foundations for the full ascendancy to come. By 1900, productive research centers already existed at Harvard, Johns Hopkins, Columbia, Michigan, Chicago, and (in physiological chemistry) Yale. There had been, since 1887, an American Physiological Society (APS) and, since 1898, an *American Journal of Physiology* (*AJP*). There was even, since 1896, an *American Textbook of Physiology*, produced under the editorship of W. H. Howell of Johns Hopkins.

Yet these attempts at national assertion in physiology could not conceal the fact that the leading international centers of research and discovery were still to be found in Germany and, increasingly, England. By 1900, to be sure, most aspiring American physiologists no longer felt the need to travel to foreign and especially German laboratories for advanced training. Among those who now stayed home was the single most important figure in twentieth-century American physiology, Walter Bradford Cannon of Harvard, who took his M.D. there in 1900 and later became famous for his contributions to the concept of homeostasis.[1] Nonetheless, for another fifteen years, until the outbreak of World War I, many of the most ambitious (and subsequently most distinguished) American physiologists continued to travel abroad—if now less often for extensive training than for concentrated exposure to the latest techniques and insight into the most exciting problems at the research front.[2] It was still rare for a European or British physiologist to undertake a journey in the opposite direction.

The Great War as Great Divide

World War I forever changed international relations in physiology as it did in politics. Like the rest of their compatriots, European and British physiologists were profoundly disrupted by the war. At least a few, including the already distinguished Keith Lucas of Cambridge University, actually lost their lives.[3] A poignant picture has been drawn of Professor John Newport Langley keeping a lonely vigil at Cambridge after 1914, working almost alone in a spacious new laboratory while his younger colleagues and students went off to war.[4] During 1915–1919, according to the best quantitative index available to us (see Table 1), the number of original research contributions by both British and German physiologists plummeted to a level barely one-third that of the prewar years 1910–1914.

Meanwhile, American physiologists, isolated from the immediate effects of the war, could happily contemplate the recent growth and foreseeable future expansion of the discipline at home. As early as November of 1915, when the war was just entering its second year, Frederic Schiller Lee of Columbia was telling a correspondent that "one of the results of the European war, I cannot help thinking, will be to increase the prominence of American science and make American men of science better known throughout the world." Alluding to "the advantages of study in American universities . . . for both Americans and foreigners," Lee had already decided that continental training was no longer of much use for aspiring American physiologists (although he did concede a continuing role for Great Britain).[5]

During part of the war, to be sure, American physiologists were diverted from their usual research into such war-related problems as aviation and submarine medicine, poisonous gases, nutrition in army troops, fatigue among

Table 1
Number of Original Contributions to Physiology in Various Countries by Five-Year Periods

	Germany	France	England	USA	Other	Unknown
1800–1804	4	2	6		2	2
1805–1809	1	3	2		2	
1810–1814	3	7	2			
1815–1819	9	8	4		1	
1820–1824	9	10	2	1	3	
1825–1829	20	7	4			1
1830–1834	21	6	5		5	2
1835–1839	25	10	4		3	1
1840–1844	38	16	7		1	1
1845–1849	53	6	6		3	1
1850–1854	52	11	5		3	4
1855–1859	74	26	3		3	
1860–1864	82	15			10	
1865–1869	89	1	2	7	1	
1870–1874	76	9	2	1	5	
1875–1879	79	5	9	1	8	3
1880–1884	49	5	10	1	15	3
1885–1889	39	5	13		13	3
1890–1894	65	7	15	3	16	4
1895–1899	54	5	18	6	15	7
1900–1904	78	2	14	11	18	4
1905–1909	59	5	28	5	11	1
1910–1914	66	6	24	9	9	4
1915–1919	20	1	9	14	4	
1920–1924	47	2	13	24	8	2

[From A. Zloczower, *Analysis of the Social Conditions of Scientific Productivity in 19th Century Germany* (Master's thesis, Jerusalem, the Hebrew University). Based on K. E. Rothschuh, *Entwicklungsgeschichte physiologischer Probleme in Tabellenform* (Munich, FRG: Urban & Schwarzenberg, 1952).]

workers in munitions factories, and above all wound shock, which Cannon investigated first hand from 1917 to 1919 while serving in France with the Harvard Hospital Unit.[6] However, original research contributions continued somehow to be made. In fact, the war scarcely impeded the flow of manuscripts to the *AJP*—somewhat to the surprise of its new editor, Donald R. Hooker, who was appointed to the post just before war broke out in Europe. In December 1918 Hooker reported that the *Journal* had experienced no dearth of manuscripts, adding that "my fear that we were to suffer in this manner during the war appears unfounded."[7]

In fact, Table 1 suggests that the number of original research contributions by American physiologists actually increased during the war years, sweeping them past their British colleagues and bringing them within range of the once-dominant German physiologists. The results would be more striking still if one corrected for the obvious pro-German bias throughout Table 1, which rests ultimately on a German compilation of physiological discoveries. If we were to divide the listed number of German contributions by two—a reasonable enough corrective given everything else we know about the history of physiology since 1800—the United States would surpass Germany during the last decade surveyed (i.e., 1915–1925). In any case, Table 1 shows a distinct upward trend in the relative international fortunes of American physiologists after 1915.

A similar but even more dramatic result emerges from Table 2, although it tabulates discoveries in all the medical sciences rather than in physiology alone.[8] Based on an American medical dictionary, it seems biased in the opposite direction from Table 1, that is, in favor of American contributions at the expense of German discoveries. Again, however, it is the trend after 1910 that it is more significant than any of the raw numbers listed. Whatever its flaws, Table 2 suggests a truly stunning reversal in the relative standing of German and American biomedical research precisely as the Great War was being waged.

In a further attempt to quantify the relative standing of physiologists in different countries after 1870, I constructed Table 3 from the entries in the second edition of John Fulton's *Selected Readings in the History of Physiology*, completed by Leonard G. Wilson after Fulton's death in 1960 and published in 1966.[9] The results, to put it mildly, fully counteract the pro-German bias in Table 1. Fulton's anthology reveals instead the Anglo-American bias of a Harvard-educated Rhodes Scholar who studied neurophysiology with Charles

Table 2
Number of Discoveries in Medical Sciences by Nations

	USA	England	France	Germany	Other	Unknown	Total
1800–1809	2	8	9	5	2	1	27
1810–1819	3	14	19	6	2	3	47
1820–1829	1	12	26	12	5	1	57
1830–1839	4	20	18	25	3	1	71
1840–1849	6	14	13	28	7		68
1850–1859	7	12	11	32	4	3	69
1860–1869	5	5	10	33	7	2	62
1870–1879	5	7	7	37	6	1	63
1880–1889	18	12	19	74	19	5	147
1890–1899	26	13	18	44	24	11	136
1900–1909	28	18	13	61	20	8	148
1910–1919	40	13	8	20	11	7	99
1920–1926	27	3	3	7	2	2	44

[From J. Ben-David, "Scientific Productivity and Academic Organization," *Am. Soc. Rev.* XXV (December 1960): 830. Based on a list of medical discoveries in F. H. Garrison, *An Introduction to the History of Medicine*, 4th ed. (Philadelphia, PA: Saunders, 1929).]

Table 3
Number of Major Contributions to Physiology in Various Countries by Decades

	USA	England	Germany	France	Other	Total
1870–1880	1	2	0	3	0	6
1880–1890	1	1	1	1	0	4
1890–1900	2	3	1	0	1	7
1900–1910	2	7	0	1	1	11
1910–1920	7	5	1	0	1	14
1920–1930	15	7	1	0	5*	28
1930–1940	7	2	0	0	2	11
1940–1950†	7	0	0	0	1	8
Totals	42	27	4	5	11	89

* Includes three Canadian entries. † Approximate ending date. From John F. Fulton and Leonard G. Wilson, eds., *Selected Readings in the History of Physiology*, 2d ed. (Springfield, IL: Thomas, 1966).

Sherrington at Oxford in the early 1920s and later became professor at Yale.[10] Of ninety excerpts for the post-1870 period in the second edition of the anthology, only four come from German physiologists and indeed only seventeen in total from physiologists outside of Great Britain, the United States, and Canada. The Anglo-American bias is so extreme that Table 3 cannot tell us much except how American and British physiologists fared vis-à-vis each other. The results confirm again the quantitative dominance of American over British physiology in the wake of World War I.

With the end of the war, the number of original contributions by British and German physiologists did rebound somewhat, at least according to the data in Table 1. By then, however, American physiologists had already established themselves on the world stage, and they took an ever larger part in the story thereafter. For all of its bias, Fulton's anthology does give some indication of just how dominant American physiology had become by 1920; from that point on, for each period listed in Table 3, the number of American entries is at least two to three times the number of British entries and even exceeds the number of entries by all other physiologists everywhere else in the world.

Needless to say, all such quantitative measures of relative national rank in research achievement are open to challenge. Even if we could eliminate the obvious national biases in Tables 1–3, who would dare to compare fifty publications on some trivial problem by some junior author in an American laboratory with even one paper by a Helmholtz, a Pavlov, or a Sherrington? From about 1900 on, such quantitative data may be particularly misleading in the case of British physiology. As I have suggested elsewhere, British physiologists, especially those trained in the Cambridge School, made fundamental contributions to modern physiology out of all proportion to their numbers or even the number of their publications. For such fundamental problems as the cause of the heartbeat, the mechanisms of muscle contraction and metabolism, nervous integration, hormones, and vitamins, British physiologists often pointed the way to fertile research programs that their American colleagues then exploited with zeal.[11]

In a letter of 1915, without a shred of quantitative evidence, Lee did not hesitate to identify British physiology as the one bright star in a fading constellation across the Atlantic:

> In physiology a rather sharp distinction can be drawn between Great Britain and the continent of Europe. In Great Britain there are still prominent leaders and some of the best of physiological research of the present day comes from the British laboratories. On the continent the number of prominent leaders has been diminishing for a considerable period of time. The famous men of the middle and latter half of the nineteenth century have practically all passed away. A few of the scientific generation immediately following . . . still remain, although most of them are lesser lights than their predecessors. Among the still younger men there are few leaders. There are several physiological laboratories in Great Britain to which I might gladly send American students [he cited specifically Cambridge, Oxford, Edinburgh, and University College, London], but there are almost no such centers on the continent.[12]

In fact, when American physiologists did visit foreign laboratories after the war, they went more often to England than anywhere else—to Cambridge, Liverpool, Oxford, or University College, London.[13]

American Physiology Triumphant: Laboratories, Publications, and Citations, 1920–1940

By the mid-1920s, however, even the most brilliant achievements of a few British physiologists could not obscure the more general shift of disciplinary hegemony to the United States. One major factor was the sheer number and increasing superiority of physiological laboratories in this country. The established centers in the eastern United States were expanding rapidly, and new laboratories were springing up throughout the country as universities and their physiologists joined the vigorous migration inland and westward.

A striking example arose in the Midwest even before World War I: the large new laboratory built at Washington University in St. Louis for future Nobel laureate Joseph Erlanger, who became chairman of the physiology department there in 1910.[14] Completed just in time for the annual meeting of the APS in St. Louis in 1914, the new laboratory attracted the admiration and envy of visiting physiologists. Lee wrote Erlanger that he had been "much impressed with the excellent arrangement of your place," which "seems to me to approach more nearly my own ideals than any other existing laboratory."[15] Perhaps Lee was struck by the fact that the new laboratory at St. Louis covered nearly three times as much floor space (22,724 sq. ft.) as his own laboratory at Columbia (8,047 sq. ft.).[16]

During the 1920s, Erlanger received at least two letters exalting his facilities in St. Louis over those at Cambridge University, the most famous center for physiology in Great Britain. One letter came in 1923 from Erlanger's collaborator (and fellow Nobel laureate) Herbert Gasser, who was then visiting

Cambridge. In a passage also quoted by Louise Marshall (in chapter XV of this book), Gasser wrote: "The laboratory here is not nearly so good as ours, I mean yours. There are many beautiful instruments for electrical investigations but they seem to belong mostly to [A.V.] Hill as the result of purchases from grants."[17] Four years later, a similar message reached Erlanger from a globe-trotting Phillipine physiologist then in Cambridge:

> The Cambridge physiological laboratory seems to me to be less well equipped than any of the three laboratories that I have seen in the U.S. [namely, those at the Universities of Chicago and Michigan and Erlanger's own laboratory at Washington University]. Owing perhaps to the present economic depression of the country, there is an atmosphere of stringency not to say stingyness in giving out materials for work. That the English should have done real pioneering work in the sciences must be credited to their ingenuity, but since modern physiology rests so much now on modern equipment, I can concur readily in Dr. Gasser's opinion expressed some time when I was still there, that the center for medical science is gradually being transferred to America.[18]

As this letter suggests, the laboratory at Washington University was just one of several superb facilities for physiological research in the United States by the mid-1920s. At Harvard, the physiological laboratories filled one wing of a huge medical building and accommodated four separate but associated departments: Physiology, Comparative Physiology, Applied Physiology, and Physical Chemistry.[19] At Chicago and Michigan, as the well-traveled Phillipine physiologist knew from personal experience, the research facilities were no less impressive. The new six-story building at Chicago, completed in 1926, dwarfed even Erlanger's laboratory in St. Louis; and although the Department of Physiology shared the building with the Departments of Physiological Chemistry and Pharmacology, its own space was now as large as the entire building the three departments had been sharing for years.[20] At Michigan, the space given to physiology in the new East Medical Building was larger still (32,000 sq. ft.). The accommodations, planned in exquisite detail by Robert Gesell upon his appointment as head of the Michigan department in 1923, included a separate laboratory for each staff member, an aquarium room, and (on the roof) extensive animal quarters with kennels for thirty-two dogs.[21] It was facilities such as these that allowed Walter J. Meek of Wisconsin to say in 1928, without fear of contradiction, that "the United States has the largest and best equipped [physiological] laboratories in the world."[22]

By then, there was another impressive if less massive institutional expression of the vitality of American physiology. As early as 1919, the APS began to lay plans for an ambitious new annual periodical to supplement its flourishing *AJP*. The new journal, *Physiological Reviews*, was designed to provide commissioned state-of-the-art review articles by leading American physiologists on the major research problems and achievements in the field. It was one way of announcing to the world that American physiologists were ready to give independent assessments of the state of research in the discipline at large. And from the time the first issue appeared in 1921, *Physiological Reviews* also served to advertise how many of the most important recent contributions had been produced in the United States.[23]

In 1929 the new international status of American physiology received a sort

Table 4
Percentage of Citations of Four Nations in Three Leading National Journals, 1938–1939

Journals Cited	AJP	J. Physiol.	Pflüger's	All
American	71%	31%	10%	37%
British	16	48	14	25
German	9	13	70	31
French	4	8	6	6
Total citations	7284	3674	2809	

From Ralph W. Gerard, *Mirror to Physiology* (1958), p. 143.

of symbolic expression when the United States served, for the first time, as host of the International Physiological Congress. As President of the Congress, W. H. Howell of Johns Hopkins had the privilege of delivering the opening address. "We look upon this meeting," he proudly observed, "as a recognition that American physiology has come of age, and has been admitted to full membership in that old world group to whom we have so long looked for guidance and inspiration."[24]

That was putting it modestly or diplomatically. In several respects, American physiologists already led the way and they gave every sign of consolidating their new position. Table 4 presents a special sort of quantitative index of the international situation as of 1938–1939, just before World War II reached Western Europe. The table obviously covers a tiny slice of historical time, but it does at least try to correct for national biases by examining citation patterns in all three of the major countries simultaneously. As Ralph Gerard noted in his brief commentary on this table, it clearly reveals "the strong position of the United States" as well as "the provincialism of each nation, least in Britain, in its professional reading."[25]

Physiologists in each country cited their compatriots much more often than foreign physiologists, almost exactly symmetrically so in the case of German and American physiologists (i.e., 70% of the citations in *Pflüger's Archiv* referred to German sources and only 10% to American sources, while 71% of the citations in the *AJP* referred to American sources and only 9% to German sources). Even the less provincial British physiologists (at least those who published in the *AJP*) cited their compatriots preferentially, but not nearly to the extent that Americans and Germans did. They therefore provide the best sense of the relative international standing of physiologists outside of Great Britain. They cited American sources more than twice as often as they referred to German publications. True, British and American physiologists shared the same language, but if we had such a table for 1910 the British column would surely reverse the relative rank of American and German physiologists. By the late 1930s, as World War II approached, British physiologists clearly believed that their American colleagues were contributing more importantly to the discipline than were German physiologists.

In short, by every available quantitative index, and by all manner of qualitative evidence, the international ascendancy of American physiology between the two World Wars is resoundingly clear. What we obviously need now is an explanation for the phenomenon. In fact, however, a large part of the story is bound up with the general ascendancy of American science during that period—and more specifically with the success of the "Flexnerian" medical reform movement, which gave the preclinical sciences a newly central role in the medical curriculum and wholly unprecedented resources to go with it. To that extent, the rise of American physiology requires no special explanation. Insofar as there was anything special about the development of physiology in the United States, it has to do with issues in the intellectual history of the discipline, including whether or not one can identify a distinctive and unusually fruitful "American style" of physiology. I believe that one can, but this is not the place to pursue the claim. To make the case persuasively would require a separate volume that focused on the conceptual and technical history of the discipline. In the meantime, let us begin by identifying the elite individuals and research centers in the history of American physiology from 1900 to 1940.

DOMESTIC ELITES

Now that we have some sense of the international ascendancy of American physiology between the World Wars, the domestic situation takes on added significance. What people and institutions in the United States were responsible for its new position in the world of physiology? How large was the community of research physiologists in the United States, and how many of them played a significant role in the discipline? How many research centers were there, and which were the most productive and most important? Where were the leading physiologists trained? Was research achievement distributed across the country or concentrated in a few major centers or regions?

The title of this section suggests most of the answers in advance: the international ascendancy of American physiology was due very largely to the institutional and intellectual leadership of a tiny portion of the larger physiological community. Even in the United States, that is to say, research achievement and international standing in physiology depended much less on a large pool of "scientific commoners" than on a very small group of people, almost exclusively men of course, at a very small number of institutions, almost all of them from the familiar list of leading research universities in this country. Democratic sensibilities notwithstanding, these leading men and research centers clearly deserve to be called the elite in American physiology.[26]

As a matter of fact, the pattern that emerges below is roughly of the sort one might expect from Alfred Lotka's "inverse square law" of scientific productivity as modified and extended by Derek J. de Solla Price twenty-five years ago.[27] Lotka's law, which dates from the 1920s, expresses the frequency distribution one finds by counting the number of papers published by each person in the cumulative index of a journal from any scientific field. According to Lotka, the number of people producing n papers is proportional to $1/n^2$. Another way of putting the same result is that the number of high producers is on the order of the square root of the total population of published scientists. Thus, in a population of 100 authors, there will be 10 high scorers—"an objective method," says Price, "for separating the major from the minor contributors."[28]

However, Price found that Lotka's law required modification for high scorers. At the top of the productivity scale, the gap between the scores is less than Lotka's law predicts. Price's modified law is essentially a quantitative expression of the qualitative fact that one scientist can do only so much. Designed to apply across the full range from lowest to highest productivity, it predicts, more or less, how many papers will be produced and in what sort of distribution by a given population of scientists in the course of their careers.[29]

No direct and systematic test of the Lotka-Price laws is possible here, partly because my data were not collected with that object in mind. Yet the results do decisively support Price's broader point that research productivity and achievement are very unevenly distributed within any broadly conclusive population of scientists. By now, that conclusion may seem to be nothing more than common sense, but perhaps it deserves emphasizing in the case of a laboratory science in "egalitarian" America.

Much of the rest of this chapter relies heavily on visual language in the form of tables and maps.[30] The approach is quantitative and populational. The activities of individual physiologists and the histories of particular institutions are almost invisible from this point of view. Eventually, of course, we will want to know much more about each of these men and institutions, including the exact character of their research and the significance of their achievements. During these very early days in the historiography of American physiology, I have thought it best to begin by providing a systematic sense of the leading people and places as American physiology became established and then moved to center stage in the disciplinary world.

Defining the Population: Physiologists and the *American Journal of Physiology*

The point of departure for all that follows was a quantitative study of articles published in the *AJP* between 1898 and 1940 (volumes 1–131).[31] One obvious question arises immediately: How representative of American physiology was the *AJP*? The answer depends on one's definition of physiology, but this much can be said at once: the *AJP* was about as representative of American physiology as was the APS itself. Indeed, the *AJP* was established under the auspices of the Society, although until 1914 it was owned and managed by William T. Porter of Harvard under formal contract to the Society. In that year, the Society assumed direct ownership and official management of the *AJP* and appointed Donald R. Hooker of Johns Hopkins as managing editor, who served in that capacity until 1946.[32]

This is not the place to pursue the complicated history of the *AJP* or its precise relationship with the APS. Suffice it to say that the Society did not exert much direct influence on the editorial policies and practices of the *AJP* before 1933, when the Board of Publications Trustees and the Editorial Board were established. Until then, Porter and Hooker effectively decided what the *AJP* would include or exclude.[33] Despite occasional hints of strain in the relationship between the APS and the managing editors of the *AJP*, the contents of the journal seem broadly representative of the interests of the Society membership.

The early volumes of the *AJP* ranged widely over the field. Among other things, the journal contained a lot of physiological chemistry and "biological" or "general" physiology.[34] But such borderland topics increasingly faded from view as alternative specialty journals were established, especially the *Journal of Biological Chemistry* (founded 1905), the *Journal of Pharmacology and Experimental Therapeutics* (1909), *Endocrinology* (1916), Jacques Loeb's *Journal of General Physiology* (1918), and the *Journal of Nutrition* (1928). From the beginning, moreover, a sizable number of American scientists and physicians who considered themselves physiologists in some sense of the word published in a wide variety of other anatomical, biological, chemical, and medical journals. Gradually, the *AJP* (like the APS itself) became identified mainly with vertebrate physiology of the sort taught to medical students. In November of 1933 the *AJP* made it official, announcing a new editorial policy that indicated its "preference [for] . . . papers from American Laboratories in the field of vertebrate physiology. . . ."[35]

Obviously, then, to restrict attention to the *AJP* is to restrict artificially the population of American physiologists, especially as one moves forward in time. However, it would be an endless and futile task to try to survey the publications of every form of physiology by everyone who ever called himself a physiologist. At least twice, the APS has actually tried to determine the total population of physiologists in the United States; both attempts immediately fell afoul of the problem of how to define the discipline and its practitioners.

The Society undertook its first systematic survey in 1945, at least partly to assess the effects of World War II on the discipline in North America.[36] The survey committee generated a list of American physiologists from a wide range of sources, including the Society's own membership roster and the names of people who listed physiology as their primary field of interest in *American Men of Science*. A list of 3,400 names was eventually compiled: "a primary file of some 2,000, identified principally through their professed adherence to the name of physiology in some connection . . . , and a secondary file of some 1,400 who seem from records of their research to deserve a place in the profession."[37] This list of 3,400 physiologists was vastly larger than the APS, which in 1945 had only 905 living members. The number of self-styled physiologists in *American Men of Science* (900) was almost precisely the same as the Society membership, but the two populations were by no means the same. In fact, only 500 of the 900 "physiologists" in *American Men of Science* belonged to the APS. The survey committee also complained that the entries in *American Men of Science* could be misleading because some scientists chose to identify themselves primarily by their subspecialty (e.g., hemodynamics, metabolism, endocrinology, nutrition, physiological chemistry) rather than the broad discipline of physiology.[38]

In 1958 the results of an updated and more ambitious survey appeared under the title *Mirror to Physiology*, edited by neurophysiologist Ralph Gerard, a past president of the APS. Generously funded by the new postwar National Science Foundation, this remarkable study bursts with information, insights, and sophisticated quantitative techniques. Yet even this survey fails to provide any clear way out of the difficulties of defining physiology or physiologist, despite an exquisitely detailed discussion of the problem. Gerard settled on an expansive estimate of "5750 physiological scientists in the United States and Canada as of October, 1953"—an estimate that encompassed one-third of all "pure" biological scientists in North America and exceeded fourfold the membership of the APS at the time (roughly 1,400).[39]

Both surveys reveal how hard it is to define and thus to count a meaningful population of American physiologists at any point in time. For current purposes, I have adopted a very simple definition and a very clearly defined population: American physiologists are those people who published at least one article in the *AJP* between 1898 and 1940. One advantage of this definition is that it permits consistent quantitative comparisons over time. More important, it links physiology with a scientist's journal of choice for a particular paper rather than his formal disciplinary identities and affiliations, which could be multiple, conflicting, and transient. An article published in the *AJP* was presumably seen as "of physiology and for physiologists" by the author at the time. If my definition seems restrictive, it nonetheless yields a total of nearly 2,000 American physiologists for the period 1922–1929 and more than 2,600 for the period 1930–1937.[40] My population of physiologists, like those in the APS surveys of 1945 and 1954, is at least three to four times as large as the membership of the APS at any given point in time.

What is ultimately most interesting about the total pool of American physiologists, however defined, is how very large it was compared with the subgroup that made a real difference in the international ascendancy of American physiology. That small subgroup—in a word that elite—is the focus of our attention hereafter. The boundaries of this elite are themselves a bit fuzzy of course, but it is a much more cohesive and more manageable unit of social analysis than "physiology" or "physiologist" writ large.

Elite Individuals: Research Productivity

As already mentioned, the research productivity of individual scientists within my total population of physiologists varied enormously. In keeping with the Lotka-Price laws, the vast majority published one or at most a few articles and then disappeared from the pages of the *AJP*. At the other end of the scale, a small percentage of the population accounted for a disproportionately large percentage of the total number of published papers. For the entire period 1898–1940, the total number of individual contributors to the *AJP* can be estimated at roughly 3,600.[41] According to Lotka's inverse square law of scientific productivity, the square root of 3600 (i.e., 60) should give a first approximation of the number of major contributors to American physiology from 1900 to 1940. Sixty is about 8% of the total membership of the APS as of 1940 (726).[42] It amounts to less than 2% of the total population of contributors to the *AJP* during the period surveyed. We are clearly talking about an elite.

Table 5 lists in rank order the sixty-four most prolific contributors to the *AJP* over the period 1898–1940 (along with the site and date of their advanced degrees and their main institutional affiliations thereafter). As it turns out, the minimum qualification for inclusion on this list is a total of 34 publications in the *AJP* during those forty years. Each of the four most prolific contributors published three times as many papers as the lowest qualifying score. The top

50% of the list contributed 64% of the articles published by all the men on it and the top 11% (7 men) contributed 20%. There was, in short, a prolific "superelite" within the highly productive group itself.

Part, but only part, of this variation within the select group can be ascribed to age. A physiologist whose active career spanned the whole period from 1898 to 1940 (and there are a few in Table 5) would reach a total productivity score of 80 by publishing two papers a year on average, while a younger colleague who published nothing in the *AJP* until 1930 would have to produce an average of eight articles per year to achieve the same high score.[43] Yet the correlation between age (or date of highest degree) and final productivity score in Table 5 is surprisingly weak. More generally and more importantly, the gap between productivity scores within this elite group is much less striking than the very uneven distribution across the total population of authors in the *AJP*. When everyone in a population is highly productive, the proportional difference between top and bottom is much less extreme. As Price's modified law predicts, the distribution of productivity within an already select group no longer approximates Lotka's inverse square law.

Needless to say, raw data about the number of papers published can be misleading—even as indicators of scientific productivity itself. For one thing, some research problems lend themselves to relatively quick and frequent publications, especially when a narrowly focused question or simple technique is exploited over a wide range of organisms or tissues. In such cases, a whole area of research can display unusually high productivity. Between 1900 and 1940, that seemed to be the case in some studies of the physiological effects of ionic solutions, drugs, and other chemical substances, or in certain kinds of nutritional and metabolic research. If physiological chemists (and especially nutritionists) were in fact consistently more prolific than mainstream physiologists, perhaps that was because of some inherent difference in the nature of their tasks—perhaps, for example, the amount of time it took the latter to master surgical techniques or to secure and maintain appropriate laboratory animals.[44]

The number of published papers can also fail to indicate actual research productivity because of systematic differences in the social organization of scientific work in different subfields, or even in different laboratories within the same research area. One example is differences in publication practices. The length of articles in the *AJP* varied widely during our period, from perhaps one page to more than forty. Some laboratories would publish nothing until they had fully resolved a research problem to their satisfaction, whereas others would submit a series of brief progress reports on a single ongoing project, a practice Hooker despised during his tenure as editor of the *AJP*.[45]

More important, individual publication scores can also be distorted by different local practices in the distribution of credit between the laboratory director and his collaborators, students, or research assistants. At one increasingly rare extreme, the director could play a major role in the conception, execution, and even the "writing up" of research work without claiming authorship. The most celebrated example in the history of physiology is the great nineteenth-century German mentor Carl Ludwig.[46] At the other extreme, the director could control all of the research and equipment in his laboratory and insist upon being listed as coauthor (sometimes even as senior

Table 5
Highly Productive American Physiologists, 1900–1940

Name and Rank	Number of Articles in *AJP*	Degree	Main site 1900–1940
1. *A. J. Carlson** (1875–1956)	123	Ph.D. Stanford 1902	Chicago
2. *A. C. Ivy* (1893–1978)	106	Ph.D. Chicago 1918 M.D. Chicago 1921	Northwestern
3. *F. C. Mann* (1887–1962)	104	M.D. Indiana 1913	Mayo Clinic
4. R. Gesell (1886–1954)	102	M.D. Wash. U. 1914	Michigan
5. W. E. Burge (1881–1948)	89	Ph.D. Hopkins 1909	Illinois
6. *W. B. Cannon** (1871–1945)	79	M.D. Harvard 1900	Harvard
7. L. B. Mendel* (1872–1935)	79	Ph.D. Yale 1993	Yale
8. *C. J. Wiggers* (1883–1963)	75	M.D. Michigan 1906	Western Reserve
9. S. W. Britton (1892–?)	71	M.D. McGill 1924	Virginia
10. O. Riddle* (1887–1968)	71	Ph.D. Chicago 1907	Carnegie Institution (Station for Experimental Evolution)
11. D. I. Macht (1882–1961)	69	M.D. Hopkins 1906	Corporate research
12. A. Rosenbleuth* (1900–1970)	69	M.D. Paris 1927	Harvard
13. G. H. Whipple*† (1878–1976)	66	M.D. Hopkins 1905	Rochester
14. *J. Erlanger**† (1874–1965)	64	M.D. Hopkins 1900	Washington (St. Louis)
15. H. S. Gasser*† (1888–1963)	60	M.D. Hopkins 1915	Washington (St. Louis)
16. *W. J. Meek* (1878–1963)	58	Ph.D. Chicago 1909	Wisconsin
17. G. N. Stewart* (1860–1931)	57	Sc.D. Edinburgh 1987 M.D. Edinburgh 1990	Western Reserve
18. *L. N. Katz* (1897–1973)	56	M.D. West. Res. 1921	Western Reserve and Chicago
19. *W. F. Hamilton* (1893–1964)	55	Ph.D. Cal. Berk. 1921	Georgia
20. J. M. Rogoff (1883–1966)	55	M.D. Ohio Wesl. 1908	Western Reserve
21. *H. E. Essex* (1893–1978)	54	Ph.D. Illinois 1927	Mayo Clinic
22. A. Forbes* (1882–1965)	52	M.D. Harvard 1910	Harvard
23. F. G. Benedict* (1870–1957)	50	Ph.D. Heidelberg 1995	Carnegie Institution (Nutrition Laboratory)
24. J. A. E. Eyster (1881–1960)	50	M.D. Hopkins 1905	Wisconsin
25. *A. B. Luckhardt* (1885–1957)	50	Ph.D. Chicago 1911 M.D. Chicago 1912	Chicago
26. E. C. Schneider (1874–1954)	50	Ph.D. Yale 1901	Wesleyan
27. Y. Henderson* (1873–1944)	49	M.D. Yale 1998	Yale
28. G. H. Bishop (1889–1973)	48	Ph.D. Wisconsin 1920	Washington (St. Louis)
29. T. H. Sollman (1874–1965)	47	M.D. West. Res. 1996	Western Reserve
30. J. L. Bollman (1896–?)	44	M.D. Illinois 1923	Mayo Clin
31. H. E. Himwich (1894–1975)	44	M.D. Cornell 1919	Yale et al.
32. *J. J. R. Macleod**† (1876–1935)	44	M.B. Aberdeen 1998	Western Reserve and Toronto
33. J. Markowitz (1901–?)	44	M.D. Toronto 1923 Ph.D. Toronto 1926	Toronto
34. E. G. Martin (1876–1934)	44	Ph.D. Hopkins 1904	Harvard and Stanford
35. G. R. Cowgill (1893–1973)	43	Ph.D. Yale 1921	Yale

Table 5—Continued

Name and Rank	Number of Articles in *AJP*	Degree	Main site 1900–1940
36. *R. W. Gerard* (1900–1974)	42	Ph.D. Chicago 1921 M.D. Chicago 1925	Chicago
37. T. B. Osborne (1859–1929)	42	Ph.D. Yale 1985	Connecticut Experimental Station
38. C. D. Snyder (1871–1973) [sic]	41	Ph.D. Cal. Berkeley 1905	Hopkins
39. F. A. Hartman (1883–1971)	40	Ph.D. U. Wash. 1914	Buffalo
40. H. Silvette (1907–?)	40	Ph.D. Virginia 1934	Virginia
41. *H. C. Bazett* (1885–1950)	39	M.D. Oxford 1919	Pennsylvania
42. *W. E. Garrey* (1874–1951)	39	Ph.D. Chicago 1900 M.D. Chicago 1909	Washington and Tulane
43. *E. F. Adolph* (1895–?)	38	Ph.D. Harvard 1920	Rochester
44. W. C. Alvarez (1884–1978)	38	M.D. Cooper Med. 1905	California Hooper and Mayo Clinic
45. E. J. Baldes (1898–1975)	38	Ph.D. Harvard 1924	Mayo Clinic
46. L. R. Dragstedt (1893–1975)	38	Ph.D. Chicago 1920 M.D. Chicago 1921	Chicago
47. E. Gellhorn (1893–1973)	38	Ph.D. Münster 1919 M.D. Heidelberg 1919	Oregon and Illinois
48. W. M. Boothby (1880–1953)	37	M.D. Harvard 1906	Mayo Clinic
49. F. P. Underhill (1877–1932)	37	Ph.D. Yale 1903	Yale
50. *W. O. Fenn* (1893–1971)	36	Ph.D. Harvard 1919	Rochester
51. J. R. Murlin (1874–1960)	36	Ph.D. Penn. 1901	Cornell Medical School and Rochester
52. F. H. Pike (1876–1953)	36	Ph.D. Chicago 1907	Columbia
53. *C. W. Greene* (1866–1947)	35	Ph.D. Hopkins 1998	Missouri
54. A. Grollman (1901–1980)	35	Ph.D. Hopkins 1923 M.D. Hopkins 1930	Johns Hopkins
55. S. W. Ranson (1880–1942)	35	Ph.D. Chicago 1906 M.D. Chicago 1907	Northwestern
56. C. I. Reed (1887–1967)	35	Ph.D. Chicago 1925	Illinois Medical School
57. B. O. Barnes (1906–?)	34	Ph.D. Chicago 1931 M.D. Chicago 1937	Chicago
58. *H. Davis* (1896–?)	34	M.D. Harvard 1922	Harvard
59. McK. Cattell (1891–1983)	34	Ph.D. Harvard 1920 M.D. Harvard 1924	Harvard
60. C. K. Drinker (1887–1956)	34	M.D. Penn. 1913	Harvard
61. D. R. Hooker (1876–1946)	34	M.D. Hopkins 1905	Hopkins
62. L. B. Nice (1882–1974)	34	Ph.D. Clark 1911	Oklahoma and Ohio State
63. J. M. D. Olmsted (1886–1956)	34	Ph.D. Harvard 1920	Toronto and California (Berkeley)
64. T. L. Patterson (1884–1966)	34	Ph.D. Chicago 1920	Detroit Medical School

Based on total number of articles published in the *American Journal of Physiology*, 1898–1940 (vols. 1–131). * Has entry in *Dictionary of Scientific Biography*. † Won Nobel prize in physiology or medicine. Names in italics, Presidents of the American Physiological Society.

Table 6
Percentage of Coauthored Articles in *American Journal of Physiology*

Dates	Volume Numbers	% Coauthored
1898–1906	1–15	27
1906–1912	16–30	42
1912–1918	31–45	42
1918–1922	46–60	52
1925*	71–75	64
1930*	91–95	63
1935*	110–114	70
1940*	128–131	67

* Volumes published from January through December of the sample year listed.

author) of papers in which he played little or no direct role. To judge from evidence cited elsewhere in this volume, Joseph Erlanger might qualify as one example of such a laboratory "chief" from our own elite group of sixty-four.[47] Perhaps A. C. Ivy would be another, for it is otherwise hard to imagine how his bibliography eventually included more than 1,500 publications![48]

Finally, the practice of coauthorship could inflate everyone's "productivity." Table 6 presents data on coauthorship in the *AJP* during the first forty years of this century. It reveals a marked trend toward coauthorship over time, moving from roughly 25% of all articles at the turn of the century to 50% in 1920 and apparently peaking at 70% in 1935. By 1949, however, the percentage of coauthored papers in the *AJP* had reached 76%. In addition, the number of different authors listed on coauthored papers increased at an even faster rate than coauthorship itself. Thus triple authorship, virtually unheard of in 1900, occurred in 32% of the articles published in 1949, while another 12% had more than three authors.[49] The increase in coauthorship is so striking that one is tempted to apply some sort of deflation factor to later productivity scores in order to make them more comparable to "real" productivity as measured in, say, 1900.

Elite Individuals: Quality in Quantity

What about quality in all this? When discussing international comparisons in the first part of this chapter, I warned that quantitative indices should not be taken as automatic measures of the quality, importance, or influence of physiological research in different countries. That caveat applies fully to the domestic situation as well. Yet, whatever their flaws, quantitative publication scores actually do bear some relation to the quality of scientific research. Citing a small but suggestive body of empirical evidence, Derek Price concluded: "Flagrant violations there may be, but on the whole there is, whether we like it or not, a reasonably good correlation between the eminence of a scientist and his productivity of papers . . . the list of high scorers contains a large proportion of names that are not only well known but even honored."[50]

Table 5 offers additional data in support of Price's conclusion. For the most part, the sixty-four names listed there are precisely those most familiar to us from the qualitative history of American physiology—precisely those that will be found, for example, in the official histories of the APS. Admittedly, there are two problems. First, perhaps a dozen of the names in Table 5 will be unfamiliar even to well-informed students of American physiology. A few men made their way onto the list as junior coauthors of a large number of articles published jointly with a more distinguished physiologist. Several others simply failed to make a profound impression on the discipline despite high and genuine productivity, perhaps because they bounced around from institution to institution, worked outside the leading research universities, or spent much of their career in medical practice or industrial research outside the academic network entirely.[51]

The second problem is arguably more serious because it concerns notable exclusions rather than dubious inclusions. Here, too, age is one obvious factor. By the time the first volume of the *AJP* appeared in 1898, some of the famous founding fathers of American physiology had little or no chance to demonstrate their productivity as measured here. (For example, H. N. Martin of Johns Hopkins was already dead, Henry Sewall of Michigan had gone to Denver for his health, and H. P. Bowditch of Harvard had long been preoccupied with his duties as Dean of the Medical School.) Other important figures in the early history of American physiology simply did not publish as many articles in the *AJP* as one might have presumed (e.g., Lee of Columbia and W. T. Porter of Harvard, although Porter fell just four papers short of the qualifying score for inclusion in Table 5).

The most astonishing omission of this sort—or indeed of any sort—is William Henry Howell of Johns Hopkins.[52] A powerful intellectual and institutional force in the discipline from the mid-1880s into the 1930s, Howell nonetheless published only 26 articles in the *AJP* and therefore does not appear among the elite group of sixty-four men in Table 5.

With the exception of a few such early leaders—and of course those future leaders at the opposite end of our chronological scale, whose careers had scarcely begun—Table 5 seems to include nearly all of the leading American workers in mainstream physiology (e.g., neuromuscular, cardiovascular, gastric, renal, or respiratory physiology) between 1900 and 1940. Outside of this mainstream, to be sure, there were major American contributors to physiology broadly conceived who did not publish in the *AJP* often enough to qualify for inclusion in Table 5. At least a few of these excluded scientists worked on exciting and fundamental research problems in which they and other Americans often led the way, notably the study of nutrition, vitamins, hormones, and reproduction. E. V. McCollum of vitamin fame comes readily to mind, as do the endocrinologists George Corner, Herbert M. Evans, and Gregory Pincus.[53] Equally conspicuous by their absence are the famous general physiologists Lawrence J. Henderson of Harvard and peripatetic, quixotic Jacques Loeb.[54] Important Canadian scientists are also missing; indeed, only one of the four men associated with the discovery of insulin at Toronto in 1922 (namely, J. J. R. Macleod) is included among the highly productive group in Table 5.[55]

If such exclusions are unfortunate in a way—at least for those who would like to claim as many eminent names as possible for American physiology—they are not merely arbitrary. All of the important scientists just named did in fact publish in the *AJP* from time to time (indeed, Evans did so twenty-one times). On those occasions, they joined the pool of American physiologists as I define them. However, they more often chose to publish their work in the specialized journals of such subdisciplines as biochemistry, pharmacology, endocrinology, and general physiology (or in the case of the Canadians, in British or Canadian journals). In doing so, they effectively excluded themselves from the American physiological elite, sometimes deliberately so. The least we can do is honor the decisions they made at the time as to whether, how often, and how much they wished to be identified with physiology, as that discipline fragmented into a variety of specialties. Their absence from Table 5 is one index of those decisions.

For those who were clear and consistent about their identity as physiologists, the highly productive list in Table 5 correlates very well, although not perfectly, with other indications of scientific achievement. Many of the sixty-four men who appear on that list received honors of the sort one might expect of those renowned for the quality as well as the quantity of their research. As the asterisks in Table 5 indicate, four won the Nobel Prize in physiology or medicine (Erlanger, Herbert Gasser, Macleod, and George Whipple). Thirteen (20%) have already won a place in the *Dictionary of Scientific Biography* and several others will surely join them in the supplementary volumes yet to come. All but one of the sixty-four men were elected to membership in the APS and no less than nineteen of them (30%) eventually served as presidents of the Society.[56]

In short, most of the physiologists in Table 5 were major contributors to the discipline in the United States by any and every standard. Those sixty-four men supplied much of the intellectual and institutional leadership required to bring about the ascendancy of American physiology after World War I. As Price would have predicted, there is indeed some correlation between research productivity and disciplinary eminence in American physiology. Nor does that correlation operate only at the level of individual physiologists. It also applies at the level of research schools and training centers, as we shall now see.

Elite Institutions: Research

Except for its policy of rejecting most papers from foreign laboratories, the *AJP* displayed no pattern of discrimination against institutions during the period 1898 to 1940. In fact, it displayed little discrimination of any sort before the 1920s. Rejections were rare, although revisions were occasionally demanded.[57] During and after the 1920s, papers on human or nonvertebrate physiology were discouraged. Even so, submissions came from an astonishing number and range of institutions.

In a sample survey of thirty-two (37%) of the eighty-six volumes that appeared between April 1918 and December 1940, the *AJP* published articles from 182 different institutions (including twenty-three foreign laboratories, despite the stricture against them). In a sense, the number of different

Table 7
Productivity by Research Centers in *American Journal of Physiology*, 1918–1940

Papers Per Institution	Number of Centers	Total Papers	Papers Per Institution	Number of Centers	Total Papers
1	73	73	25	1	25
2	28	56	27	1	27
3	14	42	30	1	30
4	13	42	31	1	31
5	8	40	34	1	34
6	4	24	36	1	36
7	3	21	38	1	38
8	4	32	39	1	39
9	3	27	41	1	41
10	3	30	42	1	42
11	4	44	48	1	48
12	1	12	55	1	55
13	3	39	58	1	58
14	1	14	64	1	64
15	1	15	90	1	90
17	1	17	115	1	115
18	1	18	130	1	130
20	1	20			
24	1	24	Total	182	1527

Volumes surveyed: (46–56) Apr. 1918 through July 1921; (71–75) Jan. 1925 through Dec. 1925; (91–95) Jan. 1930 through Dec. 1930; (110–114) Jan. 1935 through Dec. 1935; (126–131) Jan. 1940 through Dec. 1940. Average papers/center: 1527/182 = 8.4; median papers/center: <2.

institutions was even higher than that, for articles often came from two or more separate departments or laboratories within the same university or other parent body (e.g., the Mayo Clinic). All such articles have been assigned to the parent body here.

The research productivity of individual institutions, like that of individual authors, varied enormously. Table 7 displays the full range of institutional productivity in the sampled volumes of the *AJP* during the period 1918–1940. It shows that seventy-three of the 182 institutions (40%) were represented by a single article, and twenty-eight others (15%) by only two articles. In other words, more than half (55%) of the institutions never reached their third article. And 136 (70%) never made it past their fifth article.

To begin from the other end of the scale, the most productive institution (the University of Chicago) was responsible for almost as many articles (130) as all 101 of the least productive institutions put together (133)! The top twenty institutions (9%) produced 63% of the total number of articles surveyed during the period 1918–1940. For the earlier twenty-year period, 1898–1918, no such systematic data were gathered on institutional productivity across the whole range, but a very similar pattern was found among a much

Table 8
The Ten Leading Institutional Contributors to *American Journal of Physiology*, 1898–1918

Rank	Institution	Number of Articles
1	University of Chicago	139
2	Yale University	137
3	Harvard University	120
4	Cornell University Medical School	67
5	Johns Hopkins University	64
6	Western Reserve University	62
7	University of Michigan	33
8	University of Pennsylvania	30
9	St. Louis University	28
10	Washington University (St. Louis)	21

smaller number of institutions. During that period, the top ten institutions clearly accounted for more than 50% of the articles published. Tables 8 and 9 list the most productive institutions for each of these two periods. (For the period 1918–1940, the results for the top ten institutions have been further broken down into five-year periods and displayed in Figure 1).

When Tables 8 and 9 are compared, the most striking result is the long-term staying power of highly productive institutions. Eight of the ten institutions in Table 8 reappear in Table 9 (the exceptions being the medical schools at Cornell and St. Louis Universities), and four institutions are among the top five over the entire period from 1898 to 1940: Chicago, Harvard, Yale, and Johns Hopkins. In both tables, Chicago ranks first, Western Reserve sixth, and Pennsylvania eighth.

Despite the general stability, however, four dramatic examples of upward mobility do emerge: Washington University, which in the meantime had absorbed the medical school at St. Louis University,[58] leaped from tenth to fourth place, while the Mayo Clinic, the University of Rochester, and Northwestern all came out of nowhere to appear among the top ten in the 1918–1940 period (ranking seventh, ninth, and tenth respectively). Meanwhile, as Cornell Medical School disappeared from the top twenty entirely, Michigan dropped from seventh to fifteenth place in the more crowded and competitive field.

As might be expected, West Coast institutions first appeared among the highly productive centers during the 1918–1940 period (Stanford at eleventh, Berkeley at fourteenth, and UCSF at sixteenth), by which time the Midwest could claim to have more leading institutions than the two coastal regions combined. If Western Reserve in Cleveland can be considered midwestern at the time, the region harbored ten of the twenty most productive institutions during this second period—four traditional midwestern leaders (Chicago, Michigan, Western Reserve, and Washington University) and six new arrivals: the Mayo Clinic (in Rochester, Minnesota), Northwestern, Wisconsin, Min-

Table 9
The Twenty Leading Institutional Contributors to *American Journal of Physiology*, 1918–1940

Rank	Institution	Number of Articles
1	University of Chicago	130
2	Harvard University	115
3	Johns Hopkins University	90
4	Washington University, St. Louis	64
5	Yale University	58
6	Western Reserve University	55
7	Mayo Clinic	48
8	University of Pennsylvania	42
9	University of Rochester	41
10	Northwestern University	39
11	Stanford University	38
12	University of Wisconsin	36
13	University of Minnesota	34
14	University of California, Berkeley	31
15	University of Michigan	30
16	University of California, San Francisco	27
17	Columbia University	25
18	University of Toronto	24
19	University of Illinois, Chicago	20
20	University of Illinois, Urbana	18

Volumes surveyed: (46–56) Apr. 1918 through July 1921; (71–75) Jan. 1925 through Dec. 1925; (91–94) Jan. 1930 through Dec. 1930; (110–114) Jan. 1935 through Dec. 1935; (126–131) Jan. 1940 through Dec. 1940.

nesota, and both the University of Illinois in Urbana and the University of Illinois Medical School in Chicago. Also during the 1918–1940 period, the University of Toronto became the first Canadian entry among the highly productive research centers.

There is perhaps nothing very astonishing about these results, but some readers may be surprised to see how quickly Chicago took the lead from the historic sites of experimental physiology in the United States: Harvard, Hopkins, and Yale. Perhaps others will be surprised that Columbia, another early entrant in the field, is absent from the list of leading institutions for 1898–1918 and only seventeenth for 1918–1940. If so, they should consider Alejandro Laszlo's discussion in chapter III of this book of the relative failure of Lee's "biological" program for physiology at Columbia.[59] Finally, it is worth emphasizing just how utterly dominant the top three institutions were in each of the two periods. In fact, if one discounts the highly productive physiological chemists at Yale during 1898–1918, and concentrates instead on classic (vertebrate) physiology, the entire period from 1900 to 1940 is marked by the overwhelming hegemony of three institutions: Chicago, Harvard, and Johns Hopkins.

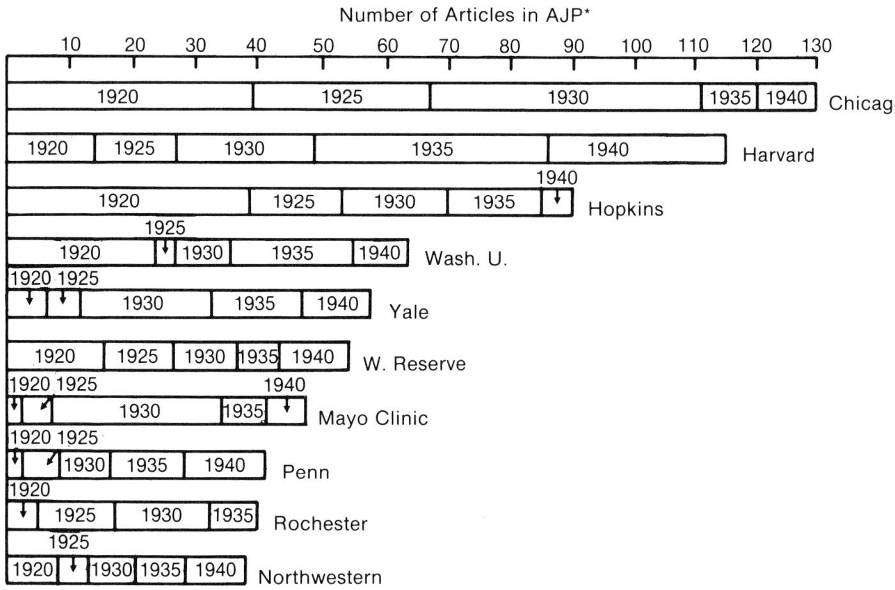

Figure 1: *Research productivity at leading centers of physiology in the United States, 1920–1940. *Based on the following sample years only of the American Journal of Physiology: 1920, 1925, 1930, 1935, 1940.*

Elite Institutions: Graduate Training

The dominant centers of research in physiology were also the leading centers of graduate training by a wide margin. (Perhaps one should add "or vice versa," since graduate students often contributed significantly to institutional productivity.) There are, to my knowledge, no systematic data on the number of doctoral degrees in physiology awarded by each American university for the period 1900–1940. We do know that by 1920, 169 members of the APS—roughly fifty percent of the total membership at the time—had received advanced degrees at one of the three leading centers. John Hopkins had trained sixty-four of them; Harvard, fifty-seven; and Chicago, forty-eight.[60] A decade later, Chicago had awarded a total of seventy-one Ph.D.s in physiology,[61] but I have as yet no comparable data for other universities.

In any case, John Hopkins, Harvard, and Chicago clearly produced more physiologists than any other American university from 1900 to 1940. They also trained most of the best. Consider, in this connection, the list of highly productive physiologists in Table 5. Those sixty-four men received their first highest degree (M.D. or Ph.D.) from twenty-seven different institutions, including six from across the Atlantic and two in Canada (McGill and Toronto). Yet nineteen of these twenty-seven universities had trained only one member of the elite group, while another four institutions had trained but two each. Yale alumni account for another six of the men in Table 5. That leaves thirty-one men, or nearly half of the highly productive group, to be distributed among the remaining three centers as follows: Chicago, twelve; Hopkins, ten; and Harvard, nine. Not so incidentally, Hopkins trained three of the four Nobel laureates in Table 5 (Erlanger, Gasser, and Whipple; the fourth, Macleod, received his highest degree from the University of Aberdeen in Scotland).

Where did the leading American centers find their own faculty? For the most part, they hired and promoted from within, i.e., from among their own graduate alumni. Such inbreeding was especially striking at Yale. Of the six highly productive Yale alumni in Table 5, four joined the faculty at their alma mater; even the other two stayed in Connecticut. Throughout the period 1880–1940, all five of the leading physiologists at Yale (Chittenden, Yandell Henderson, Mendel, Underhill, and Cowgill) held Yale Ph.D.s and, except for Henderson, all in physiological chemistry.[62]

At first glance, the situation looks quite similar in the physiology departments at Hopkins, Harvard, and Chicago. There too nearly all of the top faculty came from within, especially after 1920. By then, five of the seven leading faculty at Harvard were Harvard men, while at both Chicago and Hopkins only one of the faculty leaders came from outside (A. J. Carlson at Chicago and C. D. Snyder at Hopkins). Interestingly, none of the leading faculty members at any of these three major centers had been trained at either of the other two. Only the Chicago department was chaired by an outsider, the awesomely productive Carlson, who had taken his Ph.D. at Stanford in 1902.[63]

The graduate programs at Hopkins, Harvard, and Chicago trained many more physiologists than they could hire. Where did these surplus students go? Figures 2–4 begin to provide an answer by charting the main career paths of ten leading physiologists who received advanced degrees from each of these three centers between 1880 and 1925. Most, but not all, of the men in Figures 2–4 are also included among the highly productive group in Table 5. Insofar as the two populations differ, it is because Figures 2–4 place a higher premium on a man's institutional role and location than does Table 5, which relies solely on the number of papers published in the *AJP* between 1898 and 1940. In a few cases (most glaringly, the Chicago alumnus E. P. Lyon) Figures 2–4 include men whose productivity in the *AJP* was modest, despite their association with highly productive institutions. In these cases, it could perhaps be said that "the institution made the man."

As might be expected from its more general leadership in American graduate education, Johns Hopkins was the most active intellectual colonizer.[64] Seven of the ten Hopkins men included in Figure 2 spent the most important part of their careers elsewhere—at various universities in the East, Midwest, and even California (E. G. Martin at Stanford, after a ten-year stint at Harvard). The most celebrated of these Hopkins products, future Nobel laureate Erlanger, went to the University of Wisconsin for four years (1906–1910) before moving to Washington University in St. Louis, where he was joined in 1916 by another Hopkins man, Herbert Gasser. Meanwhile, in 1911, Wisconsin appointed another product of the Hopkins department, J. A. E. Eyster. The Hopkins-Wisconsin connection extended well beyond their departments of physiology. As Erlanger pointed out in a letter of 1913, all but one of the early professorships in the medical school at Wisconsin went to Hopkins men.[65]

Half of the ten Harvard alumni listed in Figure 3 spent their careers at their alma mater. Two of the other five (W. O. Fenn and E. F. Adolph) went to the University of Rochester and a third, W. M. Boothby, to the Mayo Clinic (in Rochester, Minnesota), which later hired other Harvard men to join him. Meanwhile, as Figure 4 suggests, Chicago had become a "feeder school" for younger Midwestern centers that perforce began by hiring products of the older physiology programs. Nine of the ten Chicago alumni in Figure 4 spent the most important part of their careers in the Midwest, four of them at the University of Chicago itself. Two others (S. W. Ranson and A. C. Ivy) moved just a few miles north to Northwestern University in Evanston. To the south, in St. Louis, Washington University hired W. E. Garrey, and the medical school at St. Louis University appointed E. P. Lyon, who barely appeared in the pages of the *AJP* but nonetheless led active research groups at St. Louis and later at the University of Minnesota.

Figures 5–9, which are quite literally maps of the leading men and institutions in American physiology from 1880 into the 1930s, try to capture in a few images much of what has already been said about domestic elites. Collectively, they display sixty-two of the most important faculty members at twenty-one leading research centers for each of the decades between 1880 and 1930. The institutions are more important than the men. In fact, there is a nearly perfect overlap between the twenty-one research centers included in these maps and the twenty-two most productive institutions listed in Tables 8 and 9. In the case of individuals, by contrast, only 71% of the men in Figures 5–9 appear among the list of highly productive physiologists in Table 5 above, and only 69% of the men in Table 5 appear in the maps. The two populations are thus by no means identical, even though their number is almost precisely the same (i.e., 64 men in Table 5 and 62 men in Figures 5–9). Of the relatively unproductive men included on the maps, but excluded from Table 5, several were founding fathers of the disipline at one or another American university, while several others barely missed qualifying for inclusion among the high producers in Table 5. The maps in Figures 5–9 were thus constructed according to a looser set of criteria than Table 5 and should be regarded as somewhat preliminary. They are nonetheless meant to be useful even in their current form. To my knowledge, nothing quite like them exists.

Figures 5–9 are designed to include nearly all of the leading institutions and individuals in American physiology up to 1940. But we should not end this discussion of training centers without drawing attention to the important role played by other, unlisted figures at these institutions. To begin with, it should be remembered that candidates for the M.D. or Ph.D. degree were not the only physiologists in the making. There were also various levels of research or teaching assistants and postdoctoral fellows. The number and significance of such positions increased immensely after 1920 as support became available for postdoctoral students through the Guggenheim and Rockefeller Foundations and especially the National Research Council. At Harvard between 1930 and 1938, Cannon trained thirty-two research and teaching fellows, of whom ten were supported by the National Research Council, five by the Rockefeller Foundation, two by the Guggenheim Foundation, and fifteen by various other organizations.[66] At Chicago during the

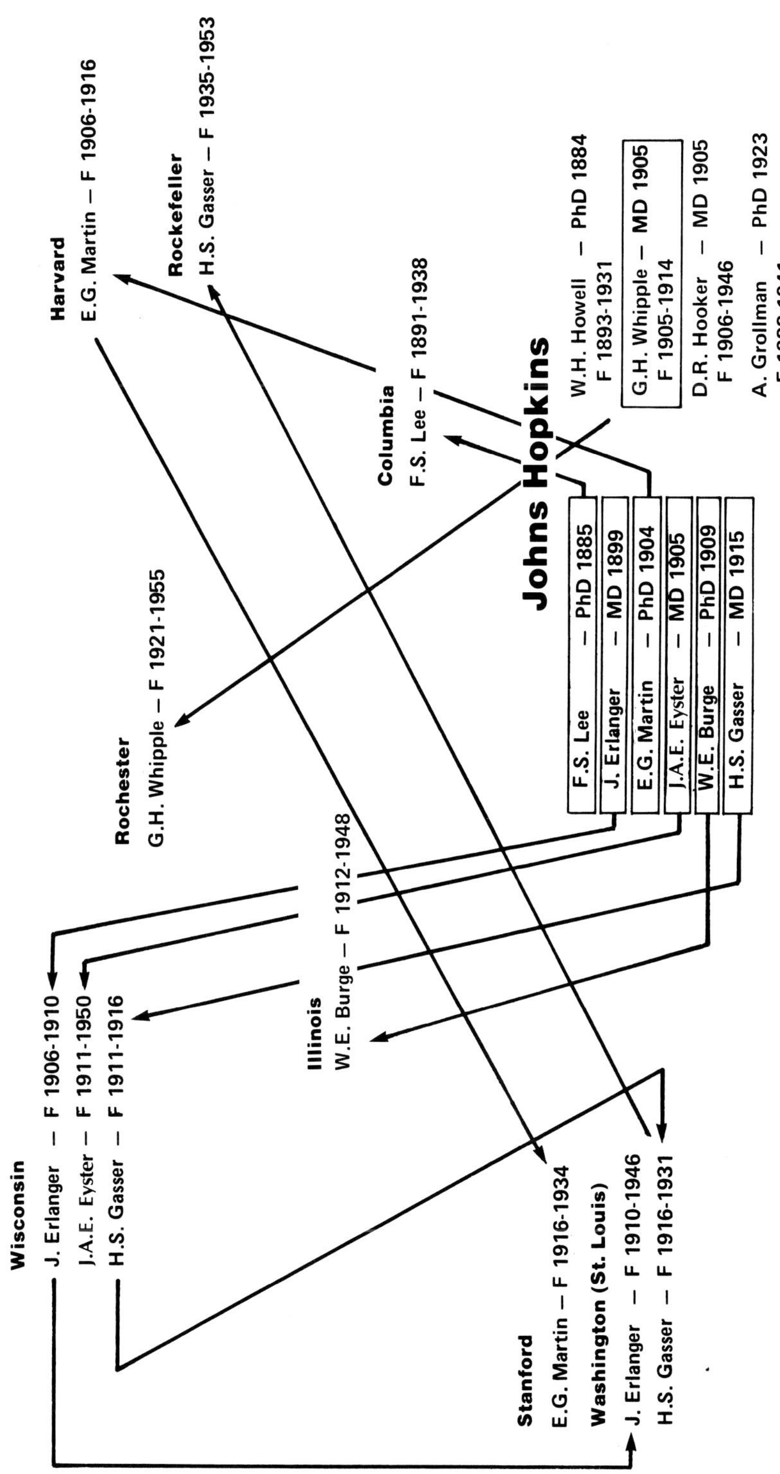

Figure 2: *Leading physiologists trained at Johns Hopkins,* ***1880–1925****. Individuals in boxes spent the most important part of their careers elsewhere. F, faculty member.*

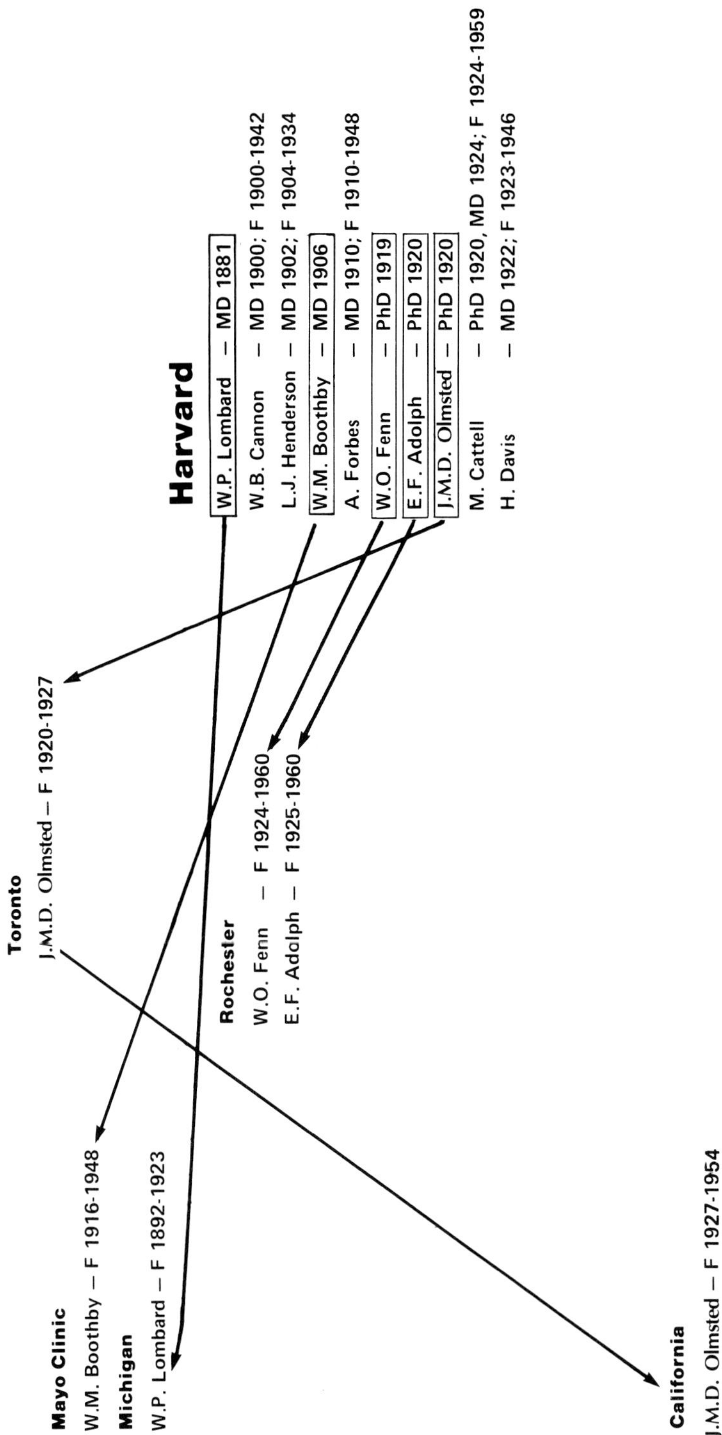

Figure 3: *Leading physiologists trained at Harvard, **1880–1925**. Individuals in boxes spent the most important part of their careers elsewhere. F, faculty member.*

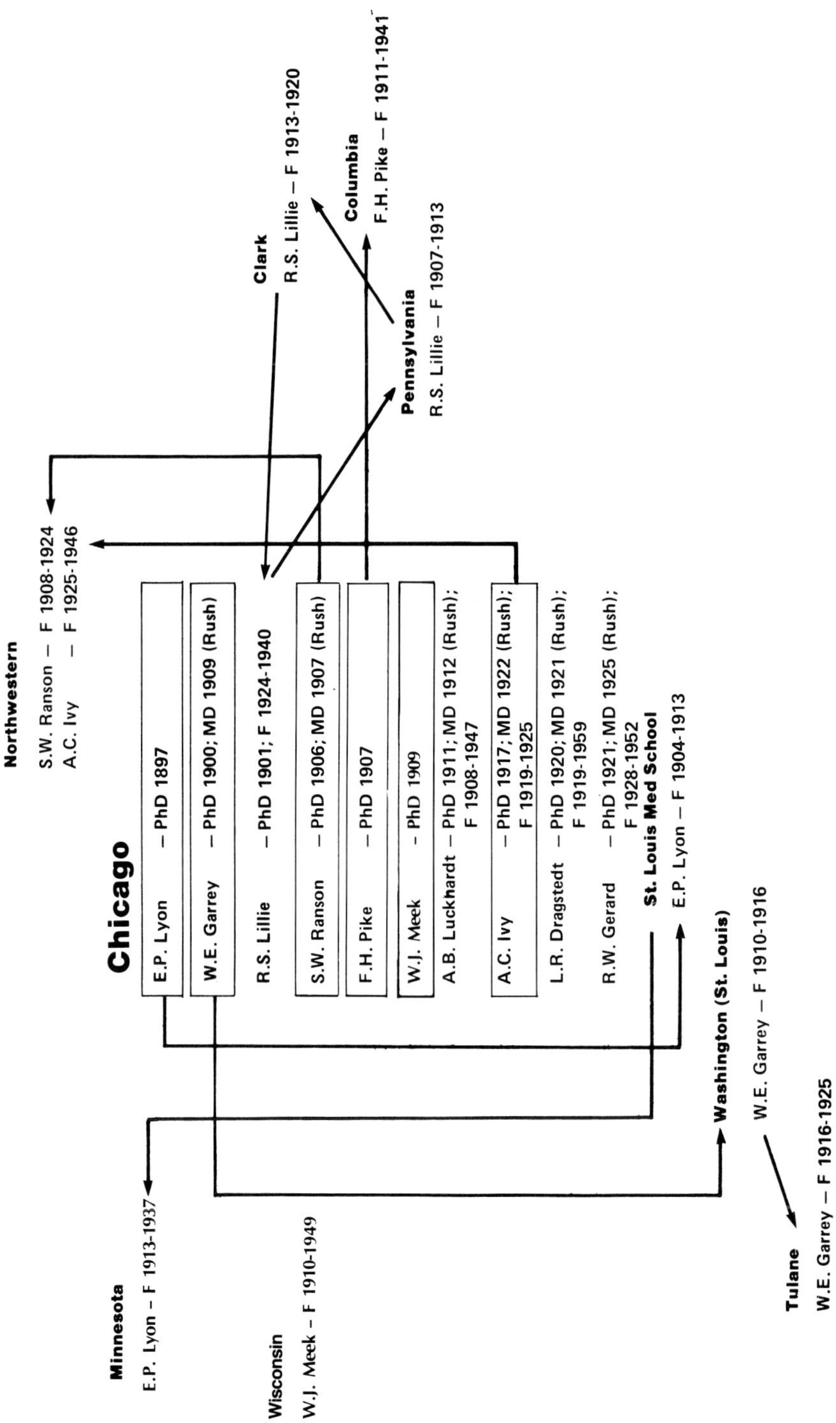

Figure 4: *Leading physiologists trained at Chicago,* ***1880–1925****. Individuals in boxes spent the most important part of their careers elsewhere. F, faculty member.*

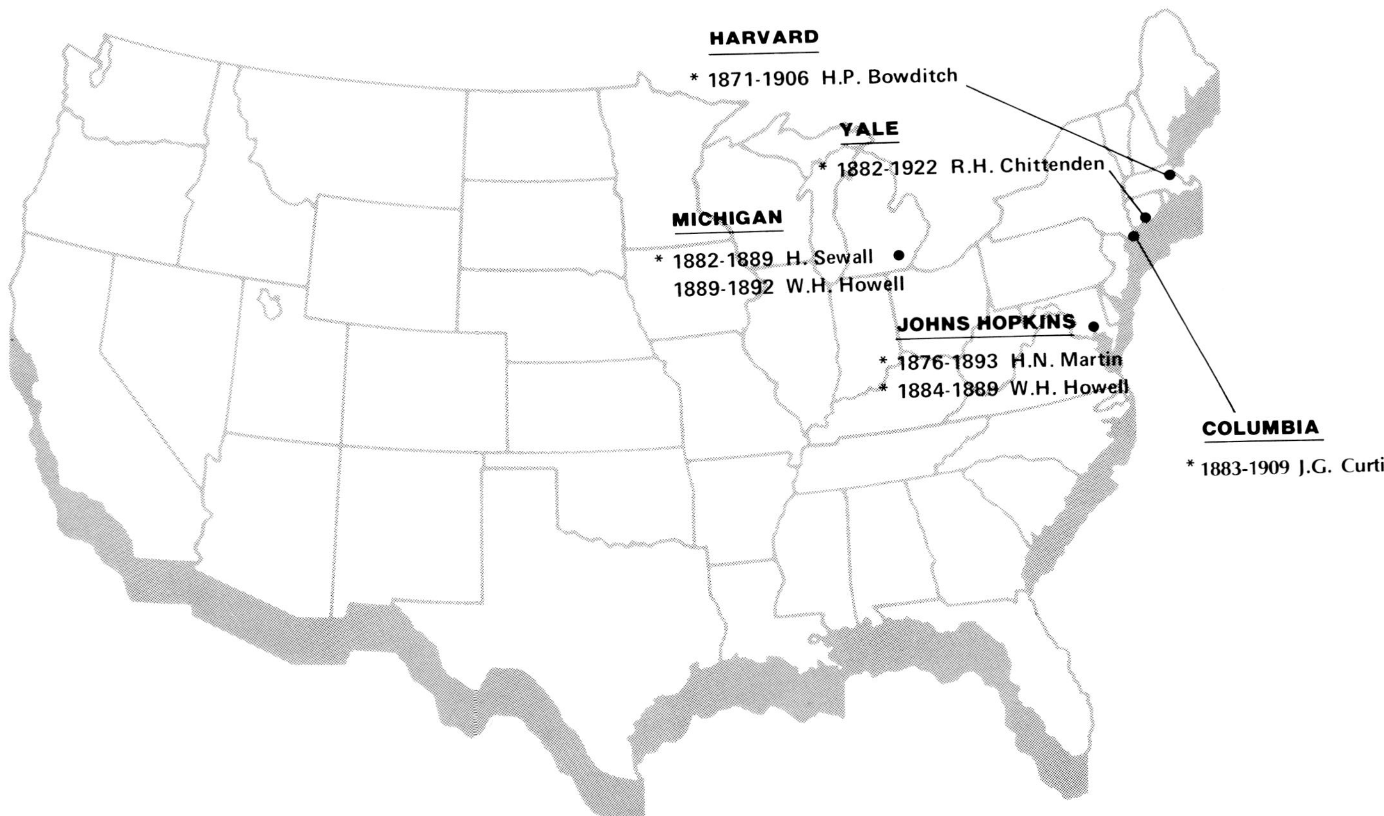

Figure 5: *Leading faculty and centers in American physiology,* ***1880–1890****. *Individual's first appearance in Figures 5–9; capitalization, institution's first appearance in Figures 5–9.*

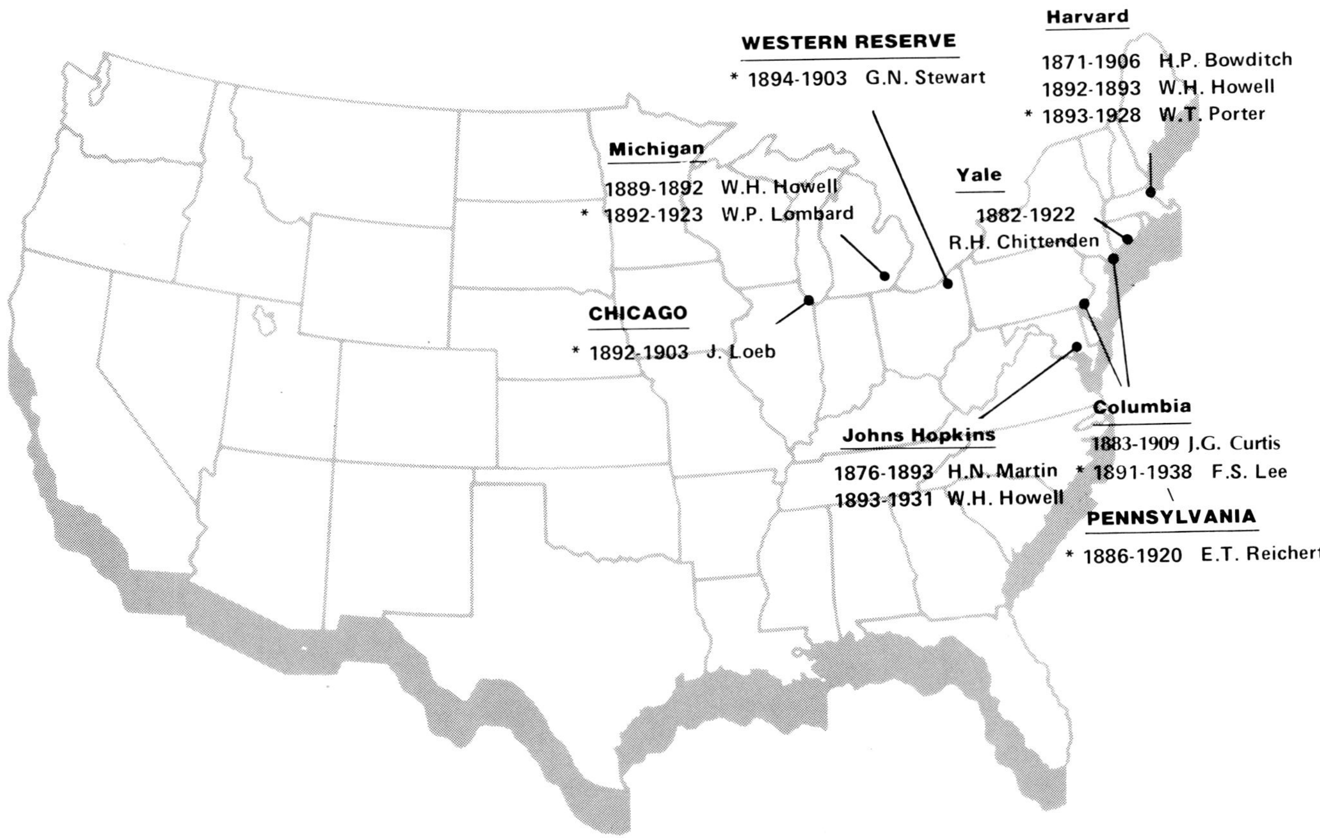

Figure 6: *Leading faculty and centers in American physiology,* ***1890–1900****. *Individual's first appearance in Figures 5–9; capitalization, institution's first appearance in Figures 5–9.*

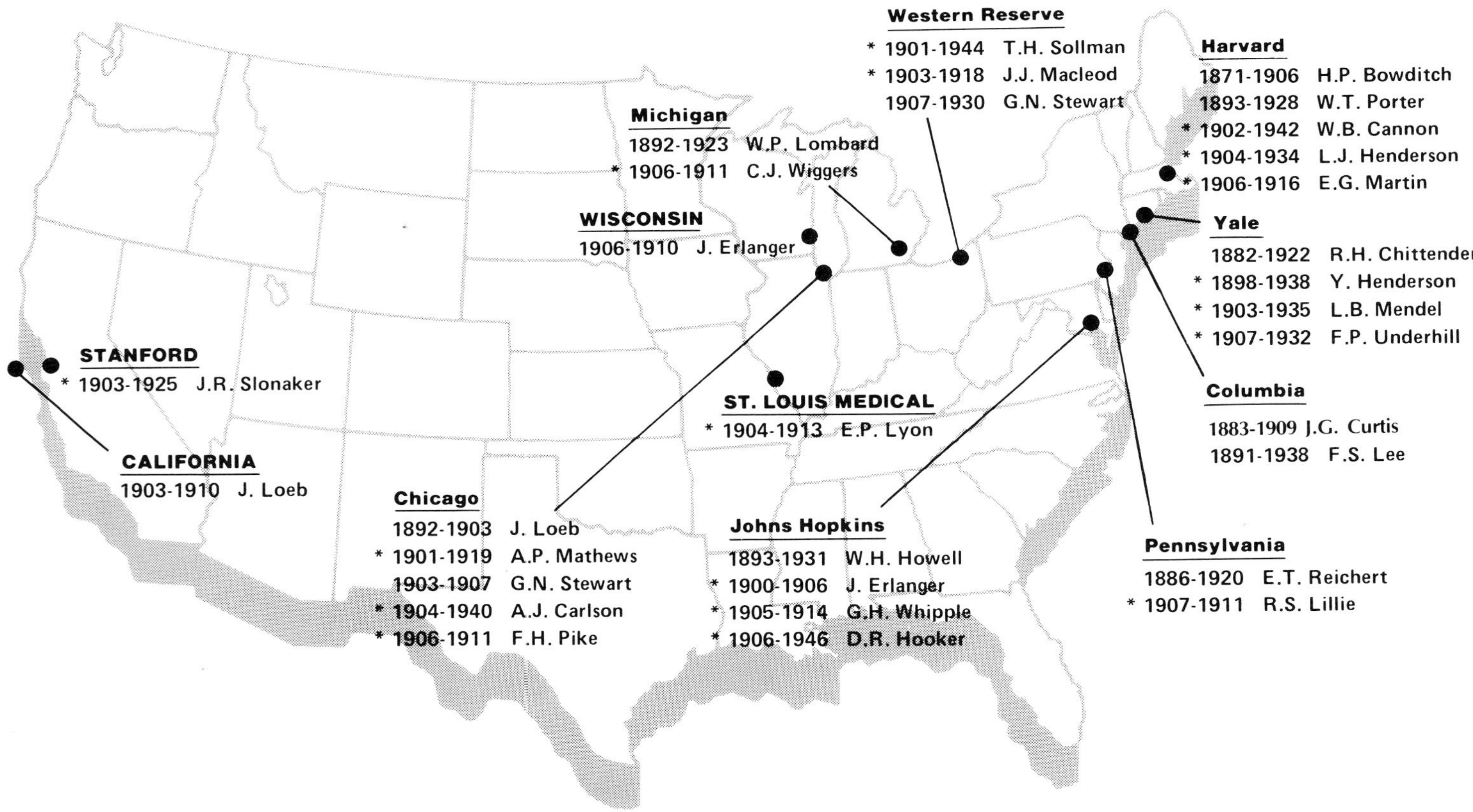

Figure 7: *Leading faculty and centers in American physiology,* ***1900–1910****. *Individual's first appearance in Figures 5–9; capitalization, institution's first appearance in Figures 5–9.*

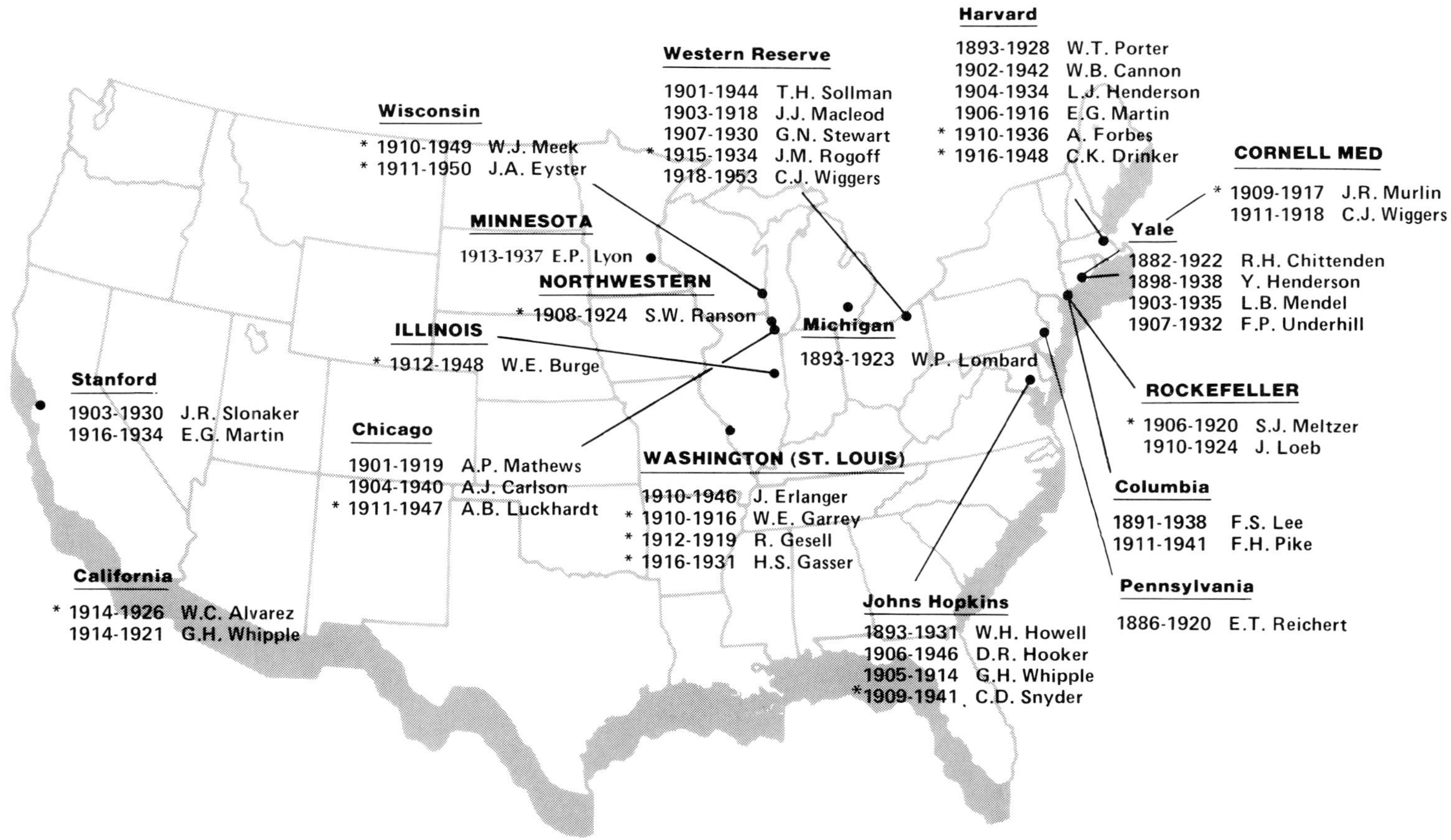

Figure 8: *Leading faculty and centers in American physiology,* ***1910–1920****. *Individual's first appearance in Figures 5–9; capitalization, institution's first appearance in Figures 5–9.*

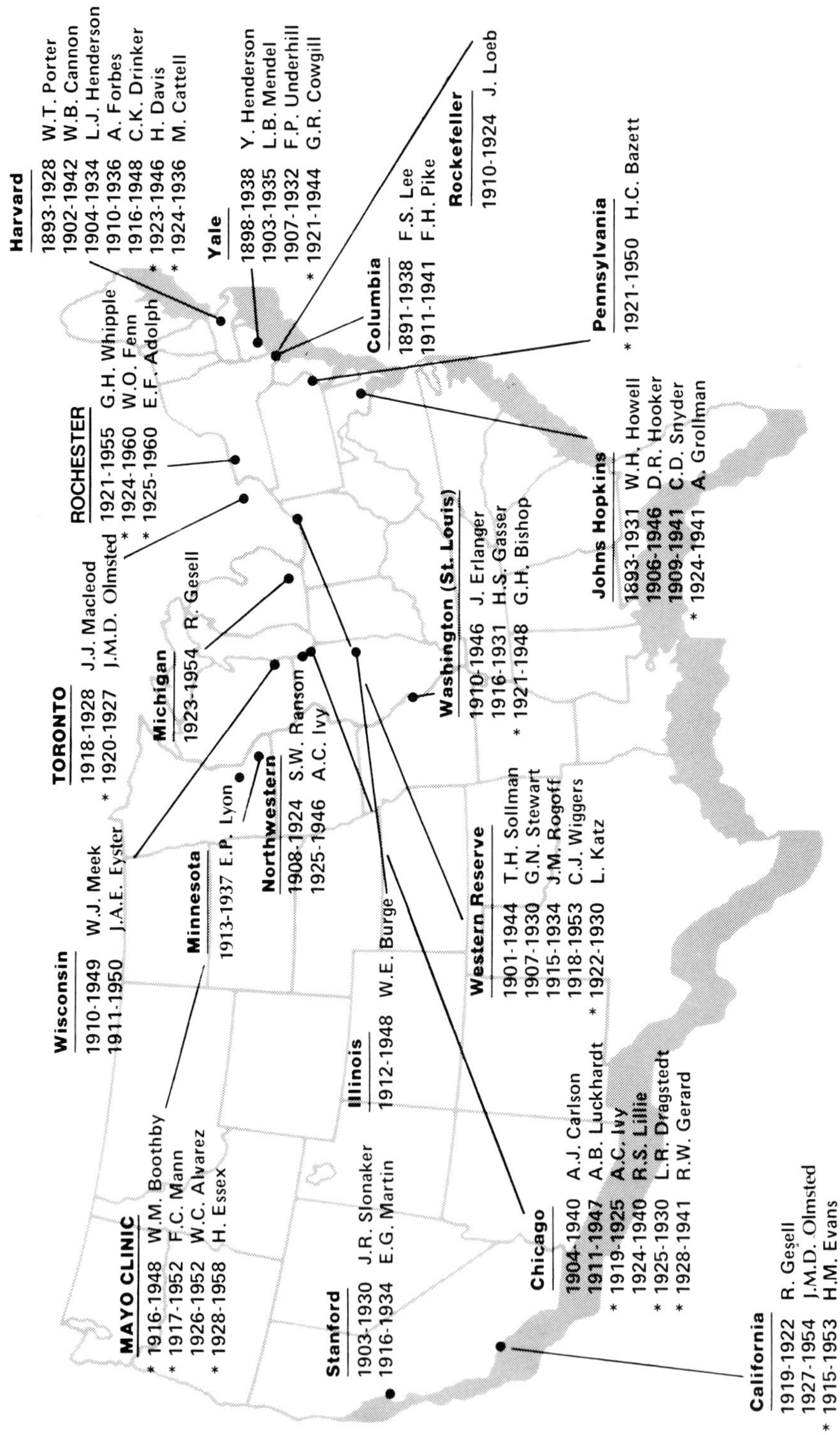

Figure 9: *Leading faculty and centers in American physiology,* ***1920–1930.*** **Individual's first appearance in Figures 5–9; capitalization, institution's first appearance in Figures 5–9.*

same period, Carlson trained an almost identical number of fellows (31) of various sorts, although the sources of their fiscal support is not yet so clear. For the two decades 1920–1940, the number of research and teaching fellows in the Chicago department reached a total of fifty-one.[67] Those who joined Cannon at Harvard or Carlson at Chicago after graduate training elsewhere must have found it difficult to say that they had been trained at any one place.

More generally, this evidence begins to alert us to the full size, complexity, and fluidity of the subprofessorial staff in leading physiology departments. Especially in elite departments, a lot of research, teaching, and training was carried out by assistants, associates, and fellows, often out of sight of and beyond the direct control of the department chairman or his full professorial colleagues. The productivity and reputation of leading centers clearly depended to no small degree on the quality of this body of staff and trainees.

A preliminary study of the department of the University of Chicago by George Anastoplo provides some sense of the scale and significance of such subprofessorial positions.[68] Between 1900 and 1940, the research and teaching staff at Chicago occupied no fewer than nine different kinds of positions, divided into four broad categories: (1) faculty; i.e., professor, associate professor, assistant professor, and instructor; (2) research associates and assistants; (3) unspecified—probably teaching—associates and assistants; and (4) fellows of various sorts, including those supported by the National Research Council. Except during the mid-1930s (presumably because of the Depression), the number of personnel at the lower levels often matched the size of the official faculty. In 1919, for example, the Chicago staff included 5 faculty members (only one of them a full professor) and four assistants, associates, and fellows, for a total staff of nine; in 1929, seven faculty members (3 of them full professors) were joined by five assistants of various sorts for a total staff of twelve; in 1939, thirteen faculty members (still only 3 full professors) and ten assistants made up the total staff of twenty-three. Such data make it clear that until we know vastly more about the role and contributions of personnel at subprofessorial levels, we will have a very incomplete picture of the process of training and its relation to teaching and research in American physiology between 1900 and 1940.

OF MEN AND INSTITUTIONS: CONCLUDING REFLECTIONS

This chapter has drawn special attention to the very uneven distribution of productivity and achievement in American physiology from 1900 to 1940. I have suggested that these "inequalities" can be detected at every level of analysis: national, regional, departmental, and individual. The argument, which is mainly quantitative in form, makes informal use of some of the techniques and insights that Price developed in his quantitative analyses of science twenty-five years ago. To some degree, this chapter could be seen as a confirmation in the case of American physiology of Price's broader conclusions about the differential distribution of productivity and achievement in science.

One crucial issue remains: why should scientific productivity and achievement be so unevenly distributed between the elite and the rank-and-file in my population of physiologists—or, so Price claims, in any other broadly inclusive population of scientists? Obviously, as he points out, the result says something about the nature of the product being measured: scientific papers do not rain down from heaven at random. Associating himself with the Galtonian tradition, Price seems to ascribe this pattern mainly to individual differences in intelligence and scientific talent.[69] Insofar as this explanation amounts to saying that gifted individuals play a crucial role in science, it seems indisputable to me.

I have already noted how very few men it took to lead American physiology to its ascendancy between 1915 and 1940—i.e., an elite group of sixty or so. The operative number may have been smaller still, for a single man often energized an entire research center into vigorous and sustained activity. In fact, one could cover an astonishingly large portion of the history of American physiology during the first half of this century by focusing on just four men and their research schools: W. H. Howell of Johns Hopkins, Cannon of Harvard, A. J. Carlson of Chicago, and Joseph Erlanger of Washington University in St. Louis.[70] They dominated the field, both intellectually and institutionally, from 1900 to 1940. Carlson, for example, was directly responsible for the surge of productivity at Chicago during the early years of this century. His prolific output of papers on cardiac and gastric physiology placed him at the very top of our individual list. In fact, he would have ranked fourth, by himself, among the leading "institutions" for the period 1898–1918.[71] Similarly, one can explain many of the intellectual developments and productivity shifts at Hopkins, Harvard, or Washington University solely by examining the activities of Howell, Cannon, and Erlanger.

The role of other important, if less dominant, individuals should begin to emerge from an examination of the maps in Figures 5–9. If one follows there the path of leading faculty within or between research centers over time, one often finds a detectable shift in productivity as particular men arrive or depart. For the second part of our period (i.e., 1918–1940), such hypotheses can be refined through the use of the bar graph in Figure 1, which allows us to follow shifts in institutional productivity at shorter intervals of time and thus to relate them more closely to the activities of leading individuals.

Yet I want to conclude, almost paradoxically, with a plea that we recognize the role of social and institutional factors in the production of such elites. Like individual talents, social circumstances and opportunities are themselves differentially distributed. One need only recall that productivity and achievement in physiology varied not only between individuals, but also between different countries and between different institutions within the United States. Indeed, as Table 7 suggests, the distribution of research productivity among centers of physiology in the United States followed a pattern very similar to that found in the case of individual physiologists.

In both cases, productivity was distributed very unevenly across the whole spectrum from top to bottom. At the top of the scale in both cases, the gap between productivity scores narrowed. When Price detected this trend at the top of high individual productivity scores, he proposed a modified version of Lotka's law to take account of the fact that one man can do only so much. Given the same basic trend in the case of institutional productivity scores for American centers of physiology, it follows that one institution can do only so much, as is indeed obvious as soon as one thinks about it.

A much more important, indeed crucial, implication flows from the similarity in the pattern of individual and institutional productivity. For if, in the case of individuals, one can ascribe such a pattern to innate differences in intellect, talent, etc., how is one to explain the same pattern in the case of institutions? Even if we wanted to include some "hereditary" component in our account of differential institutional success in American physiology, we would surely do so in terms compatible with Robert K. Merton's famous "Matthew effect," whereby the rich get richer and the poor get poorer. For we would surely find ourselves speaking in terms of the cumulative advantages that come with institutional prestige and traditions, large endowments or other fiscal resources, as well as all the other environmental forces that simultaneously attract and reproduce an elite. The elite physiologists previously discussed were themselves shaped by such external forces, with the site and nature of their graduate training being arguably the chief unifying or distinguishing feature between them. Those who emerged from the elite centers clearly had a head start in the quest for a place on the privileged side of the Matthew effect.

Yet it was precisely those privileged few—those who were smart, talented, industrious, and very lucky—who managed to effect a change in the pattern of institutional relations and mental constructs they had inherited while being "reproduced." Indeed, it is only through them that one can begin to assess in detail the relative importance of internal and external forces. Whether as agents or vehicles, they are the markers of change, the historically interesting actors in our story. In the end, this quantitative and populational exercise will have failed in its purpose if it does not lead us toward a much closer look at those very few men whose research and ideas made a real difference in the rise of American physiology.

NOTES

1. On Walter B. Cannon, see his splendid autobiography, *The Way of an Investigator: A Scientist's Experiences in Medical Research* (New York: Norton, 1945); Donald Fleming, "Walter B. Cannon and Homeostasis," *Soc. Res.* 51 (1984): 609–640; Saul Benison, A. Clifford Barger, and Elin L. Wolfe, "Changing of the Guard," *Harv. Med. Alum. Bull.* 56 (1982): 34–41; and Benison and Barger, "Walter Bradford Cannon," *Dictionary of Scientific Biography*, XV, suppl. 1, ed. C. C. Gillispie (New York: Scribner, 1970) 71–77. (Hereafter, *DSB*.)
2. See Robert G. Frank, Jr., "American Physiologists in German Laboratories, 1850–1914," chapter I in this book, pp. 11–46. esp. 19–23.
3. See Gerald L. Geison, "Keith Lucas," *DSB,* VIII, 532–535.
4. See Harvey Cushing, *The Life of Sir William Osler*, 2 vols. (Oxford, 1925), vol. 2, 413–414, esp. n. 1.
5. Frederic Schiller Lee to Prof. [Edward] Ellery, Union College, 30 November 1915, Lee Papers, f. El. The Lee Papers are deposited at the College of Physicians and Surgeons, Columbia University, Health Sciences Library, Special Collections, New York City. (Here-

after, Lee Papers.) This collection was not yet fully organized and catalogued when I examined it in the Spring of 1984.

6. Cannon, "A Parenthesis of War," in *Way of An Investigator*, pp. 130–145. On fatigue, see Richard Gillespie, "Industrial Fatigue and the Discipline of Physiology," chapter X in this book, pp. 237–262, esp. 244–249. On the American Physiological Society (APS) and World War I, see William H. Howell and Charles W. Greene, *History of the American Physiological Society Semicentennial, 1887–1937* (Baltimore, MD: APS, 1938), pp. 103–108. For other comments on physiology and World War I, see the autobiographical essays gathered together as *The Excitement and Fascination of Science* (Palo Alto, CA: Annual Reviews, 1965), passim, including pp. 87, 103, 255, 288, 367, 375, 558, 563.
7. D. R. Hooker to APS Council, 6 December 1918, Council Minutes, APS vol. 2 (1915–1923), p. 65a, American Physiological Society Archives, Bethesda, MD. (Hereafter, APS Council Minutes.)
8. I have reproduced Tables 1 and 2 as they are conveniently found in Joseph Ben-David, *The Scientist's Role in Society: A Comparative Study* (Englewood Cliffs, NJ: Prentice-Hall, 1971), pp. 188–189.
9. John F. Fulton and Leonard G. Wilson, eds., *Selected Readings in the History of Physiology*, 2d ed. (Springfield, IL: Thomas, 1966).
10. On Fulton, see A. Earl Walker, "John Farquhar Fulton," *DSB*, V, 207–209.
11. Geison, *Michael Foster and the Cambridge School of Physiology: The Scientific Enterprise in Late Victorian Society* (Princeton, NJ: Princeton Univ. Press, 1978), esp. p. 341.
12. Lee to Ellery, 30 November 1915, Lee Papers, f. El.
13. During a visit to England in 1923, Herbert Gasser of Washington University even complained that "There are also too many Americans [here]." Gasser to Joseph Erlanger, 3 October 1923, Erlanger Papers, Washington University School of Medicine, St. Louis, Special Collections, Box 2, f. 27. (Hereafter, Erlanger Papers.) For a superb introduction to and description of this collection, see Frank, "The Joseph Erlanger Collection at Washington University School of Medicine, St. Louis," *J. Hist. Biol.* 12 (1979): 193–201. For more on American neurophysiologists in English laboratories after World War I, see Louise Marshall, "Instruments, Techniques, and Social Units in American Neurophysiology, 1870–1940," chapter XV in this book, pp. 351–369, esp. 351–353. For autobiographical accounts by other American physiologists who visited English laboratories, see *Excitement and Fascination of Science.*
14. On Erlanger at Washington University, see Frank, "Joseph Erlanger Collection;" and Louise Marshall, "The Fecundity of Aggregates: The Axonologists at Washington University, 1922–1942," *Perspect. Biol. Med.* 26 (1983): 613–636.
15. Lee to Erlanger, 9 June 1915, Lee Papers, f. Erlanger. Lee asked Erlanger for his blueprints and total floor space because "the time has come for us here at Columbia to formulate plans for a new physiological laboratory—plans which may, however, never be realized." He did not elaborate.
16. See Erlanger to Lee, 12 June 1915, Lee Papers, f. Erlanger. Erlanger added that of the total floor space of 22,724 sq. ft., "the department of Pharmacology has the privilege of using about 3,000 sq. ft." In the same box in the Lee Papers is a loose unsigned drawing labeled "Floor space, department of physiology, January, 1914," indicating a total floor space of 8,047 sq. ft. This drawing presumably depicts Lee's laboratory at Columbia.
17. Gasser to Erlanger, 3 October 1923, Erlanger Papers.
18. Naciso San Louis Cordero to Erlanger, 30 October 1927, Erlanger Papers, f. 10; and Erlanger to Cordero, 26 November 1927, Erlanger Papers, f. 14. For more on Cordero, who was a Rockefeller Foundation Fellow, see Clifford W. Wells to Erlanger, 17 August 1925, Erlanger Papers, f. 10.
19. On the Harvard Laboratories of Physiology as of 1920, including floor plans, see Alejandra Laszlo, "Physiology of the Future: Institutional Styles at Columbia and Harvard," chapt. III in this book, pp. 67–96, on 85–88.
20. Anton J. Carlson, "University of Chicago Division of Biological Sciences, Department of Physiology," in *Methods and Problems of Medical Education*, 19th ser. (New York: Rockefeller Foundation, Division of Medical Education, 1931), pp. 37–41, esp. 39. For a more complete description of the Hull Biological Laboratories that this new building replaced, see *The Annual Register for the Year Ending June 30, 1925* (Chicago: University of Chicago, 1925), p.

38. My thanks to George Anastoplo for this reference. I have converted the exterior dimensions of the buildings into square feet and divided the space among the departments as indicated in the sources just cited.

21. Robert Gesell, "Department of Physiology, University of Michigan," in *Methods and Problems of Medical Education*, 18th ser. (New York: Rockefeller Foundation, 1930), pp. 1–16; and Horace W. Davenport, *Fifty Years of Medicine at the University of Michigan, 1891–1941* (Ann Arbor, MI: Univ. of Michigan Medical School, 1987), chapt. 6. My thanks to Dr. Davenport for allowing me to see and cite this chapter of his manuscript.
22. W. J. Meek, "The Beginnings of American Physiology," *Ann. Med. Hist.* 10 (1928): 111–125, esp. 124.
23. On *Physiological Reviews*, see APS Council Minutes, vol. 2, pp. 80–81, 109a–109g, and vol. 6 (1936), passim; Howell and Greene, *History of APS*, pp. 108–109; and Toby A. Appel, "The Second Quarter Century," in J. R. Brobeck, O. E. Reynolds, and T. A. Appel, eds., *History of the American Physiological Society: The First Century* (Bethesda, MD: American Physiological Society, 1987).
24. Howell, "Address of Welcome . . . ," 19 August 1929, in Howell and Greene, *History of APS*, pp. 138–140, quote on 138.
25. R. W. Gerard, *Mirror to Physiology: A Self-Survey of Physiological Science* (Washington, DC: American Physiological Society, 1958), p. 140.
26. The word "elite" can arouse surprising passion among historians of American science. I use the word here in what I take to be a commonsensical way, perhaps roughly equivalent to Daniel J. Kevles's "best-science elitism." On the tension between American democratic ideals and "best-science elitism," see Kevles, *The Physicists: The History of a Scientific Community in Modern America* (New York: Knopf, 1978).

    A closely related issue is the alleged indifference of Americans to basic research, an allegation that Nathan Reingold has repeatedly challenged in recent years. For convenient access to the pertinent literature, see Albert E. Moyer, "History of Physics," in *Historical Writing on American Science*, ed. Sally Gregory Kohlstedt and Margaret W. Rossiter *Osiris*, 2d ser., 1 (1985): 163–182, esp. 173–176, n. 20–22.
27. Derek J. de Solla Price, *Little Science, Big Science* (New York: Columbia Univ. Press, 1963), pp. 33–61.
28. Ibid., p. 46.
29. Ibid., esp. pp. 46–49.
30. For a plea that visual language be used more often by historians of science, see Martin J. S. Rudwick, "The Emergence of a Visual Language for Geological Science, 1760–1840," *Hist. Sci.* 14 (1976): 149–195.
31. As noted in the preface to this book, I am indebted to a series of talented part-time research assistants for the data on which much of the following discussion is based. For a somewhat similar body of data for three other disciplines in the United States, see Kevles, "The Physics, Mathematics, and Chemistry Communities: A Comparative Analysis," in Alexandra Oleson and John Voss, eds., *The Organization of Knowledge in Modern America, 1860–1920* (Baltimore: Johns Hopkins, 1979), pp. 139–172, esp. 161–172. See also Arnold Thackray, Jeffrey L. Sturchio, P. Thomas Carroll, and Robert F. Bud, *Chemistry in America, 1876–1976: Historical Indicators* (Boston, MA: Reidel, 1985).
32. See APS Council Minutes, vol. 1, pp. 166–170 on the founding of the *American Journal of Physiology* with copy of printed agreement between Porter and the Society; and vol. 2, pp. 382–385 on the Society's assuming direct ownership of the journal. More generally, see the sources in n. 33.
33. On the editorial policies and practices of the *American Journal of Physiology*, see Howell and Greene, *History of APS*, pp. 78–83, 97–98; and Appel's chapters on the first seventy-five years of the APS in *History of the American Physiological Society*.
34. On "biological" and "general" physiology respectively, see Jane Maienschein, "Physiology, Biology, and the Advent of Physiological Morphology," chapt. VII, this book, pp. 177–193; and Philip J. Pauly, "General Physiology and the Discipline of Physiology, 1890–1935," chapt. VIII, this book, pp. 195–207.
35. *Am. J. Physiol.* 106 (1 November 1933, no. 2), p. 250 (inside cover), quote from item 2.
36. E. F. Adolph et al., "Physiology in North America, 1945: Survey by a Committee of the American Physiological Society," *Proc. Fed. Soc. Exp. Biol.* 5 (1946): 407–436.

37. Ibid., p. 422.
38. Ibid., p. 420; the committee even included genetics among these presumed subdisciplines of physiology.
39. Gerard, *Mirror*, p. 32. This expansive estimate included, e.g., plant and bacterial physiologists, many of whom might well have named botany as their discipline if forced to choose.
40. The total number of individual authors for each of the thirty-volume units from volumes 1–120 (1898–1937) was as follows: volumes 1–30, 552; volumes 31–60, 836; volumes 61–90, 1,990; volumes 91–120, 2,633.
41. The estimate of 3,600 total individual contributors begins with the data in n. 40. Using Price's estimate (*Little Science*, pp. 87–90) that 40 percent of new authors will produce a second paper or more, one can calculate that volumes 90–120 had 1,935 new authors, while 1,750 authors dropped out before volume 90 appeared. Thus, the total number of individual authors would be 1,935 + 1,750 = 3,685, or, roughly speaking, 3,600.
42. For data on APS membership, see Joseph F. Saunders and Aubrey E. Taylor, "Membership," in *The First Century*.
43. Put another way, annual research productivity varied quite widely within the highly productive group listed in Table 5. Among the more extreme examples of the actual range are Yandell Henderson, with an average annual publication rate of less than 1.5 articles, and Arturo Rosenbleuth, who published more than 7 articles per year on average.
44. On the time-consuming task of securing and maintaining appropriate laboratory materials and animals, see Adele E. Clarke, "Research Materials and Reproductive Science in the United States, 1910–1940," chapt. XIV, this book, pp. 323–350.
45. Wallace O. Fenn, *History of the American Physiological Society: The Third Quarter Century, 1937–1962* (Washington, DC: American Physiological Society, 1963), p. 12, mentions Hooker's "very firm" attitude that "Papers were not to be given serial numbers with general titles for the whole series and separate subtitles for each paper." See also item 2g of the "Publication Policy of the *American Journal of Physiology*," dated 15 June 1930, printed and distributed with the suggestion that it "be framed and hung in your laboratory so that every worker may have access to it at all times." According to item 2g, special scrutiny would be given to "Papers which appear to be of the nature of progress reports the publication of which might be properly withheld until the research has progressed to the completion of at least a significant phase of the general problem." A printed copy of this policy statement is deposited in the APS Archives, Bethesda.
46. See Frank, "American Physiologists," pp. 35–36.
47. See Marshall, "Instruments, Techniques, and Social Units," this volume, pp. 360–361.
48. See Fenn, *History of the APS, 1937–1962*, p. 6: "Dr. Ivy is reported to have published 1,500 scientific articles, . . . which is something of a record." Perhaps Ivy does indeed hold the record; the most prolific author found by Price in his quantitative studies of science was the nineteenth-century British mathematician Arthur Cayley, who produced a mere 995 papers. Price, *Little Science*, p. 49.
49. Gerard, "The Organization of Science," in *Excitement and Fascination of Science*, pp. 149–160, on 155, n. 1. More generally on the practice of multiple authorship in twentieth-century science, see Price, *Little Science*, pp. 87–90.
50. Price, *Little Science*, pp. 40–41. Price elsewhere suggests that frequency of citation might be a better index of quality than frequency of publication. Ibid., pp. 77f.
51. For basic biographical information on the men in Table 5, see their entries in *American Men of Science*.
52. Howell still awaits a full-scale study. In the meantime, see Anne Clark Rodman, "William Henry Howell," *DSB*, VI, 525–527; and especially Erlanger's account in *Biogr. Mem. Nat. Acad. Sci.* 26 (1951): 153–180.
53. On E. V. McCollum, see Stanley L. Becker, "Elmer Verner McCollum," *DSB*, VIII, 590–591; and McCollum, *From Kansas Farm Boy to Scientist* (Lawrence, KS: Univ. of Kansas, 1964). On George W. Corner, see his delightful autobiography, Corner, *Seven Ages of a Medical Scientist: An Autobiography* (Philadelphia, PA: Univ. of Pennsylvania Press, 1981). On H. M. Evans, see E. C. Amoroso and Corner, "Herbert McLean Evans, 1882–1971," *Biogr. Mem. Fell. R. Soc. Lond.* 18, (1972): 83–186; Corner, "Herbert McLean Evans," *Biogr. Mem. Nat. Acad. Sci.* 45 (1974): 153–192; and I. D. Raacke, "The Appointment of Herbert McLean Evans as Head of Anatomy at Berkeley," *J. Hist. Biol.* 9 (1976): 301–322. On Gregory Pincus,

see Dwight J. Ingle, "Gregory Pincus," *Biogr. Mem. Nat. Acad. Sci.*, 42 (1969): 229–270; M. C. Chang, "Gregory Goodwin Pincus," *DSB*, X, 610–611; and James Reed, *The Birth Control Movement and American Society: From Private Vice to Public Virtue* (Princeton, NJ: Princeton Univ. Press 1984), pp. 317–333.

54. On L. J. Henderson, see Cannon, "Lawrence Joseph Henderson, 1878–1942," *Biogr. Mem. Nat. Acad. Sci.* 23 (1943): 52–58; John Parascandola, "Lawrence Joseph Henderson," *DSB*, VI, 260–262; Parascandola, *Lawrence J. Henderson and the Concept of Organized Systems* (Ph.D. diss., Univ. of Wisconsin, 1968); Parascandola, "Organismic and Holistic Concepts in the Thought of L. J. Henderson," *J. Hist. Biol.* 4 (1971): 63–113; and Stephen J. Cross and William R. Albury, "Walter B. Cannon, L. J. Henderson, and the Organic Analogy" (forthcoming in *Osiris*, 2d ser., vol. 3). On Loeb, see Donald Fleming, "Introduction," in Jacques Loeb, *The Mechanistic Conception of Life*, reprint ed. (Cambridge, MA: Harvard Univ. Press, 1964), vii–xli; Fleming, "Jacques Loeb," *DSB*, VIII, 445–447; and especially Philip J. Pauly, *Controlling Life: Jacques Loeb, Experimental Biology, and the Engineering Ideal* (New York: Oxford Univ. Press, 1987).
55. On J. J. R. Macleod, see Lloyd G. Stevenson, "John James Rickard Macleod," *DSB*, VIII, 614–615. For an excellent account of the Toronto discovery, see Michael Bliss, *The Discovery of Insulin* (Toronto; University of Chicago Press, 1982), which gives extensive biographical attention to the other three main participants, Frederick G. Banting, Charles H. Best, and J. B. Collip. On Banting, see also Lloyd G. Stevenson, "Frederick Grant Banting," *DSB*, I, 440–443; and Bliss, *Banting: A Biography* (Toronto: University of Chicago Press, 1984). On J. B. Collip, see also R. L. Noble, "James Bertram Collip," *DSB*, III, 351–354, C. H. Best, thus far the most neglected of the quartet, was an active member of the American Physiological Society (including its Council) and published twenty-one papers in the *American Journal of Physiology* before 1940. During the same period, Collip published twenty-four papers in the journal, while Banting published but nine.
56. The nineteen APS Presidents in Table 5 are indicated in italic print. Biographical sketches of all nineteen can be found in Howell and Greene, *History of APS*; and Fenn, *History of APS, 1937–1962.*
57. See Appel, "Second Quarter Century," chapt. 3.
58. The historical relationship of Washington University to both St. Louis Medical College and the rival Missouri Medical College is complex and deserves further study. Among other things, the latter two institutional names continued to be used in some contexts even after both schools were officially merged with Washington University in the 1890s. A first step toward sorting out these relationships is the pamphlet, "Special Collections," Library of the Washington University School of Medicine, St. Louis, esp. pp. 20–23.
59. Laszlo, "Physiology of the Future," pp. 70–78.
60. I owe these numbers to Toby Appel, historian/archivist at the American Physiological Society, who pulled them from the APS computer databases.
61. See Carlson, "University of Chicago," p. 38.
62. On physiological chemistry at Yale, see Robert E. Kohler, *From Medical Chemistry to Biochemistry: The Making of a Biomedical Discipline* (Cambridge, England: Cambridge Univ. Press, 1982), pp. 289–295. On similar inbreeding at the University of Michigan see Davenport, *Fifty Years*, chapt. 6.
63. Carlson has been woefully neglected by biographers and historians of physiology. For basic information, see Lester R. Dragstedt, "Anton Julius Carlson," *Biogr. Mem. Nat. Acad. Sci.* 35 (1961): 1 12; Maurice B. Visscher, "Anton Julius Carlson," *DSB*, III, 68–70; and Ingle, "Anton J. Carlson: A Biographical Sketch," *Perspect. Biol. Med.* 22 (1979): S114–S136.
64. For quantitative data on the leadership of Johns Hopkins in chemistry, physics, and mathematics, see Kevles, "Physics, Mathematics, and Chemistry," pp. 161–172. More generally, see Hugh Hawkins, *Pioneer: A History of the Johns Hopkins University, 1874–1889* (Ithaca, NY: Cornell Univ. Press, 1960); and Laurence R. Vesey, *The Emergence of the American University* (Chicago, IL: Univ. of Chicago Press, 1965).
65. Erlanger to Cannon, 10 March 1913, Erlanger Papers, box 2, f. 12.
66. Laszlo, "Physiology of the Future," pp. 88.
67. See George Anastoplo, "Physiology at the University of Chicago," unpublished manuscript, Princeton University, Program in History of Science, pp. 25–41.

68. Ibid.
69. Price, *Little Science*, esp. pp. 40–46, 49–51. Yet despite such Galtonian passages, Price elsewhere seemed to recognize that the "built-in *undemocracy*" of scientific productivity also depended on institutional factors and social opportunities. See, e.g., ibid., pp. 54–59. A few years later he made the point more explicitly in Price, "The Scientific Foundations of Science Policy," *Nature Lond.* 206 (1965): 233–238, esp. 235–236. Cf. Robert K. Merton, "The Matthew Effect in Science" [1968], in Merton, *The Sociology of Science: Theoretical and Empirical Investigations*, ed. Norman W. Storer (Chicago, IL: Univ. of Chicago Press, 1973), p. 457, n. 45.
70. On research schools and their advantages as a unit of historical analysis, see Geison, "Scientific Change, Emerging Specialties, and Research Schools," in *History of Science*, 19 (1981): 20–40.
71. Between 1898 and 1918, Carlson published ninety-three papers in the *American Journal of Physiology*. Had he been an "institution," that level of productivity would have placed him between Harvard and Cornell Medical School on Table 8.
72. See Merton, "The Matthew Effect in Science," pp. 439–459, esp. 457–459.

VI

Biological and Medical Societies and the Founding of the American Physiological Society

TOBY A. APPEL

> The American Physiological Society. . .from its inception has struggled with ambivalent attachment to the two worlds, of biological science and medical practice.
>
> Ralph Gerard, 1958[1]

Since its emergence as a discipline in the nineteenth century, physiology has occupied a special position as a bridge between biology and medicine. Because it stands at the juncture, its boundaries have been difficult to define and spokesmen for the discipline have long worried that it was in danger of being pulled apart, on one hand by the various clinical specialties and on the other by branches of biology.[2] This "ambivalent attachment" to "two worlds" is vividly illustrated in the founding and early history of the American Physiological Society (APS). Organized on 30 December 1887, the society had its immediate origin in the medical tradition. It was a product of a conflict within the medical community between the older American Medical Association (AMA) and the more recently established and more exclusive medical specialty societies. At the same time it was the firstborn of a series of biological specialty societies, offshoots of the American Society of Naturalists, itself a specialist offshoot of the American Association for the Advancement of Science. In its foundation and first twenty-five years, APS was shaped by its uneasy alliances with these two very different groups of organizations.

In past histories, the founding of the APS has been treated as a relatively simple and unproblematic event. William Henry Howell's history of the Society merely suggested that the time was ripe, for there were by 1887 three flourishing laboratories of physiology: Henry Pickering Bowditch's at Harvard Medical School, Henry Newell Martin's at Johns Hopkins, and Russell Chittenden's at Yale's Sheffield Scientific School. These laboratories so stimulated experimental physiology in America that "within a decade the number of workers and the output in research had reached a stage which made it desirable to form a national society of physiologists." From Howell's account, based on the minutes of the Society, we learn only that Bowditch,

Martin, and the dean of American physiologists, S. Weir Mitchell, sent a letter on 10 November 1887 to an undisclosed number of physiologists inviting them to attend an organizational meeting in New York, and that a second letter dated 2 December enclosed a copy of the proposed constitution.[3] The impression was given that although the fledgling society sometimes met with other groups, it usually met alone as a tiny band of modern experimental physiologists. Other existing accounts tell a similar unencumbered story.[4]

However, when the Society is placed in the context of science and medicine in America and in particular of other medical and biological societies, several quite interesting historical issues emerge. First, one discovers that the APS was not only the earliest disciplinary society in the biomedical sciences but also one of the earliest national disciplinary societies among the sciences in general. The physiologists preceded the anatomists, zoologists, botanists, psychologists, geologists, mathematicians, physicists, and astronomers in forming a society. Even the chemists, who formed a society in 1876, were primarily a local New York institution before the 1890s. Why were the physiologists so early? Second, the APS appears to have been the first national society in America to require publication of original research as a prerequisite to membership. Why did the Society deliberately choose to restrict itself in this way instead of attempting to acquire as large a membership as possible by admitting anyone with an interest in the subject? Who was this restriction designed to leave out?

This latter point takes us back to the question of defining a discipline. Robert Kohler has suggested that a major function of professional societies is to define fields—to map out boundaries with respect to other disciplines, or more colloquially, to stake out intellectual turf.[5] Physiology provides a particularly interesting case study in disciplinary definition, not only because of its dual ties to medicine and biology, but also because the term "physiology" was used in so varied a manner in the nineteenth century. Even within the medical community it was by no means synonymous with experimental physiology as represented by the founders of the Society. To the layman of the nineteenth century, physiology was often equated with hygiene. In fact, fifty years earlier, in 1837, an American Physiological Society had been founded in Boston by William Alcott and Sylvester Graham based on totally different principles from the later APS. The first American Physiological Society was a large lay organization of men and women whose central concern was diet.[6]

By looking at the milieu of science and medicine in the 1880s, and especially the Society's early links to other biological and medical societies, one can account for why physiologists united so early and why the founders chose to be so restrictive regarding entry. In fact, one can plausibly explain why the APS happened to be formed in New York City on 30 December 1887. Though some historians have suggested that the society nearly expired in the 1890s,[7] one can show that it was able to function tolerably well in this period even though it remained very small. Finally, one can see, in terms of professional organizations, how physiology gradually separated itself from both medical and biological specialties and instead became part of a new grouping of distinctively biomedical sciences, meeting together as part of the Federation of American Societies for Experimental Biology (the Federation).

In the broadest sense, the 1880s were a ripe time for the formation of specialist societies in America. From the American Historical Association and the Modern Language Association to the American Institute of Electrical Engineers and the American Ornithological Union, a remarkable number of societies were formed in this decade. According to Ralph S. Bates in his survey of scientific societies in America, the period from the end of the Civil War to the end of World War I was the great era of formation of national disciplinary societies. The 1880s, Bates claims, was the decade when the process of society formation got fully under way.[8]

Other considerations, however, lead one to ask whether the time was indeed ripe for forming a national physiological society. In 1887, there were still very few active experimental physiologists in America. The founders of the Society spread their net widely by including representatives of every variety of physiology among the charter members. Medical school physiologists, who from the beginning formed the mainstay of the Society, were represented of course, but so were physiological chemists and pharmacologists, physicians who experimented in private laboratories, physiological psychologists, bacteriologists, and hygienists, and even physiological botanists. Because it was decided to elect to membership only those who published original experimental researches, however, the pool of members was extremely limited. Only seventeen men attended the organizational meeting and only twenty-eight were named charter members.[9] Of the latter, it was later estimated that only 43 percent, less than half, were primarily physiologists, hardly enough to support a flourishing organization.[10] The number of members grew slowly in the early years, reaching 100 only in 1901. Through the 1890s fewer than thirty members attended the annual meetings.[11]

By the time the American Society of Biological Chemists was founded with eighty-one charter members in 1906 and the American Society for Pharmacology and Experimental Therapeutics in 1908,[12] the medical reform movement in America was well under way. As Kohler's study of biochemistry in America has shown, the radical transformation of medical schools created an unprecedented opportunity for expansion of the preclinical sciences. The formation of the two new societies at this time, both offshoots of the APS, therefore did much to consolidate the disciplines of biochemistry and pharmacology and to encourage the establishment of chairs and separate departments.[13] In 1887 reform of medical schools had hardly begun; opportunities for well-trained experimental physiologists were exceedingly limited and remained so for the next decade. The physiologists would certainly have formed a society eventually. That they did so in 1887 was due to a more immediate impetus.

Whereas the establishment of laboratories based on European models and of societies, such as the American Society of Naturalists and the Physiological Society in Britain, were important precedents for the founding of the APS, they do not account for the timing. The founding of a disciplinary society is often crystallized by a particular outside event. In the case of the APS this event was the first meeting of a federation of medical societies known as the Congress of American Physicians and Surgeons, which was to take place in Washington, D.C. in 1888.

A TALE OF TWO MEDICAL CONGRESSES

The APS was a by-product of an era of deep dissension in the medical community between the research-oriented elite, primarily associated with northeastern medical schools, and the rank and file of small-town medical practitioners. This friction came to a head in the organization of two rival medical congresses in Washington, D.C. in 1887 and 1888. The organizers of the two congresses represented very different interests and constituencies in American medicine. The Ninth International Medical Congress (IMC) of 1887, under the control of the AMA, reflected the AMA's large membership of practicing physicians across the country who were members of state and county societies and upholders of the AMA code of medical ethics, forbidding consultation with any but "regular" physicians. The Congress of American Physicians and Surgeons, conceived in 1886 as an alternative congress, consisted of members of the recently formed specialty societies and stood for the promotion of the highest quality medical research by the best-trained medical scientists.

Beginning in 1867 the IMC held meetings every two or three years in various European cities. They were attended by the greatest physicians and medical scientists of the day. Physiologists were prominently represented.[14] In May 1884 Austin Flint, Sr. proposed in his presidential address to the AMA that the next IMC be held in America. In response the AMA passed a resolution giving Flint power to appoint a committee of seven (later enlarged to eight) to extend the invitation officially at the Eighth IMC about to meet in Copenhagen. Flint named to the committee some of the most prestigious members of the medical establishment. The committee's secretary and most active member was John Shaw Billings, Librarian of the Surgeon General's Library and founder of *Index Medicus*, who traveled to Copenhagen and personally extended the invitation for the next IMC to meet in Washington, D.C. Although Rudolf Virchow offered the hospitality of Berlin, the American invitation was accepted.[15] An international meeting of physicians in Washington, D.C. promised to be a high point in the history of American medicine. However, instead of proving to Europeans that American medicine had come of age and was the equal of old world medicine, the IMC turned into a source of supreme embarrassment for the American medical community.

In late 1884 the AMA committee proceeded according to its broad mandate to enlarge itself into an executive committee and began organizing the IMC. Seventeen new members were added, a committee on finance was formed, and sections complete with officers were set up. In early 1885 a set of rules and a provisional list of officers was printed in pamphlet form and widely circulated in America and abroad. Among the many people given places of honor in the proposed organization of the Ninth IMC were four of the five men considered by Howell to have been the founders of the APS, namely S.

Weir Mitchell, Henry Pickering Bowditch, H. Newell Martin, and John Green Curtis.[16] Particularly prominent in the initial organization of the IMC was Mitchell, who was appointed to the enlarged executive committee in October and to the five-member standing committee on finance (chaired by John Call Dalton) in December. A Section of Physiology was organized with Dalton as president, Curtis as secretary, and Martin and Mitchell as members of the Council. Mitchell and Bowditch were to be presidents of the Sections of Neurology and Medical Education, respectively. Among other charter members of the APS given offices as presidents or secretaries of sections were William Henry Welch, Horatio C. Wood, and Edward T. Reichert.[17]

The organizational committee was dominated by the eastern, reform-minded elite of the AMA and it chose to fill most of the major positions of the congress organization with scientifically-oriented physicians, largely from eastern institutions. The AMA, with over 4,000 members from all over the country, was becoming increasingly organized along state and local lines, yet the organizers of the IMC gave geographic distribution a low priority in choosing officers for the congress. Geography was considered only in naming vice-presidents and council members of the sections. Billings had to defend even this much solicitude against those among the organizers who wished to reserve the vice-presidencies for distinguished foreigners. The vice-presidents, he wrote, "are mainly for the purpose of recognizing certain gentlemen, especially in the south and west, whom it is desirable to honor; but who are too remote from the great medical centers to be able to render much, if any, practical assistance in the work of organization."[18] For physiology, the proposed vice-presidents were the clinicians James F. Hibberd of Richmond and Middleton Michel of Charleston, neither of whom ever became members of the APS.[19] The initial organization of the IMC, then, was heavily weighted toward the elite of medical researchers, including many of the charter members of the APS.

In April 1885, at the annual meeting of the AMA, this carefully devised plan of organization was torn asunder by a disruptive controversy. The meeting, held in New Orleans, was heavily attended by men from the South and Southwest. When the committee on the organization of the IMC (the original eight members had since enlarged their number to twenty-five) presented its report, serious objections were quickly raised by those who felt left out. In response to the opposition to the choice of officers of the IMC, the AMA in New Orleans passed new resolutions that essentially overturned their resolution of the previous year. As the *Boston Medical and Surgical Journal* reported, the AMA was "we will not say led, but stampeded into taking the extraordinary action . . . of appointing a new Committee."[20] It was resolved on 30 April that the initial committee appointed in 1884 "be enlarged by the addition of thirty-eight members, one from each State and Territory, the District of Columbia, and the Army, the Navy and Marine Hospital Service . . . and that the Committee, thus enlarged, shall proceed at once to review, alter, and amend the motions of the present Committee as it may deem best." This meant that all additions to the original eight-member committee, among them Mitchell, were removed from power and that all appointees were also subject to replacement.[21]

The reasons for this precipitate action were twofold. First, as AMA loyalists claimed, the committee had high-handedly selected as officers of the IMC "a limited representation of the profession in the United States."[22] Large areas of the country were ignored and a number of men of importance in state and local medical politics felt slighted. Second, the imbroglio over the IMC became enmeshed in the then-raging controversy over the AMA's code of medical ethics. The AMA required of its members adherence to the code, which forbade physicians to consult with homeopaths. In 1885 the AMA was in the midst of a battle with the New York Medical Society, which on its own voted in 1882 to liberalize the code. The New York Society's delegates were thenceforth denied registration at the AMA meetings, and an alternate and rival New York society had since been formed. The first committee had sinned by placing a few of the New York dissidents in offices of the IMC. Because the large city specialists were more likely to wish to act as consultants to homeopaths than were country doctors, the conflict was sometimes perceived as one between arrogant city specialists and small-town general practitioners.[23] The IMC affair was thus a classic confrontation between the claims of "best" science and those of democratic science of the kind that has occurred innumerable times in the history of American science and medicine.[24]

The controversy left exceedingly bitter feelings in the medical community that were vented through the pages of the *Journal of the American Medical Association* and other medical periodicals. When, in June of 1885, it appeared that no compromise seemed possible, the original executive committee passed a resolution to be returned to power and adjourned.[25] Billings and several other members of the original eight-member committee resigned in disgust. Many more resignations of officers and committee members followed. Journals such as the *Boston Medical and Surgical Journal* published frequent accounts of who was resigning and printed letters from foreigners suggesting that Europeans would be highly reluctant to attend an IMC under such a shadow.[26] Among the casualties were Bowditch, Mitchell, Martin, Curtis, and Dalton. They and a great many of the most prominent medical men in America would have nothing further to do with the IMC.[27]

The Ninth IMC met as planned in September 1887 with Nathan Smith Davis, a founder of the AMA and editor of its journal, as president. By some accounts it was a great success. The attendance of over 2,500 was said to have been larger than all but one previous IMC, but according to some of the major eastern medical journals, the scientific value of the IMC was much below that of previous congresses, the number of foreign delegates—fewer than 150—was embarrassingly low, and the leaders of the medical profession in America were conspicuously absent.[28] A correspondent for the *Boston Medical and Surgical Journal* wrote that "the gentlemen who came from abroad had a right to expect to come to a representative American gathering; and this they failed to find. Sectional as regards residence, scant in men of recognized professional eminence, the American contingent of the Congress was a body of very worthy family doctors."[29]

The program of the Section of Physiology was strange indeed. Although the IMC took place only three months before the founding of the APS, not one future member of the Society was represented among the officers or speakers. Chaired by John H. Callender, M.D., an "alienist" (i.e., a student of insanity) from Nashville, Tennessee, the sessions included a few foreign visitors of note—Alexander McWilliam, professor of physiology at Aberdeen, Richard Caton, professor of physiology at Victoria University, Manchester, England, and William Dobinson Halliburton, assistant professor of physiology at University College, London—who presented creditable experimental papers. Most of the speakers, however, were obscure American and Canadian practitioners. Several of them, like the chairman, were professionally engaged in the treatment of the insane. Several of their papers were in the form of speculative discourses, including "The Vital Organism—Its Nutrition—Hereditament" by A. Wetmore, M.D. of Waterloo, Illinois; "The Basal Ganglia of the Brain as Psychic Centers" by Daniel Clark, M.D. of Toronto; and "On the Cerebral Function—Transmutation of Organic, or Life-Force, Into Psychic-Force—the Phenomena of Consciousness Sequential, etc., etc." by Orpheus Everts, M.D. of Cincinnati, Ohio.[30] The official history of the IMC written by the chairman of the new executive committee, Henry Hollingsworth Smith, took a defensive tone as it attempted to explain away the difficulties of organizing the IMC and to celebrate the virtues of democratic science. It concluded: "All must now recognize the fact that medical science is not the exclusive property of, or only cultivated by a few, no matter what their local greatness may be, but that the rank and file of our Profession are everywhere its chief basis of support." The work of the medical profession could be found "outside the cliques and circumscribed influences of a capital, and more abundant in the highways and by-ways of medical life; in towns, rather than in the sumptuous dwellings of the Dives [rich men] of large cities."[31]

Those who opposed the 1887 IMC did more than simply stay away. A group led by the surgeon Claudius H. Mastin of Mobile, Alabama, a leading member of the elite American Surgical Association, was determined to organize an alternative medical congress in Washington, D.C. that would be restricted in its membership and that would bypass the AMA. The initial plan, communicated by Mastin to the American Surgical Association meeting in April 1886, called for a congress to meet in June 1887, thus forestalling the IMC.[32] The proposed new organization, to be called the Congress of American Physicians and Surgeons, would be controlled not by a large heterogeneous medical organization such as the AMA but by the smaller and highly selective societies of medical specialists. The societies in question were all products of the past two decades and were formed in part as a result of dissatisfaction with the AMA. They were the ophthalmologists (1864), otologists (1868), neurologists (1875), gynecologists (1876), dermatologists (1876), laryngologists (1878), surgeons (1880), and climatologists (1883). Yet another participating society was to be the Association of American Physicians, founded in 1886 with a broad-sounding name, but with membership limited to a select 100.

In effect Mastin hoped to create a national medical congress by convincing the specialist societies to agree to hold their annual meetings at a common time in Washington, D.C. every year. The societies invited to participate responded favorably but cautiously to Mastin's scheme. Because of the concern of the various societies with retaining their autonomy, the representatives of the societies in September 1886 rejected the plan for an annual joint meeting in favor of one every three years to which foreign guests would also be invited. The first congress was planned for September 1888. An executive committee of representatives of each of the societies was formed to handle arrangements. Among the representatives were several of the men who were originally to have taken part in the Ninth IMC. In October 1887 John Shaw Billings was elected president of the new Congress of American Physicians and Surgeons.[33]

FOUNDING OF THE AMERICAN PHYSIOLOGICAL SOCIETY

At the IMC in 1887 physiology was one of the sections; at the Congress of American Physicians and Surgeons of 1888 specialized societies took the place of the sections. If the physiologists wanted to take part in this new national organization, created by the elite of the medical profession and dedicated to "scientific medicine," they had no choice but to form their own specialized society and apply for admission to the congress. Indeed, this was the immediate occasion of the formation of the Society, as is made clear by a circular letter dated 23 May 1887, recently donated to the APS. It was written on stationery of the Physiological Laboratory of the Harvard Medical School and, like the two later letters published by Howell, was signed by Mitchell, Martin, and Bowditch. It read:

> Dear Sir,
>
> The assembling of the various societies of specialists in Washington in 1888 with a view of forming a national medical association, offers a favourable opportunity for bringing together those who are interested in the pursuit of Physiology, and formation of a Physiological Society which shall, by its annual meetings, afford facilities for professional and social intercourse and supply that mutual aid and encouragement which are so valuable to those who have a common interest in any department of science. In order that the society, if formed, may be ready to join with other societies of specialists in the summer of 1888, it seems desirable that a meeting for organization be held during the coming summer. You are therefore requested to reply to the following questions. 1) Are you in favour of the formation of a National Physiological Society? 2) Are you willing to attend during the coming summer a preliminary meeting for the discussion of plans of organization? 3) At what time and place can such a meeting most conveniently be held?[34]

Both Howell and Henry H. Donaldson said they believed (but were not certain) that the initial impetus for the founding of the APS came from Mitchell.[35] Given the background just described, this supposition is quite likely despite the fact that Mitchell, at this time preoccupied with his neurological clinic and novel writing, would have been less concerned with the creation of a scientific discipline than either Bowditch or Martin. Mitchell, the most medically oriented of the founders of the APS, had close ties with both Billings and Bowditch and would have recognized the opportunity for physiology that the proposed congress afforded. Although Mitchell was abroad for much of 1888,[36] and therefore did not attend the first congress (and probably for that reason also declined to become the first president of the APS), he was greatly admired by the organizers of the new congress and was made president of the second congress in 1891.

The organizational meeting of the physiologists took place on 30 December 1887, just in time to assure participation of the new society in the new congress. On 23 January 1888, the APS was accepted for admission by the congress and Bowditch, the first president of the new Society, took his place along with the other society representatives on the executive committee.[37] The first "special meeting" of the APS was therefore held in Washington, D.C. from 18 to 20 September 1888, as one of twelve member societies of the Congress of American Physicians and Surgeons.[38] The new congress was attended by some 380 Americans and 23 foreign guests.[39] The organizers were so intent on devoting the congress to research papers alone that the meeting proved exceedingly exhausting for the participants. Scientific sessions of the individual societies during the day were followed by plenary symposia in the evening. The high point of the meeting was a plenary session on cerebral localization in which David Ferrier and Victor Horsley, both eminent British neurophysiologists, participated. Twelve members of the newly formed APS registered, though several members of the other societies no doubt also attended the physiological sessions held in the Army Medical Museum on the Mall in Washington, D.C.[40]

By forming a society, selecting charter members, and applying for membership in the congress, physiologists were making an important statement—that they and the physiology they represented wanted to be a part of the best medical schools of the future. The leaders of the clinical societies, for their part, welcomed the physiologists as representatives of the medical science that ought be the basis of medical practice. Representatives of the other basic science discipline commonly found in medical schools, anatomy, did not want to be left out. The anatomists therefore also organized their own professional society (now known as the American Association of Anatomists) during the congress and held their first scientific meeting as an unofficial part of the congress.[41] Thus, because of the congress of 1888, physiologists and anatomists were among the earliest scientific disciplines to form national professional societies.

The Congress of American Physicians and Surgeons stood for values antithetical to those of the Ninth IMC in 1887. Best science and selectivity in membership became goals to be zealously pursued. According to spokesmen for the congress, its purpose was the advancement of medical research to the exclusion of all politics. Even drug, instrument, and book vendors were barred from the congress. William Pepper, in his opening address, hoped that each society would be so stimulated by the meeting that "membership in any of these bodies will come to be regarded as more and more an honor, and in time the scientific qualifications of the candidates will be more and more strictly scrutinized."[42] Thus it was to be expected that a physiological society formed in order to participate in such a congress would exercise special care in formulating its requirements for membership.

BIOLOGICAL CONNECTION: AMERICAN SOCIETY OF NATURALISTS AND ITS OFFSHOOTS

Before considering in detail the kind of society the physiologists formed, we must turn to the organization of the biological sciences in America in the

1880s. From the beginning, the APS was medically oriented, yet in its early years it had equally strong ties to biology and biological societies. For the biological specialties the American Association for the Advancement of Science (AAAS), formed in 1848 with a large and unrestricted membership, played a role analogous to that of the AMA for medical specialties.

One of the first groups to break away from the AAAS was the naturalists. The Society of Naturalists of the Eastern United States was founded in 1883 with 109 original members (including both Bowditch and Martin) and became known after 1886 as the American Society of Naturalists. It and the Physiological Society, founded in England in 1876, served as the chief models for the APS. The impetus to found the American Society of Naturalists came from Samuel Clarke, a former student in the Department of Biology at Johns Hopkins, who had become professor of biology at Williams College, Williamstown, Massachusetts. As Donaldson recalled, the society was formed "as a protest against conditions existing in the natural history section of the American Association for the Advancement of Science," which was "over-diluted with non-scientific members" and, moreover, met in midsummer at a time inconvenient to naturalists.[43]

According to Charles Sedgwick Minot, a member of the naturalists and, like Donaldson, a charter member of the APS, the naturalists were pioneers in their restriction of membership to professionals, in their selection of the Christmas academic recess as a time to meet, and in their spawning of and subsequent affiliation with other scientific societies.[44] The constitution of the society stipulated: "Membership in this society shall be limited to Instructors in Natural History, Officers of Museums and other Scientific Institutions, Physicians and other persons professionally engaged in some branch of Natural History." New members were to be proposed by existing members, approved by the executive council, and then approved by the majority of members at any meeting. Thus constituted, the society excluded amateurs but otherwise membership was relatively broadly defined. The APS was considered by Donaldson and later historians of the naturalists to have been an early offshoot society. By its date of founding it could, in fact, be considered the first of many offshoots of the naturalists. This assertion is a matter open to some debate, as only three of the five founders (Bowditch, Martin, and Chittenden) and half of the twenty-eight charter members of the APS were members of the naturalists in 1887. Yet the APS did to a large extent model its constitution on that of the older society, although its membership policy was considerably more restrictive.[45]

In another recently discovered letter, the reason for the delay of the organizational meeting of the physiologists from summer until December is revealed. Bowditch wrote to Mitchell on 1 June 1887:

> There seems to be a good deal of difference of opinion about the best time for a preliminary meeting. It has been suggested that we postpone the meeting till the Christmas recess & then have it either in New Haven or New York just before or just after the meeting of the Soc. of Naturalists. How does this strike you? I think it is very important to have the meeting when both you & Martin can be present.

Mitchell annotated the letter, "Suit me very well."[46]

Many, if not most, organizational meetings of new societies have been tacked onto the meetings of existing societies. In 1887, the naturalists were to meet between 27 and 29 December at Yale. Martin was on the program scheduled to present an exhibit of "newly devised Physiological Apparatus."[47] Both Chittenden at Yale and Curtis at Columbia were consulted in the arrangements.[48] New York was probably chosen for the 30 December meeting of the physiologists because in September 1887, the College of Physicians and Surgeons, Columbia University, had just moved from 23rd Street to a new site on 59th Street. Curtis's successor in the Columbia chair of physiology, Frederic Schiller Lee, noted that this move "allowed the installation for the first time of adequate physiological laboratories, which were provided with an unusually ample outfit of apparatus."[49] Curtis had gone abroad to visit foreign laboratories and to purchase the instruments. Thus the physiologists, by meeting in New York, would have the opportunity to see Curtis's new quarters.

Other specialized biological societies, also considered by the naturalists to be offshoots, soon followed the APS. The American Association of Anatomists was founded in 1888, the Geological Society of America in 1888, the American Morphological Society (a forerunner of the American Society of Zoologists) in 1890, the American Psychological Association in 1892, and the American Society of Plant Morphologists and Physiologists (forerunner to the Botanical Society of America) in 1896.[50] All of these new societies had more or less restrictive membership policies. One consequence of this proliferation of societies was to reduce the initial breadth of membership of the APS.

MEMBERSHIP, AIMS, AND ACTIVITIES OF AMERICAN PHYSIOLOGICAL SOCIETY, 1887–1912

In one important respect the APS went beyond all previous societies (and served as a model for future societies) by requiring publication of original research of its members. The constitution stipulated: "Any person who has conducted and published an original research in Physiology or Histology (including Pathology and experimental Therapeutics and experimental research in Hygiene), or who has promoted and encouraged Physiological research, and who is a resident of North America, shall be eligible for election as an Ordinary member of the Society."[51] (The clause allowing promoters of physiology to be elected was not used and was eventually eliminated.) Bowditch recalled in 1896 that standards for membership were due to Martin: "It was always Dr. Martin's contention that a candidate for admission to our ranks should be required to demonstrate his power to enlarge the bounds of our chosen science, and not merely to display an interest in the subject and an ability to teach text-book physiology to medical students."[52] Because there were many medical schools of widely different quality in which clinicians taught physiology but did little or no physiological research,[53] it was necessary to be more restrictive of membership than the naturalists. Had membership been open to those with a professional position related to physiology, the society could have been overrun with members representing very different interests than the founders.

The twenty-eight charter members of the APS were an elite, exclusive, and remarkably cohesive group. One of the most striking features of these men was their youth. Twenty-four were under forty-five and over half (eighteen) under forty. The youngest member, Joseph Jastrow, was twenty-four. All twenty-eight charter members had either a Ph.D. or an M.D. or both and twenty-one had spent time abroad in laboratories in France, Britain, or especially in Germany. Except for Victor Vaughan and Henry Sewall of the University of Michigan, all resided in the East. All except Vaughan had at one time studied or taught in the institutions associated with one or more of the five founders: i.e., Hopkins (Martin), Harvard (Bowditch), Jefferson Medical College and University of Pennsylvania in Philadelphia (Mitchell), Columbia (Curtis), and Yale (Chittenden). Five of the twenty-eight had earned Ph.D.s at Hopkins under Martin: Henry Gustav Beyer, Henry Herbert Donaldson, William Henry Howell, Sewall, and William T. Sedgwick. Five others were also associated with Martin or Johns Hopkins in some way: Isaac Ott (a fellow under Martin in 1878–1879), Thomas Wesley Mills (Osler's successor at McGill, who spent several months at Johns Hopkins with Martin), Jastrow, G. Stanley Hall, and William Henry Welch. Seven of the twenty-eight charter members had come into Mitchell's orbit in Philadelphia: Henry Cadwalader Chapman, Hobart Amory Hare, William Osler, Ott, Edward T. Reichert, Robert Meade Smith, and Horatio C. Wood. Six had studied or worked with Bowditch at Harvard: Frederick Ellis, Hall, Warren P. Lombard, Minot, Ott, and Joseph Weatherhead Warren. Another charter member, George Lincoln Goodale, was professor of botany at Harvard. Two were linked to Curtis and Columbia: John Call Dalton and Lombard, the first as mentor and the second as assistant. Chittenden's many students in physiological chemistry at Yale's Sheffield Scientific School came somewhat later and filled the membership rolls of the Society in the 1890s and the first decade of the twentieth century.[54]

Left out of this select group of charter members were a great many older men who taught physiology in medical schools all over the country. Some of them wrote scientific papers in physiology and identified themselves as physiologists. The most obvious omission was Austin Flint, Jr., professor of physiology at Bellevue Hospital Medical College and author of a textbook of physiology. It has sometimes been wondered why Flint was not included.[55] A member of the original council of the section of physiology for the IMC of 1887, Flint did not abandon the Congress that his father had initiated; it was he who delivered the keynote address on fevers. A plausible explanation for why he never became a member of the APS was that he fell victim to the ill will generated by the Ninth IMC.

The founders of the APS were in effect boldly appropriating the term "physiology" for themselves. For most of the nineteenth century, physiology was a very broadly defined term, by no means restricted to the tradition of experimental physiology imported from France, England, and Germany. Although a young and exclusive group, the APS was readily accepted by both the specialty medical societies and by the naturalists as the proper representatives of the science. The new society by its membership policy, programs, and journal helped to define the discipline, at least in the early years, as experimental, medically-oriented animal physiology, neither too zoological nor too clinical.

Through the 1890s, the APS remained a small society. Its membership, though it grew rapidly by percentages, remained less than 100. Attendance at annual meetings in the 1890s reached a high of twenty-nine in 1898 and went as low as nine and ten. The Society had very limited means and correspondingly limited functions. Its resources consisted entirely of dues, set each year at one or two dollars according to need.[56] Yet it was able to function well enough, even though small and poor, because it operated within a framework determined by other scientific and medical societies. Its chief activities were the annual election of new members, the holding of scientific programs at which physiologists could share with their colleagues the latest developments in ideas and techniques and, after 1898, the sponsoring of a journal. The Society as an entity did not generally concern itself with combating antivivisection, reforming medical education, or other broad social and professional issues. It did not need to do so because there were other societies available to handle such problems.

By its choice of new members each year the Society served an important function in shaping the discipline. Candidates for membership were to be proposed in writing by two ordinary members of the Society and nominated for election by the Council. The name of the nominated candidates, their sponsors, and their qualifications were to be posted and voted on by the members attending the meeting. From the beginning, the Society took the annual election of new members very seriously. As early as 1888, a candidate was "postponed" and not elected until 1890. In 1895 two candidates were rejected. In the period from 1888–1912, eight people were rejected and fourteen others deferred, sometimes for a period of several years.[57] Though the grounds were rarely given, it appears that candidates were rejected because they did not yet have appropriate degrees (with very few exceptions, all members had either an M.D. or Ph.D.), because they had not yet published enough, or because the publications were not sufficiently "physiological." For an individual to be elected a member of the Society was an honor that also conveyed status to the institution where he was employed. Membership brought with it the right to present papers at each meeting of the Society and (after 1895) to have the abstracts published.[58] By its very existence the Society encouraged schools and other institutions to hire well-trained researchers rather than mere teachers of the subject. A school whose faculty included members of such societies as the APS was marked as a progressive institution that promoted the modern research ideal.

The Society did not have a journal until 1898, eleven years after its founding. Even then the journal was not fully owned by the Society nor was subscription required of members. An American journal was not an immediate necessity because the British *Journal of Physiology*, founded by Michael Foster in 1878, also served American physiology. Half or more of the editorial board were Americans.[59] Discussions on whether the Society should have a journal began in the mid-1890s, when an ad hoc committee was formed, but nothing was decided for several years. Societies such as the APS simply did not have the resources to take on the financial responsibility of a journal. The *American Journal of Physiology* was founded in 1897 only because a member of the Society's committee, William Porter, assistant professor of physiology at Harvard, took matters in his own hands and personally assumed full financial and editorial responsibility. He edited the journal, which he owned, under contract to the Society, from 1898 through the early part of 1914. In 1914 Porter decided to cease editing the journal and, after considerable negotiation, agreed to let the Society take full ownership. By contracting with Porter and providing an editorial board, the APS had more control of its associated journal than did most similar societies.[60] Most small disciplinary societies had only an unofficial relation with a privately owned journal.

The chief function of the APS in its first quarter century was to hold scientific meetings. These were held in the physiological laboratories at university campuses, giving members the opportunity to compare equipment and techniques. Experimental demonstrations formed an important part of the program. Except for its first annual meeting in 1888 the Society never met on its own, but always in conjunction with other societies. Before the founding of the Federation in 1912, the APS was loosely affiliated with two other federations of societies. On the medical side it participated in the Congress of American Physicians and Surgeons held in Washington, D.C. every third year and on the biological side it held its annual meeting each year with the American Society of Naturalists and other affiliated societies. The history of working with other medical and biological societies as part of federations provided the necessary experience for the founding of a more formal and more lasting federation of biomedical societies in 1912.

The APS held seven meetings with the Congress of American Physicians and Surgeons from 1888 through 1907. These "special" meetings were in effect regular meetings of the Society at which the Council transacted business, reports of ad hoc committees were presented, and minutes were recorded. In addition to the physiological sessions the Society participated from time to time in joint evening symposia such as "Trophic Lesions and Diseases of the Nervous System" in 1888 and "Internal Secretions" in 1897, both with the American Association of Physicians. Although most of the societies participating in the Congress of American Physicians and Surgeons represented specialties of medical practice, at least two other preclinical scientific societies—the anatomists and the biochemists—eventually joined.[61]

Though the congress was established with the intention of concentrating on science to the exclusion of all political issues, it did in fact concern itself with such issues from time to time. In the 1890s, for example, antivivisectionist agitation became a serious concern when federal legislation was proposed that would have regulated animal experimentation in the District of Columbia by setting up a system of licensing and registration. William Henry Welch, a charter member of the APS and leader of the campaign in defense of medical research, worked through the framework of the congress, of which he was president in 1897, to combat the threat and to prevent the bill from ever coming to a vote.[62] Occasionally, resolutions of a political nature were circulated and signed at the congress. At the 1907 congress, for example, the APS and the biochemists held a joint session during which the two societies approved a movement sponsored by the AAAS "to increase and coordinate the present activities of the federal government in matters pertaining to public health."[63]

The alliance of physiology, biochemistry, and anatomy with the clinical specialty societies in the Congress of American Physicians and Surgeons proved to be short-lived. In 1908 the APS decided to cease meeting with the congress.[64] Indeed, all the societies representing the preclinical sciences had withdrawn by then. Although the reasons for the physiologists' action were not recorded, they are not difficult to guess. The meetings were expensive affairs for a society like the APS because the congress required a contribution from the Society for the publication of the *Transactions* and a hefty registration fee for those attending.[65] The congress meetings were poorly attended by physiologists in comparison to the annual meetings of the Society. At the congress meetings, clinical concerns far outweighed those of the preclinical sciences. More significantly, the congress had lost much of its original rationale as an elite forum for the medical sciences. By 1908 the AMA had revised its code of ethics and reorganized for medical reform. Thus reconstituted, the AMA offered physiologists a better forum for general medical concerns than a congress that met every three years. Although the APS ceased to meet with medical societies after 1907, the latter continued to play an important role for physiology. In the first decade of the twentieth century the AMA Council on the Defense of Animal Research, chaired by Walter B. Cannon, spearheaded the battle against the antivivisectionists.[66] Problems of medical education, including the teaching of physiology in the medical curriculum, were the province of the AMA's Council on Medical Education and of the Association of American Medical Colleges. The APS did not seriously concern itself with these professional issues, important as they were for the discipline. It did not need to do so. They were much more effectively handled by APS members through the medium of large and well-funded general medical organizations rather than by a small group of physiologists.

Annual meetings of the APS were held as part of a second informal federation of societies led by the American Society of Naturalists. While societies like the American Surgical Association held their annual meetings with the Congress of American Physicians and Surgeons, it is significant that physiologists' meetings with the Congress were "special" meetings added on to annual meetings. In 1888, the APS held its first annual meeting in Philadelphia just after the Naturalists' meeting in Baltimore. That same year the Naturalists, concerned about the possible loss of membership from the recent formation of societies of physiologists, anatomists, and geologists, decided to invite the new societies to meet with them the following year at the same time and place. With only two exceptions, every year from 1889 to 1913 the APS met in affiliation with the American Society of Naturalists and shifting combinations of other biological societies.[67]

Affiliation with the Naturalists allowed the APS to survive the 1890s as a very small society. Much of the work of organizing the meetings was done by the Naturalists. The combined meetings assured an adequate attendance and the officers of the APS could concentrate their efforts on organizing the scientific sessions. The physiologists took part in common functions such as the annual banquet with the presidential address and invitations to receptions and social clubs. Financially the arrangement was beneficial because the combined societies usually provided enough members to obtain reduced fares on the railroads. Meetings were reported on in the *American Naturalist*, the official organ of the Naturalists, where first Sedgwick and then Lee served as physiology editor. Such joint meetings also allowed for the possibility of overlapping membership in the societies. Several early APS members, for example, were primarily anatomists, among them Minot, Franklin P. Mall, and Ross G. Harrison, while others such as James McKeen Cattell were primarily psychologists. Among the presidents of the Naturalists in the 1890s were a number of physiologists, including Martin, Chittenden, and Bowditch.[68]

The joint meetings of biological societies, like the medical congresses, provided a framework for dealing with professional issues. Presidential addresses reviewed scientific progress and dealt with the social role of biologists. Martin in 1890 spoke on "The Relations of Professor and Student," Chittenden in 1893 spoke on "Recent Chemico-physical Discoveries Regarding the Cell," Minot in 1894 spoke on "The Work of the Naturalist in the World," and Bowditch in 1898 spoke on "Reform in Medical Education."[69] Joint committees were sometimes appointed to deal with common issues such as the tariff on scientific and philosophical instruments in 1893.[70] In 1895 the Naturalists even set up a committee to cooperate with other societies "in taking action against the opponents of vivisection."[71] In 1898, representatives of six affiliated societies presented papers on "Advances in Methods of Teaching." Porter, one of the first physiologists to organize laboratory instruction in a medical school, presented the paper on physiology. In 1899, a joint discussion was held on "The Position That Universities Should Take in Regard to Investigation."[72] The Naturalists tried to formalize the relationship with their affiliated societies in 1893 by setting up a federation in which members of each of the constituent societies would also be members of the Naturalists but the specialty societies balked and the plan fell through.[73] Many of the provisions were later carried out by the Federation.

For the most part, affiliation with the Naturalists worked well, though the physiologists did not always feel completely at home. They objected when the meetings were held in schools without physiological laboratories,[74] such as Princeton, and they were unable to get a quorum when the meeting was held in an inconvenient location like Cornell. They did not always wish to participate in symposia on topics such as evolution, which were of little interest to them. After 1902, however, what was once a more or less pleasant grouping of a small number of similar biological societies turned into a large congress of very diverse societies. This change came about when the AAAS decided to switch from a summer gathering time to meeting at the same time as the Naturalists during Christmas vacation. Many of the professional societies that had previously met with the AAAS also changed to the December date. The week became known as "Convocation Week," the time set aside for meetings of scientific and learned societies.[75]

In 1901, probably in expectation of the inauguration of Convocation Week, the AAAS, long since divided into sections by scientific field, established "Section K, Physiology and Medicine."[76] At joint meetings of the APS and the AAAS, a special joint session of the APS and Section K was generally held. The existence of Section K provided an argument against liberalizing the membership policy of the APS because those who were not eligible for membership in the APS could attend Society meetings as members of Section

K (which had no membership restriction);[77] for societies like the APS and the Naturalists, this change created problems. The AAAS was a large national society committed to meeting in the South, Midwest, and West as well as the East. Friction arose each year over whether to meet with the AAAS or separately elsewhere at the same time. These conflicts were resolved only with the founding of the Federation.

THE AMERICAN PHYSIOLOGICAL SOCIETY AND THE FEDERATION OF AMERICAN SOCIETIES FOR EXPERIMENTAL BIOLOGY

Twenty-five years after the founding of the APS, at the annual meeting in December 1912, the Society and its two recent offshoots, the American Society of Biological Chemists and the American Society for Pharmacology and Experimental Therapeutics, united together to form a permanent organization: the Federation of American Societies for Experimental Biology. Thereafter they would hold their meetings in concert, with the programs of the societies combined in a single printed program.

The origin of the Federation is somewhat obscure. Critical documentation—the first volume of minutes of the Federation, and the papers of the first chairman, Samuel J. Meltzer—is missing. Yet some of the pieces of the story can be laid out. The major events leading up to the Federation were the founding of the two offshoot societies of the APS, the biochemists in 1906 and the pharmacologists in 1908. Both societies had large overlaps of membership with the APS as well as with each other. Once founded these two societies continued to meet at the same time and place as the physiologists and to participate in joint sessions. The Federation was a response to the concern felt by all to prevent any further separation. There were also the difficulties attendant on holding annual meetings with the AAAS and its heterogeneous group of affiliated societies and the resulting desire on the part of members of the three societies to create the option to meet separately elsewhere. By 1912, the year the Federation was formed, over 1,000 people attended the meeting of the AAAS and its affiliated societies in Cleveland. Twenty-three societies, among them the APS, the biochemists, and the pharmacologists, affiliated with AAAS for the event.[78] It is not to be wondered that small and relatively exclusive societies such as the APS would wish to extract themselves from this crowd. The Federation brought together a large enough group to hold viable meetings separate from other societies.

There was in this period a general movement among biological societies to create an American Biological Society on the order of the American Chemical Society.[79] Though no progress was made in the direction of a more general union, the Federation was conceived by some as a first step. Also, there was the problem of the increasingly crowded programs of both the APS and the biochemists, necessitating remedial action. By allowing papers to be shifted among the programs of the member societies, the formation of the Federation staved off for a while the inevitable choice between denying some members of the APS an opportunity to speak or abandoning the unity of physiology in favor of holding more than one simultaneous session.

The move to create a federation seems to have been initiated by the biochemists in 1911.[80] In 1912 each of the three societies appointed members to a conference committee that met the evening of 31 December and formally established the organization. Its statement of purpose read: "The Federation shall consist of the American Physiological, Biochemical and Pharmacological Societies and its object shall be to establish a stable connection between the three societies for the purpose of fixing the place and time of the annual meetings, to arrange joint sessions whenever possible, and in general to establish officially closer scientific and social affiliations between the sister societies while retaining their individual independence." The Federation was to be governed by an executive committee consisting of the presidents and secretaries of each of the three societies. A fourth society, the newly formed Society for Experimental Pathology, was added in 1913.[81]

The provisions for the Federation included this statement: "That a common meeting place of the Federation with the anatomists, zoologists and naturalists is desirable but not mandatory." As this provision suggests, the formation of a stable union of societies through the Federation allowed biomedical scientists to select their own meeting site and, in effect, to declare their independence from both the AAAS and the Naturalists. Despite some friction, the Federation managed to reach an accommodation with the Naturalists for the 1913 meeting. In 1914, however, the Federation set out on its own to Washington University in St. Louis, Missouri, leaving the Naturalists, zoologists, and the AAAS behind in Philadelphia, PA. Though the Federation met on a few more occasions with the Naturalists, the links with the other biological societies had been effectively severed.

In its early years the APS was heavily dependent on its ties to other disciplinary societies. Founded so that physiologists could take part in a national medical congress the APS spent its first twenty-five years as part of an informal federation of broadly biological societies. It was founded at the tail end of a series of clinical specialty societies and at the beginning of a series of clinical specialty societies and at the beginning of a series of biological societies. As the number of physiologists increased, however, they found they had less and less in common with medical specialists or with scientists in other branches of biology, who were based primarily in departments in colleges and universities rather than medical schools. Physiology's closest links henceforth were to be with other preclinical experimental sciences based in medical schools. Once the Federation was established in 1912, it operated as an autonomous entity and grew very rapidly. Its formation might be said to represent the biomedical sciences coming-of-age in America.

NOTES

1. Ralph W. Gerard, *Mirror to Physiology: A Self-Survey of Physiological Science* (Washington, DC: American Physiological Society, 1958), p. 73.
2. See, for example, Gerard, *Mirror to Physiology*; E. F. Adolph et al., "Physiology in North America, 1945," *Federation Proc.* 5 (1946): 407–436; Peter F. Hall, "Fragmentation of Physiology: Possible Academic Consequences," *Physiologist* 19 (1976): 35–39; Gerald L. Geison, "Divided We Stand: Physiologists and Clinicians in the American Context," in *The*

Therapeutic Revolution, ed. Morris J. Vogel and Charles E. Rosenberg (Philadelphia: Univ. of Pennsylvania Press, 1979), pp. 67–90; and Philip J. Pauly, "The Appearance of Academic Biology in Late Nineteenth-Century America," *J. Hist. Biol.* 17 (1984): 369–397.

3. William H. Howell and Charles W. Greene, *History of the American Physiological Society Semicentennial, 1887–1937* (Baltimore, MD: American Physiological Society, 1938), pp. 1–4. The originals of these letters and the constitution are found in Society Minutes, vol. 1 [1887–1919], American Physiological Society Archives, Bethesda, MD (hereafter, APS Society Minutes). Council meetings were recorded separately in Council Minutes, vol. 1 [1887–1919] APS Archives, Bethesda, MD (hereafter, APS Council Minutes).
4. Henry H. Donaldson, "The Early Days of the American Physiological Society," *Science* 75 (1932): 599–601; Russell H. Chittenden, *The Development of Physiological Chemistry in the United States* (New York: American Chemical Society, 1930), pp. 41–43.
5. Robert E. Kohler, *From Medical Chemistry to Biochemistry: The Making of a Biomedical Discipline* (Cambridge, UK: Cambridge Univ. Press, 1982).
6. Hebbel E. Hoff and John F. Fulton, "The Centenary of the First American Physiological Society Founded at Boston by William A. Alcott and Sylvester Graham," *Bull. Hist. Med.* 5 (1937): 687–734.
7. Kohler, *From Medical Chemistry to Biochemistry*, p. 110; Geison, "Divided We Stand," p. 73.
8. Ralph H. Bates, *Scientific Societies in the United States*, 3d ed. (Cambridge, MA: MIT Press, 1965), pp. 85–136.
9. Howell and Greene, *American Physiological Society Semicentennial*, pp. 5–6; APS Society Minutes, 30 December 1887; APS Council Minutes, 30 December 1887.
10. Adolph, "Physiology in North America, 1945," 416.
11. Membership can be determined from published annual membership directories, 1888–1913, in the APS Archives. Attendance of members at early meetings is listed in APS Society Minutes.
12. Chittenden, *The First Twenty-five Years of the American Society of Biological Chemists* (New Haven, CT: American Society of Biological Chemists, 1945); K. K. Chen, ed., *The American Society for Pharmacology and Experimental Therapeutics: The First Sixty Years, 1908–1969* (Bethesda, MD: American Society for Pharmacology and Experimental Therapeutics, 1969). Eighteen men attended the organizational meeting of the pharmacologists and thirty-four more were later named charter members.
13. Kohler, *From Medical Chemistry to Biochemistry*, pp. 121–214; Kohler, "Medical Reform and Biomedical Science: Biochemistry—A Case Study," in *Therapeutic Revolution*, ed. Vogel and Rosenberg, pp. 28–66.
14. When the Seventh International Medical Congress (IMC) was held in London in 1881, the British Physiological Society invited foreign physiologists to attend and hosted a dinner for them at the congress. Among the guests was Bowditch. As president of the Section of Physiology of the IMC, Michael Foster gave an address on "The History of Physiology in England." See Edward Sharpey-Shafer, *History of the Physiological Society During its First Fifty Years, 1876–1926* (Cambridge, UK: Cambridge Univ. Press, 1927), pp. 55–56, 60.
15. *Transactions of the International Medical Congress, Ninth Session*, 5 vols. (Washington, DC, 1887) (hereafter, *Transactions IMC*) 5: 736.
16. Howell and Greene, *American Physiological Society Semicentennial*, p. 21. The fifth founder, Chittenden, was understandably omitted because he had only become professor of physiological chemistry at Yale two years previously and he did not have a medical degree.
17. *J. Am. Med. Assoc.* 2 (1884): 565, 566, 573; "The International Medical Congress and the American Medical Association," *Boston Med. Surg. J.* 112 (1885): 485–486. Details of the initial organization of the congress can be found in John Shaw Billings to S. Weir Mitchell, 11 October 1884, 12 December 1884, 27 March 1885, S. Weir Mitchell Correspondence, College of Physicians, Philadelphia, PA (hereafter, Mitchell Correspondence); and in the letters and announcements sent by Billings to Mitchell dated 2 December 1884–20 January 1885 in the International Medical Congress Papers, 9th, Washington, DC, 1887, College of Physicians, Philadelphia, PA (hereafter, IMC Papers). See esp. Billings to Mitchell ("Circular 3"), 20 January 1885, which contains the names of the appointees.
18. Billings to Mitchell, 14 December 1884, IMC Papers.
19. Other members of the Council included Henry F. Campbell of Augusta, GA; Charles N.

Ellinwood of San Francisco; Austin Flint, Jr. of New York; W. Lee of Washington, DC; and John I. Mason of Newport, RI.

20. "The International Medical Congress and the American Medical Association," *Boston Med. Surg. J.* 112 (1885): 485–486.
21. An amendment gave the state delegations the power to select their representative and a second resolution gave the new committee power to elect its own officers and to appoint the officers of the IMC. See *Transactions IMC* 5: 736.
22. Ibid.
23. On the American Medical Association (AMA), the Ninth IMC, and medical ethics, see Lester S. King, *American Medicine Comes of Age, 1840–1920: Essays to Commemorate the Founding of the Journal of the American Medical Association, July 14, 1883* (Chicago, IL: American Medical Association, 1984), pp. 37–41. On the controversy over the AMA code of ethics, see also Paul Starr, *The Social Transformation of American Medicine* (New York: Basic, 1982), pp. 93–112.
24. Many examples of this type of conflict in science are described in Daniel J. Kevles, *The Physicists: The History of a Scientific Community in Modern America* (New York: Vintage, 1979).
25. Billings, Resolution, n.d. ("Whereas an unfortunate difference of opinion . . .), Billings Papers, New York Public Library, box 67. The changes made by the new committee when it met in June are detailed in *Boston Med. Surg. J.* 113 (1885): 41–42, 44–45.
26. Ibid.: 45–46, 117–118, 163–164, 284–286, 356, 382, 383, 427, 601–602.
27. The only charter members of APS on the list of those registered for the IMC were Henry Gustav Beyer who, as an employee of the U.S. Navy in Washington, DC, was obliged to attend, and Victor Vaughan. The list of registrants is printed in *Transactions IMC* 1: 89–151.
28. "The Meeting of the International Congress at Washington," *Boston Med. Surg. J.* 117 (1887): 269–271; "The Meeting of the Congress," *Med. News* 51 (1887): 322–324.
29. "Letter from Washington," *Boston Med. Surg. J.* 117 (1887): 271.
30. The papers were published in *Transactions IMC* 3: 231–348. Biographies of a number of the participants appear in Howard A. Kelly and Walter A. Burrage, *Dictionary of American Medical Biography* [1928] (Boston, MA: Milford House, 1971).
31. *Transactions IMC* 5: 735–761, esp. 761.
32. Mark M. Ravitch, *A Century of Surgery: The History of the American Surgical Association*, 2 vols. (Philadelphia, PA: J. B. Lippincott, 1981) 1: 76–77, 86. Membership in the American Surgical Association was limited to 100 fellows and twenty-five honorary fellows.
33. *Transactions of the Congress of American Physicians and Surgeons. First Triennial Session Held at Washington, D.C., September 18th, 19th and 20th 1888* (New Haven, CT: Congress of American Physicians and Surgeons, 1889) (hereafter, *Transactions CAPS*), pp. xiii–xxii.
34. Mitchell, H. Newell Martin, and Henry P. Bowditch to Warren P. Lombard, 23 May 1887, APS Archives, Bethesda, MD.
35. Howell and Greene, *American Physiological Society Semicentennial*, p. 5; Donaldson, "Early Days of the American Physiological Society," 600.
36. Thomas Frederick Crane to Mitchell, 27 October 1888, 22 May 1889, and 30 May 1889. Mitchell Papers, Trent Collection, Duke University, Durham, NC.
37. *Transactions CAPS*, p. xxii.
38. APS Society Minutes, 18–20 September 1888.
39. The minutes of the meeting were published in *Transactions CAPS*, pp. xxiii–xxxvi, and the list of attendees in ibid., pp. xxvii–xlvi.
40. "The Congress of American Physicians and Surgeons," *Boston Med. Surg. J.* 119 (1888): 314; "Congress of American Physicians and Surgeons," *Med. News* 53 (1888): 326–343; "Cerebral Localization at the Washington Congress," *Med. News* 53 (1888): 421–422.
41. Nicholas A. Michels, "The American Association of Anatomists: A Sketch of Its Origin, Aims and Meetings," *Anat. Rec.* 122 (1955): 685–714.
42. "Congress of American Physicians and Surgeons," 326.
43. Donaldson, "Early Days of the American Physiological Society," 600.
44. Charles Sedgwick Minot, "The Relation of the American Society of Naturalists to other Scientific Societies," *Science* 15 (1902): 241–244.
45. Martin, who appears to have had a large hand in the drafting of the constitution of the APS, was a charter member of both the American Society of Naturalists and the Physiological Society in Britain, the two chief models for the APS. The constitution and lists of members

of the Naturalists by year are found in *Am. Soc. Naturalists Records* 1 (1884–1895). In its early years, so as not to conflict with AAAS, the programs of the naturalists emphasized methodology and education rather than biological research.

46. Bowditch to Mitchell, 1 June 1887, Mitchell Correspondence.
47. "American Society of Naturalists," *Am. Nat.* 22 (1888): 91–92.
48. Howell and Greene, *American Physiological Society Semicentennial*, pp. 19–21.
49. John Call Dalton, *History of the College of Physicians and Surgeons in the City of New York: Medical Department of Columbia College* (New York: published by order of the college, 1888), pp. 165–197; Frederic Schiller Lee, "The School of Medicine," in *A History of Columbia University, 1754–1904*, ed. Brander Matthews et al. (New York: Columbia Univ. Press, 1904), pp. 307–334, esp. p. 328.
50. On the founding of these societies, see Michels, "American Association of Anatomists"; Edwin Butt Eckel, *The Geological Society of America: Life History of a Learned Society* (Boulder, CO: Geological Society of America, 1982); Brother C. Edward Quinn, "Ancestry and Beginnings: The Early History of the American Society of Zoologists," *Am. Zool.* 22 (1982): 735–748; Samuel W. Fernberger, "The American Psychological Association, 1892–1942," *Psychol. Rev.* 50 (1953): 33–60; *Botanical Society of America: Fifty Years of Botany* (New York: McGraw-Hill, 1958).
51. APS Society Minutes, 30 December 1887; Howell and Greene, *American Physiological Society Semicentennial*, p. 56.
52. APS Society Minutes, 30 December 1896.
53. John Harley Warner, "Physiology," in *The Education of American Physicians: Historical Essays*, ed. Ronald L. Numbers (Berkeley: Univ. of California Press, 1980), pp. 48–71.
54. Biographies of the charter members appear in Howell and Greene, *American Physiological Society Semicentennial*, pp. 6–51.
55. Ibid., pp. 2–3.
56. Dues were set each year by the Council. See APS Council Minutes.
57. Persons nominated, deferred, and rejected each year are found in APS Council Minutes.
58. Beginning with the 1895 meeting, abstracts were published in *Science* after the meetings. Abstracts from the 1897 meeting until the founding of *Federation Proceedings* in 1942 are found in the *American Journal of Physiology*.
59. William F. Bynum, "A Short History of the Physiological Society, 1926–1976," *J. Physiol. Lond.* 263 (1976): 23–72, esp. 58.
60. Howell and Greene, *American Physiological Society Semicentennial*, pp. 78–83, 97–98; APS Society Minutes; APS Council Minutes. After 1914, the journal was edited by Donald Russell Hooker and the APS Council took over the function of the editorial board. Regular review procedures were not initiated until 1933.
61. Programs of meetings are found in APS Society Minutes.
62. Patricia Peck Gossel, "William Henry Welch and the Antivivisection Legislation in the District of Columbia, 1896–1900," *J. Hist. Med.* 40 (1985): 397–419.
63. APS Society Minutes, 7 May 1907.
64. APS Council Minutes, 18 December 1908, and 29 December 1908.
65. For the first special meeting, in 1888, the Society was charged $72.90 for its share of the congress, leaving Martin, secretary-treasurer of APS, with a deficit of $13.63. See APS Society Minutes, Treasurer's Report, 1888–1889.
66. Saul Benison, "In Defense of Medical Research," *Harv. Alumni Med. Bull.* 44 (Jan.-Feb. 1970): 16–23.
67. E. D. Cope and J. S. Kingsley, "Editors' Table," *Am. Nat.* 23 (1889): 32–34. The years 1910 and 1911 are exceptions. The Naturalists, frustrated with the large meetings with the AAAS, decided to meet separately from the AAAS in Ithaca, NY in 1910 and Princeton, NJ in 1911. Both locations were without medical schools. The APS held its meeting with just the biochemists and pharmacologists in New Haven, CT in 1910 and with the AAAS in Baltimore/Washington in 1911.
68. E. G. Conklin, "Fifty Years of the American Society of Naturalists," *Am. Nat.* 68 (1934): 385–401. Programs of the Naturalists' meetings can be found in the American Society of Naturalists Archives, American Philosophical Society, Philadelphia, PA. Details of the meetings can also be found in *Am. Soc. Nat. Rec.* 1 (1884–1895) and 2 (1896–1913).
69. Conklin, "Fifty Years," 388–389; Bowditch, "Reform in Medical Education," *Boston Med. Surg. J.* 139 (1898): 643–646.

70. APS Society Minutes, 29 December 1893.
71. *Am. Soc. Nat. Rec.* 2, pt. 1 (1896): 32.
72. APS Society Minutes, 29 December 1898; William T. Porter, "On the Teaching of Physiology in Medical Schools," *Boston Med. Surg. J.* 139 (1898): 647–652; APS Society Minutes, 29 December 1899.
73. *Am. Soc. Nat. Rec.* 1 (1884–1895): 281–282, 304–307, 309, 335–337; Quinn, "Ancestry and Beginnings," 736; APS Society Minutes, 29 December 1893 and 27 December 1894.
74. APS Council Minutes, 28 December 1892.
75. Minot, "Relation of American Society of Naturalists."
76. *Proc. AAAS* (1901): 386.
77. Reid Hunt to Joseph Erlanger (and enclosure), 21 January 1908, Erlanger Papers, Washington University School of Medicine Archives, St. Louis, MO, folder 136.
78. H. E. Summers, "The Cleveland Meeting of the American Association for the Advancement of Science," *Science* 37 (1913): 41–46.
79. APS Society Minutes, 30 December 1908.
80. Business Minutes, 1908, and A. B. Macallum to A. N. Richards, 11 May 1912, American Society of Biological Chemists, Minute Book, American Society of Biological Chemists, Bethesda, MD.
81. APS Council Minutes, 29 and 30 December 1912, and 28 December 1913; APS Minutes, 31 December 1912, and 29 December 1913; *Science* 37 (1913): 193–194; *Science* 39 (1914): 217–218.

VII

Physiology, Biology, and the Advent of Physiological Morphology

JANE MAIENSCHEIN

Through Johns Hopkins's membership in the Society of Friends (Quakers) physiology first became a part of biology in the United States. As it turned out, physiology was only rarely accepted as a central part of biology. Although examples of this alliance remain few, they are nonetheless important and it is the earliest examples from the late nineteenth century that I shall explore in this chapter.

Johns Hopkins fell in love with his cousin; however, in keeping with Quaker doctrine of the time, his local Friends' meeting refused to allow them to marry. Because Hopkins never wanted to marry anyone else, the fortune he accumulated had to find other beneficiaries. Eventually he decided to donate his substantial resources to establish a new research university for Baltimore, with a medical school and a teaching hospital. Setting up two boards of directors to develop the university and the hospital, respectively, Johns Hopkins left it to the board members to determine how best to realize these objectives.[1] Although Baltimore already had ten colleges in the 1870s, none had anything like the healthy endowment of Johns Hopkins' new enterprises. The trustees chose Daniel Coit Gilman as president and decided to establish the university in advance of the hospital and associated medical school. As Philip Pauly has shown, this decision provided an atmosphere more conducive to the development of biology than would probably have occurred if the medical program had been established first.[2] True, the awareness of a coming medical school exerted some influence on the biological work there, but biology was nonetheless afforded the opportunity to develop on its own for several years.

At first, President Gilman had promised to build a program in natural history along with the other sciences of the day, but he knew how to consult the right people for advice and how to shift his goals when a revision was indicated. Thus he traveled to Europe and discussed his plans for this unusual new American research university with Thomas Henry Huxley, among others. Huxley convinced Gilman to design a broadly conceived program in biology rather than in natural history alone, which had different, less modern, and less scientific connotations. Gilman followed Huxley's lead in more than one sense, while also reassuring skeptics and critics back home in the Baltimore community that biology was not really something radically new or threatening, but merely a combination of old friends: natural history and physiology.[3]

Gilman was sufficiently persuaded by Huxley's views that he decided to build his own program at Hopkins along the lines of Huxley's laboratory in South Kensington, London. In particular, this meant to Gilman that the department should be divided between morphological and physiological work.[4] In his role as keynote speaker at the opening of the Johns Hopkins University in 1876, Huxley expressed his approval of Gilman's decision. Noting that the biology program was intended to introduce students to medicine as soon as the medical school opened, Huxley insisted that the medical student should "come to his medical studies with a comprehension of the great truths of morphology and physiology, with his hands trained to dissect and his eyes taught to see."[5] Huxley thought that his was an appropriate program for both biology and medical students.

Given Huxley's role in Gilman's decision, what better choice for first biology professor at Hopkins than Huxley's assistant Henry Newell Martin? Martin had received England's first D.Sci. degree in physiology at Cambridge University working with Michael Foster, so he represented physiology very nicely. Martin was also prepared to teach physiology as part of biology. Job prospects for physiologists in England did not seem terribly promising in 1875, so Martin considered the invitation with interest. Nevertheless, this offer from a new university in a foreign land raised concerns; in particular Martin worried that casting his lot in America meant forsaking scientific research, if not civilization itself. After lengthy negotiations over salary, title, and laboratory details, he did finally decide to join the original Hopkins faculty of six, all young and half of them British.[6]

At Hopkins, Martin set out to establish a program in biology, which meant that first he established a physiology laboratory to pursue his own research, begun under Foster, on the action of the mammalian heart. In addition, however, he taught courses in general biology and he and Gilman hired a second instructor for the biology program, William Keith Brooks. Initially appointed as a postdoctoral fellow under Hopkins' generous and innovative fellowship program, Brooks had already received a promotion to assistant professor by the time he began his work in Baltimore.[7] Martin insisted that Brooks should not serve as his assistant, but instead should represent the other, that is, morphological, side of biology. Martin also insisted that he needed an additional assistant in physiology and he soon obtained support to hire William Henry Howell, then a student in physiology at Hopkins. As Martin laid out in a lengthy memo to Gilman, the department should eventually include representatives of biology, comparative anatomy and zoology, physiology, human anatomy, botany, and a museum curator. Because of the central role planned for medicine at Hopkins, Martin held physiology as the first priority, with other developments to follow.[8] In setting up this colony of physiological work in a biology department, Martin had to improvise. He had stepped outside the traditional homes for physiology, such as they were in the late nineteenth century, and into a different setting with new constraints. Thus the way he set up the department and his own role provided the primary examples for later institutional leaders to emulate or reject.

Given this unusual setting for a physiological program within a more broadly conceived biology program, Charles Rosenberg's suggestion that historians might experiment with an ecological approach to research disciplines has particular appeal.[9] I can imagine Martin and the other physiologists in American biological programs as loosely analogous to the Arizona salamanders that one of my colleagues studies. Salamanders exist in Arizona in ponds or lakes, some of which dry up in the arid climate. Within a given population, some of these salamanders may exhibit a distinct cannibalistic trait; they devour their own kind. My colleague, who examines what factors influence the proportion of cannibals in the population, finds that density of individuals as well as elements of the food supply are relevant. With salamanders the population may change as individuals wander away, for example, or as heavy rains wash part of the population away to another pond, thus changing the situation in several places at once. Some ponds have no cannibals; others have many.

What is the analogy here? Surely I do not mean to suggest that physiologists are cannibals. Rather, in both the academic and amphibian cases, we have a minority form occurring within a larger, more normal population. Like the salamanders, physiologists existed in pools or academic environments, some of which dried up or changed with time. The proportion of physiologists within an environment as well as the resource supply were both relevant features in shaping the physiologists' own development. Some pools became completely zoological, others physiological. I am interested here in those biological pools with coexisting physiologists and morphologists, and particularly in the earliest such pools in the late nineteenth century.

The situation remains complex, so that defining biology generally, and physiology within it, becomes a process of untangling and understanding relations. No research program that self-consciously identified itself as biology existed before 1876 in America. Since biology as such did not exist before the arrival of physiology at Hopkins, both biology and physiology evidently arrived there at the same time. We cannot, therefore, consider the move of physiologists into a preexisting biology but only the parallel development of both.

Overall, given the ultimate lack of success of general physiologists within biology, physiology would seem to have had very limited success in colonizing the larger domain of biology. Generally, physiologists did not often achieve a balance with morphologically oriented zoologists within biology. For the most part they migrated instead into other, usually medical, pools. There, "nonbiological" physiologists, namely those without any biological training or perspective, generally proved more successful and absorbed whatever resources were available for physiologists. As a result physiology in America developed as a largely medical discipline with the exception of a few early biological efforts. My concern here is with these exceptional efforts, specifically at Hopkins, the Marine Biological Laboratory (MBL), and the University of Chicago. Only the Chicago school really established a tradition of biological work in physiology, but all three offered important possibilities in the 1890s.

JOHNS HOPKINS

To pursue the first part of this quasi-ecological study, let us return to Johns Hopkins and Henry Newell Martin. Martin did set up a laboratory and pursued some work of his own, especially studies of the mammalian heartbeat. In particular, he and his student collaborators isolated the heart and sustained its beating, then tested the effects of artificially (or experimentally) altering such conditions as temperature, arterial pressure, or ethyl alcohol concentration. Other studies examined the effects of respiration on the heartbeat.[11] A few advanced students climbed the stairs or ascended in the elevator to begin their own research careers in that small physiological laboratory at Hopkins, complete with its kymograph and characteristic smell. By 1887, in fact, Martin had produced nine Ph.D.s in physiology, at least seven of whom had continued to work in some aspect of physiological research—namely, Henry Sewall, Henry Gustav Beyer, Henry Herbert Donaldson, E. A. Hartwell, William Henry Howell, George Theophilus Kemp, Frederic Schiller Lee, William Thompson Sedgwick, and C. Sihler.[12]

These Hopkins men were trained as physiologists; in what sense were they also biologists, that is, in what sense did their institutional environment make a difference and thus reward us for pursuing an ecological approach to their history? All of them, though nominally students in biology, had received their specialized degree in physiology rather than the more general field. Indeed, at least as late as the 1890s, Johns Hopkins had awarded no degrees in biology as such, despite the label and the coexistence of course offerings in both physiology and morphology.[13] The physiologists in question did indeed take courses in morphology, meaning essentially comparative anatomy and embryology, with Martin's colleague Brooks. They were also eligible to join the summer sessions at the Chesapeake Zoological Laboratory directed by Brooks and designed to provide practical field experience in marine biology. Some actually did attend.[14] There they learned to think about the structure, function, and behavior of the whole organism. They also explored evolutionary relationships among organisms, as they would not in studying traditional physiology alone. A few of those Hopkins physiologists, including Lee, worked in biology before moving into the discipline of physiology per se. In a lecture before the biology section of the New York Academy of Sciences, Lee acknowledged that most people did not understand what physiology was and emphasized that physiology in fact went far beyond the medical study of humans.[15] At this point in his career, at least, Lee regarded the type of work he did as different from that of "biologists" in general, just as it was more than exclusively human physiology. Presumably his conviction was a result of the particular training he had undergone.[16]

No doubt individuals entered the Hopkins program and specialized in biological physiology for a variety of reasons. Those few who eventually earned Ph.D.s gained an unusual sense of what physiology could be and the plausibility of its role within biology broadly conceived. Yet nearly all migrated off to enter medical schools rather than to positions in zoology or in other fields of American biology. With the exception of Sedgwick's unusual position teaching biology and bacteriology at MIT, and Donaldson's appointment within biology in neurology at the University of Chicago, these Hopkins Ph.D.s did not receive offers from biology programs.[17]

Whereas before 1893 graduate students specializing in physiology had some opportunity to enter biology as a profession, after 1893 they did not. By then

Martin had himself largely withdrawn from his research, which had never been prolific anyway, so he could not help his students.[18] Martin's assistant, Howell, left Johns Hopkins in 1889 for the University of Michigan, there replacing his fellow Hopkins graduate Sewall, who had moved on to teach physiology in Denver. In 1892 Howell moved to Harvard, and then returned to Hopkins (after Martin's retirement and the simultaneous opening of the medical school) to assume the chair in physiology.[19] During the intervening years, physiology hit a low point at Hopkins, enough so that by 1893 Howell expressed his dismay at Hopkins' loss of prestige. Whereas "the Hopkins has been recognized as the one place in the country where graduate work in animal physiology was encouraged," Howell now lamented that the failure to appoint assistants had seriously undercut that former position.[20] After 1893, physiologists at Hopkins experienced increasing subservience to the medical goals of William Henry Welch and the general Hopkins medical ambitions.

The density of physiologists in biology in general also dropped in the late 1880s and 1890s so that a larger number of students in biology sought degrees in morphological work or zoology rather than in physiology. Individual factors including Martin's collapse from alcoholism also undercut physiological participation in biology and pushed students and researchers into the larger medical environment. True, a few new biology programs in the United States did embrace physiology, but not always of the Hopkins sort, while older programs did not rush to add physiologists at all.[21] In the late 1880s and early 1890s, "biological" physiology (as opposed to "medical" physiology) found a friendly environment only at Clark University, the University of Chicago, or the MBL, all of which centered around one man: Charles Otis Whitman.

CHARLES OTIS WHITMAN AND THE MARINE BIOLOGICAL LABORATORY

Charles Otis Whitman (1842–1910) played an intriguing role in the history of American biology. Although repeatedly only second choice for the jobs he held, Whitman used all his opportunities brilliantly. After attending Louis Agassiz's Penikese Island (Massachusetts) experiment in 1873 and again in 1874, he traveled to Europe for a Ph.D. under Rudolf Leuckart. Thereafter he applied for a fellowship at Johns Hopkins, but instead accepted an offer to teach zoology at the Imperial University of Tokyo, an opportunity that allowed him to begin crystallizing his ideas about how to train research scientists. In fact his four Tokyo students all became professional biologists, a remarkable record. Upon returning to the United States and after two years at Harvard with Alexander Agassiz, Whitman went to Milwaukee in 1886 to direct the Lake Laboratory for Edward Phelps Allis.[22] Once again recommended with several other people and not the clear-cut first choice, Whitman seized the opportunity offered in Milwaukee. His mandate there was to train Allis as a biologist and to develop a research station for morphological work. He also persuaded Allis to finance a new journal for American zoology, the *Journal of Morphology*, which began publication in 1887. Whitman insisted that Americans should not have to send their papers to Europe all the time and he convinced Allis and others of this point fairly easily.[23] If his position at

Milwaukee was less than ideal, he nonetheless used it well to make contacts, to bring other good researchers into the community (including William Patten, Howard Ayers, and Henry van Peters Wilson, as well as William Morton Wheeler, indirectly), and to continue his own research and articulation of his views about how biology should be carried out.

At one point Whitman applied for a job at Columbia, which he did not receive.[24] He had more success with other positions elsewhere, if only as the ultimately successful second choice. In 1888, for example, the Boston-based trustees of the MBL in Woods Hole, Massachusetts, offered Hopkins morphologist Brooks the post as first director of their new research and teaching laboratory. When Brooks declined the offer, Whitman accepted.[25] The next year, in 1889, Whitman received an offer to head the new program in biology at Clark University. Once again, Brooks had been offered the position first but had elected to remain in Baltimore,[26] and once again Whitman accepted and thereby joined the first string of influential leaders who defined the boundaries of American biology.

Whitman exhibited a continuing concern to articulate what biology is and should be. His convictions evolved gradually, partly in response to his changing institutional ties, as Pauly has argued. Indeed, Pauly interprets Whitman's statement of 1887 about biological instruction as an obvious job advertisement.[27] Perhaps this is so. Whitman originally presented the paper at a meeting of the American Society of Naturalists at a time when the various overlapping American biological societies (or incipient societies not yet formalized) were considering their respective roles. Whitman's interest in several of those societies may thus have stimulated his expression of what biological instruction should be like. Whatever its purpose, Whitman's paper advocated what he saw as a traditional German model for biological education, focusing on research as its chief end. A modern university should provide work in all the biological fields, he urged, ideally with separate institutes on the German model. Accordingly, botany, zoology, physiology, anatomy, and pathology should receive attention.[28] In Whitman's view, zoology included anatomy, histology, embryology, phylogeny, taxonomy, and physiology, with cytology as a more recent addition. These specialties were nothing to fear, Whitman reassured the American Society of Naturalists, but rather represented a necessary move in modern biological work. Zoology and botany remained the familiar primary divisions. In his role at the Allis Lake Laboratory in Milwaukee, Whitman had not yet been in a position to exploit those basic divisions in any practical way, yet he had clearly thought about how biology should be organized and taught.

In his first position of real national influence, at the MBL, Whitman had little chance to do much organizational work at first. Only seventeen people attended that first year, 1888; Whitman had only received his own invitation to assume the directorship in mid-May for the session opening on 17 July; and the announcements had appeared rather later as well. The first year's program at the MBL was thus limited and the atmosphere quiet, a feature that longtime investigator Cornelia Clapp recalled fondly in later more hectic years.[29] Whitman's second summer at the MBL brought more students and teachers, but followed a basically similar program, with one instructor to coordinate a course of lectures for all the students. Whitman's report of that year urged the addition of a full-time instructor in botany and of two assistants to the chief instructor. Yet instruction at the MBL continued to proceed along the same lines as had Agassiz's school at Penikese, through a series of lectures on a wide variety of topics in biology presented by a diverse group of lecturers.

Not until 1890 did the MBL present more than one series of formal lectures. Courses in marine zoology, botany, and a special course on coelenterates were now offered by Whitman's colleague from Clark, John Playfair McMurrich. Undoubtedly, Whitman's move to Clark University as head of the biology program in 1889 led him to think more deeply about the relation between work in biology at the university and the summer laboratory. The move also allowed him to attract new colleagues to the MBL for lectures and research and thereby to widen the interests and offerings.

To this point, about 1890, Whitman had assigned physiology only a minor role at best and stressed morphological work instead, in accordance with his own training. For him, biology might well include a few physiologists, but they would remain relatively uninteresting in the general scheme of things. As Whitman assumed a role as director of several national programs, however, that attitude began to change. In 1890, he articulated more fully his view that biology must proceed with a division of labor among specialists who would then work together. Specialization was indeed desirable and necessary, Whitman announced in response to criticism that there was too much specialization.[30] Yet "Specialization is not science, but merely the method of science. For the sake of greater concentration of effort, we divide the labor; but this division of labor leads to interdependence among the laborers, and makes social coördination [sic] more and more essential. This is the law of progress throughout the social as well as the organic world."[31] After 1890 what may have begun as a rhetorical explanation of the particular selection of lectures for the evening series of public lectures given at the MBL became a persistent theme in Whitman's writings and letters, an ideal to which he became committed: the ideal of specialization and coordination working together.

At first Whitman's primary division of specialties for biology at the MBL included zoology and botany. By 1891, however, he had decided that he wanted physiology as a central part of biology. However, Whitman did not have in mind the traditional physiological work, which remained too closely allied with medicine and too narrowly centered on human studies. Instead he sought to promote a "biological physiology."[32] In particular he found recent work in the experimental physiology of development, such as that by Eduard Pflüger, Wilhelm Roux, Theodor Boveri, and others, of interest. He mentioned Jacques Loeb's ideas on "physiological morphology" as especially promising. Whitman reported that the limited resources available at the MBL had hitherto precluded the development of physiological work, but he clearly intended that the situation should change and he saw the physiology of marine organisms as particularly appropriate for his purposes. Accordingly, in 1892, Whitman added physiology to the MBL program, though not yet as a separate department of instruction or investigation, by inviting Loeb.

In his annual report for 1892, after Loeb had spent one summer at the MBL, Whitman focused on the relationships between physiology and morphology. They were, he explained, two aspects of the same thing. Biologists pursue both morphological study and physiological study of adult organisms, developing individuals, and developing species. Morphology and physiology examine form and function respectively, and the two should come closer together. The physiologist must realize the significance of "non-adult" organisms or risk "having some of its most inviting fields pre-occupied and developed by morphologists." In particular, embryology should provide a meeting ground for both specialties because "The embryological series, often including free larval stages, furnishes one of the grandest fields for experimental study. Here the physiologist has an opportunity not only to study by experiment but also by direct observation and inference, and thus to join hands with the morphologist both in methods and results."[33] Through work in the physiology of development Whitman saw the prospects for a true biological physiology.

He also saw the field as underdeveloped, suffering from "the lack of interest taken in general physiology, and the difficulty experienced in securing active coöperation [sic] from physiologists. As a rule physiologists look upon marine biology as something quite remote from their field of work, and the cases are rare indeed where they have taken an active part in seaside work."[34] Thus he regarded the MBL as fortunate to have attracted Loeb "whose enthusiasm, zeal, and accomplishments in general physiology, make him a fitting director of his department."[35] Several associates to help with instruction, including Lee, who had earned his Ph.D. in physiology under Martin at Hopkins, and a few students filled out the physiological program. Thus physiology first entered the MBL in 1892, with the formal addition of an instructional program in 1894 at Whitman's invitation and initially in the person of Loeb. The work evidently followed the lines of Loeb's research in physiological morphology, discussed next in more detail.

THE BIGGER PICTURE

To appreciate fully Whitman's commitment to introducing what he called biological physiology, or what Loeb called physiological morphology, we must understand the character of the MBL and the broader field of American biology a bit better. The MBL Board of Trustees began as a local Boston-based group, with several nonscientific members of the Woman's Education Association playing an influential role through their assistance in funding the MBL's predecessor under Alpheus Hyatt in Annisquam, Massachusetts. One of the trustees, Sedgwick, had been Martin's student in physiology at Hopkins and a fellow student of Edmund Beecher Wilson, who graduated from Hopkins in morphology. Wilson also became active at the MBL.

Wilson had played an active role in helping to direct the Chesapeake Zoological Laboratory under Brooks at Hopkins, had worked at the Naples Zoological Station, and had a strong commitment to marine biology. Because his teaching position in biology at Bryn Mawr from 1885 to 1891 left him relatively little research time and resources, he began to attend the summer sessions at the MBL, a habit that continued after he moved to Columbia in 1891 and for the rest of his life. E. B. Wilson's work on cell lineage fit closely with Whitman's own embryological interests and made him an immediately attractive colleague, given the emerging MBL direction in both research and teaching. Whitman quickly saw Wilson as a kindred spirit, and by 1890 Wilson had become an MBL trustee and trusted advisor.

Meanwhile, when Wilson left Bryn Mawr for Columbia in 1891, he was succeeded at Bryn Mawr by another Hopkins graduate destined to assume a central role at the MBL, Thomas Hunt Morgan. In the same year, Loeb had let it be known that he wished to move from Germany to the United States. At the time, Whitman was settled at Clark University, Wilson at Columbia, and Morgan at Bryn Mawr. All of them knew of Loeb or his work and Whitman acknowledged him as a good biological physiologist. Pauly has suggested that both Whitman and Franklin Paine Mall, who was also then at Clark and had met Loeb in Naples, may have used their influence to secure Loeb a position at Bryn Mawr with the intention of transplanting him thereafter.[36] If so, the plan worked. Loeb moved to the United States to begin teaching and research in the biology program at Bryn Mawr in September of 1891. Here was a physiologist, trained in a medical environment in Germany and who had even practiced clinical ophthalmology (though not very happily), now in a biology program. If my analogy with the ecology of salamanders is useful, Loeb's move into the larger pools of biology on the American scene, initially if very briefly at Bryn Mawr, should have influenced both the character of his work and that of the surrounding population. There is some evidence that it did. Loeb's experience at Bryn Mawr helped to introduce him to American biological work and perhaps influenced or reinforced the particular kind of physiological work he pursued thereafter. Pauly reports that Loeb did not take gladly to the traditional embryological work that he was hired to teach at Bryn Mawr.[37] In fact Loeb found the preparation of sections and staining procedures tedious and deadly dull, an attitude that such embryological researchers as Ernest Everett Just later came to ridicule.[38] Loeb nonetheless reluctantly accepted his embryological tasks and thereby entered biology, as a general physiological morph amidst the dominant morphological and zoological forms.

WHITMAN AND CHICAGO

In 1892 the pools changed as something like a flood occurred. Biologists of all sorts were "washed away" from Clark University. Major battles with the administration at Clark led most of the biologists there (and indeed many of the scientific faculty members generally) to leave abruptly. They surely would not have migrated, some presumably to die, had no other pool with promising resources attracted them. The establishment, however, of another major research university, the University of Chicago, provided the desired new location. Mass migration from Clark to Chicago occurred, suddenly and generally, more like a flood than a purposeful migration.[39]

Whitman himself acted more carefully. After considering the advantages of making the move, he insisted that Chicago must provide an adequate modern laboratory for the biological sciences in order to make the move acceptable to him. Always interested in modern techniques and methods, Whitman considered the laboratory details crucial.[40] He had already given up on Clark, but he wanted to be absolutely certain that Chicago would prove more congenial, with a healthier supply of necessary resources to support a robust, diverse, and growing population. In fact Whitman had continually revised his ideas about what diverse specialties he wanted to have working cooperatively together under the broader rubric of "biology." Mall, convinced very early to move to Chicago, wanted Whitman to go as well. He thus worked hard to keep Chicago's President Harper positive about Whitman and to persuade Whitman that he needed Chicago and should not accept another offer he had evidently received.[41] Mall's efforts paid off, for Chicago and for Whitman, at least at first. Whitman finally allowed himself to be persuaded and accepted the call to Chicago in 1892.

In his early correspondence with Chicago's new president, William Rainey Harper, in 1891, Whitman outlined what he would want from Chicago—namely programs in zoology, botany, paleontology, and physiology, with anthropology to follow. He claimed that he already had with him at Clark one of the very best men in the country in paleontology and another in anthropology. He had others with him as well, professors and assistants. In a letter of March 1892, Whitman further expressed his desire that "Physiology, Anatomy, Histology, Embryology, Zoology, and Botany—should not fare worse than a science representing a much more limited domain."[42] He referred to chemistry, which had just received a substantial gift for its laboratories. This early correspondence between Whitman and Harper considers the biological sciences one department, presumably subdivided into specialty studies. Yet because Whitman kept referring to different specialties, he may well have confused President Harper about what he really sought for the biological sciences as Chicago.

In an undated memo, though it presumably reflects his ideas soon after arriving at Chicago in 1892, Whitman sought to clarify his views. There he referred to biology as divided into zoology and botany, with each in turn divided into morphology and physiology.[43] To reach European standards in that particular subset of biology that Whitman knew best, namely animal morphology, would require the development of several further subspecialty areas: zoology, anatomy, histology, neurology, paleontology, and pathology. Anthropology, cellular biology, and experimental biology would follow as further, more recent subspecialties. Within physiology, one finds human physiology, general physiology, physiological chemistry, and hygiene as the first subdivisions, with psychology coming later; Whitman regarded psychology as essentially physiological but as falling somewhere between physiology and neurology. Botany would consist of structural botany, physiological botany, systematic botany, and bacteriology. All botanical work, in the beginning at least, could exist together in the same organizational unit or institute, while the other subspecialties should each have their own units, following Whitman's perception of the structure of German universities.[44] He obviously knew most about zoological morphological study, about which he wrote in great detail, but Whitman's grand ambitions for biology did not stop there. He intended and probably fully expected to develop all branches, and he expected support from the administration.

Whitman thus went to Chicago with high hopes. He took most of the Clark faculty in biology with him, including Mall, Donaldson, George Baur, and Wheeler. He also attracted Loeb, as he had already done at the MBL, to develop the physiological side of biology, beginning that first year, 1892.[45] Despite Loeb's claim in a letter to a friend that he was surprised by the offer from Chicago, he had already written a letter to President Harper in June 1891 specifying his qualifications and explaining that he intended to make his home in America.[46] Having procured all this talent for Chicago, generally at lower salaries than he thought they actually deserved, Whitman felt he had the makings for the first-rate biology program he sought to establish.

President Harper and the University of Chicago, however, were not quite so forthcoming on all that they had promised or implied they would provide. The permanent laboratory buildings had to wait until after 1895, when Helen Culver gave $1 million for the development of the biological sciences at Chicago.[47] Whitman had to watch helplessly, though never patiently or quietly, as some of his leading researchers received better offers elsewhere that Harper could not, or would not, match.[48] The lack of proper funding for equipment was another persistent problem and the continued failure to establish an inland research station hampered research and caused Whitman and others great concern.[49] By 1899, Whitman had become deeply frustrated, as is clear from the following letter to President Harper written from Woods Hole:

> Within a few months zoology loses two men that are known throughout America and the whole scientific world as biologists of the first order. When such men go, everybody knows why they go, and they do not ask what is the matter with the man, but what is the matter with the University.
>
> P. S. Yours of August 30 at hand, I can make recommendations for the vacancies, but I would be glad to know first of all what can be done to put the department into working order. With no aquaria, no museum, and no laboratory service, and with appropriations cut to the lowest figure of a vegetative existence, it will not be easy to get such men as we have had. Nothing less than a thoroughgoing reorganization is possible or even worth doing.... So far as men are concerned we have had the best dept. in the country; so far as equipment is concerned we have had the *weakest* in nearly every respect. That has crippled us and led to dissatisfaction and steady decline. Your best help is needed. What do you propose?[50]

Perhaps all modern administrators will recognize the complaints and the resulting loss of morale that comes when great promises are broken and great hopes frustrated. Indeed, most of the men who had gone to Chicago with Whitman left for better situations, which disturbed Whitman personally as well as professionally. He especially lamented the loss of Shosaburo Watase, who he regarded as the very best cytologist of the day, better even than Wilson. Watase combined a physiological and a morphological approach to his work, as well as working with both plants and animals. He was therefore invaluable for the kind of program of cooperation that Whitman wished to build and it saddened him to lose Watase and others with similar talents, such as Wheeler.[51] Gradually, Whitman gave up and retreated into his work on the behavior and evolution of pigeons but only after continually fighting through the 1890s.[52]

In the end, Whitman's failure to achieve all the goals of his program can be explained partly through his unrealistic demands and ambitions. His problems with the administrators at Chicago paralleled his relations with the MBL trustees; he had also wished to expand the MBL into a major research and teaching center while the trustees resented the increasing costs that seemed inevitably to come with such expansion.[53] In 1892, when Whitman announced to the MBL trustees that he was adding a course of instruction and investigation in physiology, headed by Loeb, he urged them to provide more funds to equip the physiology laboratory.[54] For the prior two years of his directorship, Whitman had disagreed with the trustees, who envisioned a slow development for their local laboratory while he saw prospects for a first-rate facility for research and teaching in the most up-to-date biological areas. Whitman at times took no salary from the MBL or spent what he did take on what he regarded as necessary laboratory expenses, just as he spent much of his salary at Chicago on establishing and maintaining a pigeon colony for his research on animal behavior.[55] Certainly Whitman demanded no less of himself than he did of others, including the administrators of both the MBL and Chicago. Yet they could hardly be expected to share his vision and singleness of purpose even though Whitman's plans for biology—for all of biology divided up into specialties and then reunited through cooperation—provided a clearly articulated and solid base for both institutions.

By 1894 Whitman clearly had ambitions as high for the MBL as he had for Chicago. He no longer feared competition or objections from the United States Fish Commission in Woods Hole, which had decided not to develop its own research center.[56] The Board of Trustees of the MBL had begun to expand so that some of Whitman's trusted friends joined the primarily Boston-based original group; Henry Fairfield Osborn and Wilson particularly supported Whitman.[57] By 1894 Whitman wanted to expand physiology at Woods Hole as well as at Chicago, and he worked to attract people who would want to join Loeb. Loeb, with his German training, international reputation, and experience at the Naples Zoological Station, clearly appealed to Whitman, despite Loeb's difficult personality, which was to cause problems later.[58] Whitman wanted more people with Loeb's particular approach to biology.

Whether Whitman saw Loeb as extending what he already regarded as the important physiological side of biology or whether Whitman had first articulated a newly acquired sense of the value of physiology only after learning of Loeb's availability remains something of an open question. Whitman's interest in both Loeb and physiology does seem to have developed at the same time. He certainly insisted upon the importance of both morphology and physiology as the two major standpoints of biology at the same time that he realized the value of adding Loeb to his groups. For whatever reasons, Whitman became an advocate of "biological" physiology as a central, and not just a distant, part of biology. He saw Loeb's type of physiology as a particularly desirable model and tried to make both the MBL and Chicago places where physiologists and morphologists could feel comfortable.

PHYSIOLOGICAL MORPHOLOGY

What exactly was that particular form of physiological work that Whitman found attractive? In a set of papers presented as part of the series of evening lectures at the MBL, Loeb outlined the ideas he had begun to publish in 1891, under the title *Untersuchungen zur physiologischen Morphologie der Thiere.*[59] There he explained what the term "physiological morphology" meant and outlined the basis of his research program: "I have chosen the name Physiological Morphology for these investigations, inasmuch as their object has been to derive the laws of organization from the common source of all life phenomena, i.e., the chemical activity of the cell." Thus such work sought causal accounts of life phenomena in physical and chemical terms. Yet "the aim of Physiological Morphology is not alone an analytical one. It has another and higher aim, which is synthetical or constructive, that is, to form new combinations from the elements of nature, just as the physicist and chemist form new combinations from the elements of non-living nature."[60]

Pauly has focused on this second aspect of Loeb's program. Yet I suspect that only the first part of Loeb's program had any major appeal for the MBL researchers and for Whitman.[61] That MBL audience, especially Morgan, did find this part of the program intriguing. Through his various studies of regeneration and transplantation, Loeb was attacking the fundamental questions about the "physical forces that determine the formation of a new organ," and "How can these chemical forces be brought into relation with the visible changes which take place in the formation of a new organ?" The mechanics of growth would provide the answer to these basic questions of physiological morphology.[62]

Such work fit nicely into the context of emerging German studies in the physiology of development, studies that combined traditional physiological and embryological methods with basic morphological questions. In the process each specialty was changed into something new. These physiological studies had especially attracted the attention of those American biologists who spent their summers at the MBL. In particular, the work on frogs at Jena carried out by Pflüger, Gustav Born, and Roux excited interest. These researchers had picked up on questions raised by Wilhelm His—especially questions about what mechanisms direct organisms and how individual organisms develop. In addition, the suggestions by August Weismann—about the existence of a separate, inherited germ plasm that directed subsequent development of each organism—while convincing virtually none of the Americans, pointed to explanations of how development might be accomplished. Weismann asked not merely what sequence of forms the individual passes through, but how that sequence takes place and through what processes. Such questions concerned function and were thus traditionally physiological, but they also focused on embryos and were thus traditionally morphological as well.[63]

To Whitman, whose Ph.D. dissertation had embraced E. Ray Lankester's suggestion that some sort of "precocious segregation" might occur in the egg cell, such concentration on development and on the causes of differentiation represented an attack on one of the fundamental problems of biology.[64] If Whitman found none of the theories yet put forward completely convincing, he nonetheless saw the value of exploring them further. He had also concluded that development was more complex than simple precocious segregation could explain. In his view, physiologists and morphologists must cooperate to answer those fundamental questions. Like the other researchers at the MBL and like

Loeb, Whitman was more sympathetic to Hans Driesch's and Oscar Hertwig's epigenetic views than to Roux's and Weismann's predeterminism.[65]

Now we can see why Whitman found Loeb's work particularly attractive. Not caught up in detailed studies of how one or another isolated part of the body works, Loeb concentrated on the whole organism—and better yet, on the embryonic organism. Whitman consistently urged the importance of the "organismic standpoint," and it cannot be coincidental that such men as Charles Manning Child, William Ritter, William Locy, Frank Lillie, and Loeb, each of whom worried about how the whole organism works in some way, were all at Chicago while Whitman was in charge. Indeed, I would argue that all were participants in a Chicago school of biology. Whitman's vision, like the Hopkins program earlier, had an important impact on physiology and biology in the United States.

THE NEXT CENTURY

As the Hopkins program faded after Martin's fall, with the rise of the medical school, and as Whitman failed to achieve full support for his programs and turned to his pigeons, the study of biological physiology became increasingly marginal throughout the country. Those physiological morphologists who had found comfortable niches during the 1890s at the MBL and Chicago did not always fit happily elsewhere. Other historical studies, notably the valuable contributions by Robert Kohler and Kenneth Manning, have illustrated what happened to some of the "general" physiologists. Kohler and Manning have also described some of the battles for resources and position that occurred as the biological physiologists struggled to retain their places in biology. In this book, Pauly has further detailed those problems for general physiologists such as Loeb. Specialization on the whole brought divergence, with a loss of the unified perspective that the Hopkins program had preached and the coordination that Whitman had valued so highly. On balance, general and biological physiology lost out as relatively few researchers continued to pursue physiological morphological work. Only Chicago provides a possible exception, with its continuing strong tradition in physiology into this century. Despite promising beginnings, the physiological variants were largely kept out of biology in the twentieth century by scarce resources, increased population density, and the impact of such overzealous individuals as Loeb.

NOTES

1. Allen Kerr Bond, *When the Hopkins Came to Baltimore* (Baltimore, MD: Pegasus, 1927); John C. French, *A History of the University Founded by Johns Hopkins* (Baltimore, MD: Johns Hopkins Press, 1946), esp. pp. 1–26; Hugh Hawkins, *Pioneer: A History of the Johns Hopkins University* (Ithaca, NY: Cornell Univ. Press, 1960), pp. 3–20.
2. Philip Pauly, "The Appearance of Academic Biology in Late Nineteenth-Century America," *J. Hist. Biol.* 17 (1984): 369–397.
3. Daniel Coit Gilman, "The Original Faculty," in *The Launching of a University and Other Papers* (New York: Dodd, Mead, and Co., 1906), pp. 47–56, 51; *Inaugural Addresses* (Baltimore, MD: John Murray, 1876), pp. 15–64, esp. 43; Thomas Henry Huxley, "On the Study of Biology," in *Science and Education* (London: Macmillan, 1893), pp. 262–293, esp. 263.

4. Gilman, "Original Faculty," p. 52; Gilman to Henry Newell Martin, 14 March 1876, Gilman Papers, Special Collections, Milton S. Eisenhower Library, Johns Hopkins University, Baltimore, MD (hereafter, Gilman Papers).
5. Huxley, "Address on University Education," in *Science and Education*, pp. 235–261, esp. 254.
6. Huxley to Gilman, 20 February 1876, Gilman Papers, to recommend Martin; Martin to Gilman, 5 April 1876, Gilman Papers; he declined at first but by 29 May 1876 he had accepted and had outlined a program for Johns Hopkins University; French, *History*, pp. 34–40; Charles E. Rosenberg, "Henry Newell Martin," *Dictionary of Scientific Biography* vol. 9 (New York: Scribner): pp. 142–143; Russell Chittenden, "Henry Newell Martin," *Dictionary of American Biography*, pp. 337–338.
7. Martin to Gilman, n.d., Gilman Papers; Brooks to Gilman, 18 June 1876, Gilman Papers. Brooks explains that he has an offer at the University of Cincinnati but wants to be Martin's assistant; French, *History*, pp. 40–42.
8. Martin to Gilman, n.d., but responds to 29 April 1881, Gilman Papers. For more on Brooks's role see Keith Rodney Benson, "William Keith Brooks (1848–1908): A Case Study in Morphology and the Development of American Biology" (Ph.D. diss., Oregon State Univ., 1979).
9. Rosenberg, "Towards an Ecology of Knowledge: On Discipline, Context, and History," in *The Organization of Knowledge in Modern America, 1860–1920*, ed. Alexandra Oleson and John Voss (Baltimore, MD: Johns Hopkins Univ. Press, 1979), pp. 440–455.
10. Actually, salamanders are particularly appropriate here because one of the principals of this story, Charles Otis Whitman, spent many hours trying to locate the salamander *Necturus* near the Allis Lake Laboratory in Milwaukee. He wanted to study the embryological development of an interesting local species and knew that *Necturus* was alleged to exist in the lake waters of northern Wisconsin. Unfortunately the species proved elusive and, after several futile attempts to locate specimens, Whitman and his student/associate Edward Phelps Allis had just about given up. The local fisherman they had hired insisted that he knew just where to find some of these salamanders: "under almost any stone or stump in the water along the edge of the lake." However, Whitman and Allis had already spent a day overturning every rock in sight and found nothing. They concluded that they had been victims of another "fisherman's story" and that no *Necturus*, or mudpuppies, were to be found there at all. When their hired fisherman, Meyers, insisted that they meet him at 5:00 A.M. the next morning for one last try, they reluctantly agreed, even though they planned to catch a six o'clock train. As Allis later recalled, Meyers had told them: "'You'll find one under that big stone out there in the middle of the race. I see his hole.' Whitman laughed at him and told him he had better go out and find it himself. But Meyers, somewhat nettled, said he wasn't going to get wet to the middle for one of those d---ed things. So I [Allis] told Whitman it wouldn't do any harm to try it; and out he went, trying not to look foolish at allowing himself to be sent on such an errand. He waded out with great deliberation, turning over a few other stones as he went, like a child who wants to make believe that he isn't doing what he has been told to do. Meyers in the meantime had left, saying that he had something to do and couldn't hang around any longer and that if we didn't find a puppy under that stone we would under another a little further out.

"Whitman had now, in his leisurely course, reached the stone he was after and with a smile at me bent down to turn it over. He turned it, then dropped it, and standing up looked at me for a moment. 'There was one,' he said, in an awe-stricken voice. Then he picked up the stone, and without a word brought it ashore and showed me, attached to its under surface, a few large eggs.

'I can't believe it,' he said. 'Why, Mr. Allis, I would give my year's salary for those eggs.'"[10] From Dornfeld, "The Allis Lake Laboratory, 1886–1893," *Marquette Med. Rev.* 21 (1956): 115–144, esp. 126–128.
11. Martin, "Physiological Papers," in *Memoirs from the Biological Laboratory*, vol. 3 (Baltimore, MD: Johns Hopkins Press, 1895); Larry Owens, "Pure and Sound Government. Laboratories, Playing Fields, and Gymnasia in the Nineteenth-Century Search for Order," *Isis* 76 (1985): 182–194, points out the unusual nature of the Hopkins Laboratory. Students and assistants worked together with Martin so that even the major contributions came from the group together rather than from Martin alone.
12. "Class of '90," *Hopkins Medley* (Baltimore, MD: Guggenheim, Weil, 1890), pp. 56–58; these are the students listed in the files of correspondence, Gilman Papers.

13. William Keith Brooks, response in *Science* 3 (1896): 708, to Conway MacMillan, "On the Emergence of a Sham Biology in America," *Science* o.s. 21 (1893): 184.
14. Howell did, for example, as well as others not named. Brooks to Gilman, Gilman Papers.
15. Frederic Schiller Lee, "The Scope of Modern Physiology," *Am. Nat.* 28 (1894): 380–388, esp. 380, 388, and 473–482.
16. After his Ph.D., Lee went to Leipzig for a year, then taught biology at Sarah Lawrence for a year, physiology at Bryn Mawr, and physiology at Columbia from 1891; For more on Lee, see chapter III by Alejandra Laszlo and chapter X by Richard Gillespie, in this book.
17. In addition to Lee's one year at Sarah Lawrence. Henry Herbert Donaldson began in physiology, then received an M.D. degree and eventually moved into a research position at the Wistar Institute.
18. W. Bruce Fye, "H. Newell Martin—A Remarkable Career Destroyed by Neurasthenia and Alcoholism," *J. Hist. Med. Allied Sci.* 40 (1985): 133–166.
19. Allan Chesney, *The Johns Hopkins Hospital and the Johns Hopkins University School of Medicine*, vol. 1 (Baltimore, MD: Johns Hopkins Press, 1943), pp. 230–233.
20. Fye, "H. Newell Martin," p. 161; Gilman Papers reinforce the validity of Howell's convictions.
21. Robert Kohler, *From Medical Chemistry to Biochemistry* (Cambridge, UK: Cambridge Univ. Press, 1982); Chapter 11 discusses biological programs. See also chapter III by Laszlo, chapter XIII by Merriley Borell, and chapter VIII by Pauly in this book on particular programs at Harvard and Columbia in general physiology.
22. Dornfeld, "Allis Lake Laboratory," p. 120.
23. Dornfeld, "Allis Lake Laboratory," pp. 121–124, 133.
24. Whitman to Alexander Agassiz and Anton Dohrn to Whitman, Agassiz Collection, Museum of Comparative Zoology, Harvard University.
25. *MBL Minutes*, vol. 1 (Board of Trustees, 1888), MBL Archives.
26. Brooks to Gilman, 21 June 1889, Gilman Papers.
27. Pauly, "Appearance," 383.
28. Whitman, "Biological Instruction in Universities," *Am. Nat.* 21 (1887): 507–579; esp. 517–518.
29. Cornelia Clapp, "Some Recollections of the First Summer at Woods Hole, 1888," *Collecting Net* 2 (4) (1927): 3, 10; Jane Maienschein, "Agassiz, Hyatt, Whitman and the Birth of the Marine Biological Laboratory," *Biol. Bull. Woods Hole* 168 suppl.(1985): 26–34.
30. In particular by Henry Fairfield Osborn, "The Hereditary Mechanism and the Search for the Unknown Factors of Evolution," *Biol. Lect.*, 1894 (1896): 79–100; "Evolution or Heredity," *Biol. Lect.*, 1890 (1891): 130–141; Whitman to Osborn, Osborn Collection, C. O. Whitman folder, Archives, Department of Vertebrate Paleontology, American Museum of Natural History, New York.
31. Whitman, *Report of the Director*, MBL, 1890, p. 21. Elaborated in "Specialization and Organization," *Biol. Lect.* 1890 (1891): 1–26.
32. Whitman, *Report of the Director*, 1891, pp. 14–17, esp. 15.
33. Whitman, "General Physiology in its Relation to Morphology," *Am. Nat.* 27 (1893): 802–807, esp. 804–805.
34. Whitman, *Report of the Director*, 1892, pp. 29–36, esp. 35.
35. Whitman, *Report of the Director*, 1892, p. 35.
36. Pauly, "Jacques Loeb and the Control of Life: An Experimental Biologist in Germany and America, 1859–1924" (Ph.D. diss., Johns Hopkins Univ., 1980), pp. 127–128.
37. Pauly, "Loeb," pp. 128–129.
38. Kenneth R. Manning, *Black Apollo of Science: The Life of Ernest Everett Just* (New York: Oxford Univ. Press, 1983).
39. Lincoln C. Blake, "The Concept and Development of Science at the University of Chicago, 1890–1905" (Ph.D. diss., Univ. of Chicago, 1966), esp. "C. O. Whitman: The Adaptation of Science," chapt. 4, pp. 122–150.
40. Whitman published *Methods of Research in Microscopical Anatomy and Embryology* (Boston, MA: S. A. Cassine, 1885) following an earlier series of articles about method.
41. Franklin Paine Mall to Harper, 8 February, 7 April, and 9 April 1892, Whitman Collection, University of Chicago Archives, Joseph Regenstein Library, Chicago, IL (hereafter, Whitman Collection, Chicago).
42. Whitman to Harper, 19 December 1891, for example, Whitman Collection, Chicago.

43. Whitman to Harper, memo, n.d., Whitman Collection, Chicago.
44. Whitman looked to Germany for models and for ammunition for experiments, but he also called for particularly American work as well.
45. Whitman to Harper, n.d., Whitman Collection, Chicago.
46. Pauly, "Loeb," p. 130; Loeb to Harper, 30 June 1891, Harper Collection, University of Chicago Archives, Joseph Regenstein Library, Chicago (hereafter, Harper Collection, Chicago).
47. Whitman Collection, Chicago; Whitman Collection, MBL Archives. Whitman had hoped that the endowment would be used for the MBL and an inland research station as well as the University of Chicago. He was bitterly disappointed that his inland station was postponed yet again.
48. Whitman to Harper, letters in Chicago and MBL Archives, throughout the 1890s document his frustration. He got Watase, Wheeler, Loeb, and others in succession.
49. Whitman to Harper, letters in Chicago such as that of 4 September 1899, document this concern. Also, biographical notes about Whitman gathered by Frank Lillie discuss Whitman's growing dismay about the Chicago situation. His letter of 13 March 1896 laments the use of endowment funds to meet basic costs rather than extending resources for biology.
50. Whitman to Harper, 4 September 1899, Whitman Collection, Chicago, in response to a letter from Harper. Quoted with permission from Chicago Archives.
51. Whitman to Harper, 3 May 1899, Whitman Collection, Chicago. Regrets the loss of Wheeler earlier and of impending loss of Watase.
52. Oscar Riddle, ed., *Posthumous Works of Charles Otis Whitman. I. Orthogenetic Evolution in Pigeons*; *II. Inheritance, Fertility, and the Dominance of Sex and Color in Hybrids of Wild Species of Pigeons*; Harvey S. Carr, ed., *III. The Behavior of Pigeons* (Washington, DC: Carnegie Institution, 1919).
53. Lillie, *The Woods Hole Marine Biological Laboratory* (Chicago, IL: Univ. of Chicago Press, 1944); Maienschein, "Early Struggles at the Marine Biological Laboratory Over Mission and Money," *Biol. Bull. Woods Hole* 168 suppl. (1985): 192–196.
54. Whitman, *Report of the Director*, 1892, pp. 12–13.
55. Whitman to Edwin Grant Conklin, Whitman Collection, MBL. His former Japanese student Chiyomatsu Ishikawa reported, in a biographical sketch translated in the Whitman Papers, Chicago, that Whitman had turned his entire attention to his pigeons, pp. 17–18.
56. Whitman to Conklin, n.d., Whitman Collection, MBL; Brooks to Gilman, Gilman Papers.
57. Whitman to Osborn, Osborn Collection; Whitman to Wilson, Whitman Collection, MBL.
58. At Chicago, Albert Prescott Mathews and Loeb battled over questions of priority, for example, Harper and Whitman Collections, Chicago.
59. Jacques Loeb, *Untersuchungen zur Physiologischen Morphologie der Thiere. I. Heteromorphosis. II. Organbilding und Wachsthum* (Würzburg, Germany: Hertz, 1891, 1892).
60. Loeb, "On Some Facts and Principles of Physiological Morphology," *Biol. Lect.* 1893 (1894): 37–61, esp. 60–61; Pauly, "Loeb," provides a useful summary of those MBL lectures.
61. Pauly, "Loeb," and his book on Loeb with Oxford University Press, in press.
62. Loeb, "On Some Facts," 45.
63. E. S. Russell, *Form and Function* (London: John Murray, 1916), discusses the physiological and morphological traditions.
64. Whitman, "History of the Egg of Clepsine Previous to Cleavage," *Q. J. Microsc. Sci.* 71 (1978): 215–315, esp. 300.
65. Maienschein, "Preformation or New Formation—or Neither or Both?" in *A History of Embryology*, ed. Timothy Horder, et al. (Cambridge, UK: Cambridge Univ. Press, 1985).

VIII

General Physiology and the Discipline of Physiology, 1890–1935

PHILIP J. PAULY

Anyone beginning a survey of the literature on the history of general physiology finds a peculiar situation. Thomas S. Hall's two-volume *History of General Physiology* provides solid coverage of the intellectual developments up to 1900 and indicates that general physiology has been a fundamental subject that has engaged the interest of leading figures from the beginnings of scientific and medical thought.[1] However, general physiology after 1900 has received no such attention. Robert E. Kohler's *From Medical Chemistry to Biochemistry*—the only work to discuss early twentieth-century general physiology in a systematic fashion—explains the reasons for this contrast. For Kohler, American general physiologists were less significant as followers of a great intellectual tradition than as a small group of chemical biologists, largely independent of physiology per se, who had only middling academic success. In spite of a charismatic leader—Jacques Loeb—and considerable foundation interest in the 1920s, general physiology was a minor specialty, never realizing the potential that chemical biology—reappearing as molecular biology—eventually possessed.[2]

Kohler points to two reasons for this lack of success. The first, and most important, revolves around the "ecology" of scientific disciplines. Caught between physiology, biochemistry, biology, and chemistry, general physiology was never able to gain a significant institutional niche. The second factor was that general physiologists, especially Loeb, were inept and even self-destructive entrepreneurs, refusing to seize opportunities and fighting among themselves rather than establishing a common cause for their discipline.[3]

Disciplinary ecology can indeed go a long way in accounting for the peculiar history of general physiology, and this chapter builds on Kohler's discussion. First, it addresses the question why American general physiologists, and not only Loeb, were such indifferent entrepreneurs and why they were so contentious. Second, it considers the changing relations between general physiologists and mainline physiologists; it is a story of a specialty in many ways at war with its putative parent discipline. Third, it examines the persistent interest in building a discipline of general physiology when opposition was substantial and when the project was in a certain sense inherently implausible. The original title of Hall's book, *Ideas of Life and Matter*, was not a description of a likely specialized discipline, with journal, society, and professorships; there was no comparable development of general chemistry, physics, psychology, or history. Scientists publicly committed to specializing in generalities were faced with unique problems as well as unusual opportunities. Finally, it assesses the significance of general physiology's history in America up to the 1940s. A small specialized field did eventually stabilize, but the significance of general physiology is not solely or even primarily as a failed or only partly successful disciplinary program. Rather, it was an early instance of a significant feature in the history of scientific disciplines in America, namely, their roles as temporary loci for unsocial individualists. General physiology was an early instance of what can be called an "undiscipline."

In the middle of the nineteenth century the term "general physiology" was bound up with "teleomechanism," as discussed by Timothy Lenoir.[4] The Göttingen professor Rudolph Wagner, for example, explained that general physiology was "the theory of the life process," the search for "the laws which lie at the base of the combined activity of the organ-complex." As a result of the exigencies of medical teaching, however, such issues were separated from "special physiology" and disappeared from physiology texts after the 1850s; after short introductions, most writers launched immediately into discussions of vertebrate metabolism. With the decline of teleomechanism, general physiology came increasingly to savor of the mythical age of *Naturphilosophie.* A few iconoclasts, such as Eduard Pflüger in Bonn and Wilhelm Preyer in Jena, resisted this specializing tendency. Claude Bernard held a chair of general physiology at the Sorbonne and in his last years, with an appointment at the Muséum d'Histoire Naturelle, he assembled his *Leçons sur les Phénomènes de la Vie Communs aux Animaux et aux Végétaux.* Both the German and French efforts were fragmentary, however, and had little institutional impact.[5]

The physiologists involved in creating the American Physiological Society (APS) in the 1880s likewise had little interest in the general part of physiology. Their concern, even more than their German predecessors, was in building a medical science validated through rigorous standards of experimental work. The stimulus for general physiology came from outside physiology—from the small group of young scholars who were shaping what would become academic biology. The Johns Hopkins University, from its founding in 1876 to the long-delayed opening of its medical school in 1893, contained a unified department of biology. Its faculty, physiologist Henry Newell Martin and morphologist William Keith Brooks, were satisfied to pursue their fields in traditional fashion but their students had more expansive goals and soon had opportunities to realize them.[6]

Brooks's student Edmund Beecher Wilson, for example, became professor of biology at Bryn Mawr College in 1885. He sought to replicate his Hopkins experience by hiring a young physiologist. Yet a collegiate environment, especially where the students were female, was very different from medical schools, or even Hopkins itself. On one hand, academic questions about the nature of life took on greater significance; on the other, a male professor could teach physiology only if he omitted both mammalian vivisection and discussion of sex. In such circumstances the general part of physiology took on new significance. Frederic Schiller Lee, a Hopkins Ph.D. who had worked on muscle with Carl Ludwig, assisted Wilson at Bryn Mawr. Wilson supported

Lee's effort to define and teach a general, or biological, physiology. When Wilson moved to Columbia University in 1891, Lee followed, obtaining a position devoted in part to general physiology, which suspended him between the academic Faculty of Pure Science and Columbia's recently absorbed medical school, the College of Physicians and Surgeons.[7]

Similar considerations motivated Charles Otis Whitman, the organizer of the University of Chicago division of biology and director of the Marine Biological Laboratory at Woods Hole, Massachusetts, in the early 1890s. Whitman was a follower of Herbert Spencer, both in his belief that evolutionary progress was "an integration of matter and concomitant dissipation of motion," and in his faith that the growth of knowledge came through specialists pursuing diverse disciplines in an environment that assured balance and coordinated understanding. When offered the position of head professor of biology at the University of Chicago in 1892, Whitman bombarded President William Rainey Harper with organizational charts describing the relations among the life sciences. Whitman was especially concerned to sponsor a coeducational physiology that was not limited to "the practical needs of medicine." The nature of this "biological" or "general zoological" physiology was vague and fluctuating but Whitman, as a student of teleomechanist Rudolph Leuckart, emphasized the "economy of organisms" rather than the "economy of organs," focused on developmental study of invertebrates, and put his discussion in an explicitly evolutionary framework.[8]

The person Whitman chose to develop "biological physiology" at both Chicago and the Marine Biological Laboratory was the thirty-three-year-old Loeb, a German Jewish scientist who had come to America the year before and had succeeded Lee at Bryn Mawr. Loeb seemed to fit Whitman's specifications precisely. As an M.D. who had studied in the early 1880s at Strassburg with the neurophysiologist Friedrich Goltz, his physiological credentials were solid. He had recently become interested in invertebrates and comparative methods, and Whitman would have been especially impressed by a paper he had written with an English zoologist, Theodore Groom, on the role of tropisms in the depth migrations of the pelagic fauna—a good example of Whitman's "economy of organisms." Whitman brought Loeb to Chicago in 1892 and in the same year put him in charge of the physiology course at Woods Hole.[9]

In fact Loeb did not fit well into the physiology slot in Whitman's organizational chart. He had never studied Spencer and he was uninterested in the economy of organisms; the idea for the "pelagic fauna" paper was probably Groom's. Loeb's passion was to solve problems that would lead to scientific control over organisms. He was antagonistic toward attempts to explain the nature of life; he lumped evolutionism with other aspects of "metaphysics" and *Naturphilosophie*. While he accepted appointment as assistant professor of physiology (with assurance that no other more senior physiologist would be hired), he called himself an "experimental biologist" and considered his territory biology as a whole, opposing attempts to divide the broad subject either intellectually or professionally.[10]

Loeb recognized that his unwillingness to integrate himself into Whitman's plans limited his career. In 1894, with the combination of self-deprecation and sarcasm that distinguished him from his earnest American colleagues, he wrote his former advisor Nathan Zuntz that scientists succeeded in America "only through 'Pull.' In order to have 'pull' one must be a diplomat; but since none of my ancestors was employed in the Foreign Office—Palestinian times excepted, but there the documents are missing—it is not surprising that I have little diplomatic talent." He was, however, more interested in having freedom to work as he wished than in acquiring influence in the American academic world. Whitman took advantage of Loeb's predilections, limiting his role in biology at Chicago by giving him institutional autonomy in return for his accepting an identity as a physiologist.[11]

By the late 1890s, as he accumulated a laboratory, graduate students, medical courses, and a place on the editorial board of the new *American Journal of Physiology*, Loeb began to see the value of Whitman's use of the term "general physiology." He was not interested in building a specialty by that name, however; as with experimental biology a few years earlier, he hoped to use it as a rubric for remaking physiology in general. In 1897 Loeb announced that physiologists' focus on vertebrates was "a waste of time" with respect to such fundamental problems as secretion, oxidation, tropisms, and sleep. Physiology, he argued, could only be developed on a comparative, general basis.[12]

What, then, was general physiology? Loeb was clear about what it was not. The German zoologist-physiologist Max Verworn had recently published a semipopular survey with that title; Loeb insisted that Verworn's Haeckelian amalgam of protozoology, cell theory, and evolutionism was pointless—it was absurd to build a general physiology on "a few amoeba observations" and the "scientifically worthless" phrase of the "cell-state."[13] On the other hand, Loeb's own views on the subject and framework of general physiology were inconsistent and unsystematic. In one year, 1897, he defined general physiology's subject variously as "the phenomena of life which are common to all animals and plants," as "the energetics of life phenomena," and as "the constitution of living matter."[14] He pointed to the course he offered at Chicago as evidence that the subject could be organized and taught, but his lectures from this period, transcribed in the hope that they would provide the basis for a treatise (on "experimental biology"), were notably lacking in organization and he soon abandoned the project.[15] Almost a decade later, when he finally did publish *The Dynamics of Living Matter*, he retained the diffuse "lecture" format.[16]

American physiologists interested in protozoa and cell theory (Lee), sleep (William Henry Howell), oxidation (Russell Chittenden), and behavior (Henry H. Donaldson, Loeb's dean) found Loeb's attitude offensive and apparently considered the 1897 outline of his position, with its claim that much of their work was a scientific "waste of time," a "personal attack."[17] Although Loeb was a major contributor to the *American Journal of Physiology* during its first years, his relations with the leaders of the APS deteriorated rapidly from 1898 on. By 1902, even as Loeb's work on artificial parthenogenesis was leading to nominations for the Nobel Prize and administrative offers of faculty positions free of undergraduate teaching at both Chicago and the University of California, the leading American physiologists were pushing him out of their professional community. Franklin Mall reported to Simon Flexner that "the latest discovery in physiology is the Loeb-reaction." He gave directions for eliciting the response: "Next time you meet a physiologist or anything of the kind say

'Loeb' to him & you will understand what I mean. I hide now whenever anyone says 'Loeb' by mistake, for you may rest assured that no one says it intentionally."[18] When Loeb moved to California in 1903, the APS Council, led by Chittenden, removed him from the editorial board of the *American Journal of Physiology* on the grounds that he was no longer affiliated with a major institution. More generally, physiologists cast their lot increasingly with medicine rather than with Loeb's "biological" conception of the field.[19]

Loeb did not care about maintaining an identity as a physiologist. He refused to rejoin the editorial board of the *American Journal of Physiology* when the leadership changed. While he was professor of physiology at the University of California from 1903 to 1910, he did not teach medical students, and when free to choose he spoke of his work as "biological" research.[20] He spent much of the year at a small laboratory on the Monterey Peninsula studying artificial parthenogenesis, fertilization, and heredity in marine invertebrates. While most of his students from the University of Chicago (including Samuel S. Maxwell, Walter E. Garrey, Elias P. Lyon, and Martin Fischer) identified themselves as physiologists and went on to medical school appointments, albeit sometimes with difficulty, in California his students and assistants included the anatomist John B. MacCallum, zoologists William F. Allen, Frank W. Bancroft, and Wolfgang Ostwald, botanist W. J. V. Osterhout, physiologist T. Brailsford Robertson, and chemist Hardolph Wasteneys. The environment was individualistic, even bohemian, as Loeb and his students migrated back and forth between Berkeley and the oceanside. Loeb was roused to defend his identity as a physiologist only when challenged to do so by Berkeley President Benjamin Ide Wheeler or the San Francisco clinicians operating the university's medical school.[21]

On returning East to the Rockefeller Institute for Medical Research in New York City in 1910—after being passed over for the physiology chair at Columbia's College of Physicians and Surgeons (which went to the more medically oriented Lee)—Loeb immediately established a department of experimental biology. His main concern in negotiations regarding this post was retention of his "freedom," and he continued to work on subjects ranging from mutations and the duration of life to tropisms and brain waves.[22]

Loeb became a scientific entrepreneur only in 1917, at age fifty-eight, as part of his wartime concern to free American scientists from German influence.[23] Yet he was uncertain about the nature and scope of the discipline he was seeking to build. He initiated a book series, *Monographs in Experimental Biology*, which was designed to demonstrate that "experimental biology and general physiology should be clearly recognized as one science, covered by one chair, and not as two different sciences."[24] At the same time, with the advice and support of the head of the Rockefeller Institute, Simon Flexner, he organized the much more specialized *Journal of General Physiology*.[25] Although the monographs were varied in content and influence, Loeb identified himself increasingly through his editorial role in the *Journal of General Physiology*. He worked, largely through Simon and Abraham Flexner, to establish a discipline of general physiology in the years up to his death in 1924.[26]

Interest in general physiology, as Kohler has noted, increased dramatically in the 1920s. Many leading biologists had long adopted Whitman's use of the term to describe comparative physiology within a framework of arts and sciences. Edwin Grant Conklin had hired E. Newton Harvey at Princeton in 1910 and Herbert Spencer Jennings supported Samuel O. Mast at Hopkins a year later. George Howard Parker at Harvard drew Osterhout from California at the same time, and replaced him in 1925 with William J. Crozier. At that time Thomas Hunt Morgan and E. B. Wilson found a position for Selig Hecht at Columbia, and Ralph S. Lillie gained a chair at Chicago. Major medical schools were also interested in general physiology. Hopkins sought to lure A. V. Hill from England in 1923. At Harvard, Lawrence J. Henderson, after years in limbo, began to develop the Physical Chemistry Laboratory with Edwin J. Cohn. Rockefeller philanthropy was crucial. The Rockefeller Institute subsidized Loeb's journal and provided positions for him, John Howard Northrop, and Osterhout. Rockefeller organizations provided funding for Hecht, Crozier, and Cohn, among others.[27]

In spite of this interest, general physiology never really took off as a separate discipline. Interest on the part of administrators and leaders of other scientific fields was shallow. General physiologists were unable to acquire the stable service roles necessary for long-term academic autonomy. Other fields, ranging from academic chemistry, physics, and biology, through medical school physiology and biochemistry, were well-established competitors that limited the institutional influence of general physiologists during the best of times, and eliminated them as a significant independent force in the Depression.[28] While general physiologists gained some autonomy at nonmedical universities such as Princeton, Stanford, and the California Institute of Technology, as well as at the Rockefeller Institute, they experienced continuous problems at Hopkins, Columbia, and Harvard, the universities with both major biological and medical programs. The fundamental organizational problem was duplication of effort: if one person claimed general physiology, it was likely that two would do so.

The protozoologist Jennings considered himself a general physiologist in Whitman's sense as late as 1909, when he was negotiating inconclusively for positions at Princeton and the Rockefeller Institute.[29] In 1911, established as the Hopkins professor of zoology and moving into genetics, he hired Mast to take over this half of biology. Jennings's department was separate from the university's medical school, and the lack of cooperation was a disappointment to him. He was pleased when the new Rockefeller-funded School of Public Health hired his former student Raymond Pearl, who was interested in both genetics and "the biology of death," as professor of biometry in 1918. Jennings assumed that Pearl's appointment would mean a substantial increase in common interest between the two faculties. His enthusiasm rapidly waned in the next few years, however, as the university's medical leaders gave Pearl the title of professor of biology, and at the same time made clear their interest in supporting work in general physiology within their confines. In 1926 the Rockefeller General Education Board rejected a plan to increase the research emphasis of the university's academic departments because they were too small and undistinguished. At the same time the medical school was unwilling to provide a good permanent position for the general physiologist Leonor Michaelis, who left for the Rockefeller Institute. Only in 1930 did the university seek combined funding for zoology, botany, physiology, and Pearl's Institute for Biological Research; however, given the number of established programs and the constricted resources of the Depression years, general physiology played no role in these developments.[30]

At Columbia the issue revolved around the name "general physiology" itself. While physiologists at the College of Physicians and Surgeons had only minor research interest in general physiology, they continued to offer courses and claim the territory. T. H. Morgan, with the help of Flexner and Osterhout, was able to obtain a Rockefeller grant to support Hecht in the zoology department at Columbia's main campus in 1925; however, he was concerned that it not appear that Hecht was encroaching on the medical school's expertise. As Crozier put it, the "method of dodging trouble with P. & S. 'physiologists'" was to call Hecht's position a professorship in "biophysics."[31]

Perhaps the most interesting situation was the one at Harvard. Henderson and Cohn worked in the medical school on blood as a system and on the physicochemical properties of proteins. At the same time Osterhout was professor of botany, publishing on the permeability of algal membranes and offering a laboratory course in what many considered "biochemistry." When Osterhout left Harvard to take Loeb's place at the Rockefeller Institute in 1925, he was replaced by Crozier, as "professor of general physiology." Crozier and his medical school colleagues were immediately at loggerheads; when the General Education Board provided $3 million to consolidate zoology, botany, and general physiology into a new Institute of Biology, Crozier made sure that Henderson and Cohn would not be involved. He vowed that he was "quite determined upon" one thing—that he would not "work in the same building with L. J. H.; that stuff is medical school business and I do not care to be associated with it."[32]

Yale was the exception that proved the rule. Foundation support there was directed toward a cooperative program in "human science" involving the medical school, law school, and the social sciences. The zoology department, under the leadership of the embryologist Ross Harrison, was anatomical in orientation. Within such an environment, John Fulton, the newly appointed professor of physiology, took a strong position against general physiology. In 1931 he described general physiology as a research program "centered upon fluids and substances derived from living things, rather than upon the living organism itself"; he contrasted general physiologists with the "larger group of physiologists who conscientiously feel that the true objective of their science is the human body, the highest product of the Great Artificer of Nature," and who wished "to comprehend *its* functions rather than those of frogs or cats or apes." Fulton argued that physiology was not a big enough discipline for both groups, and that "it were perhaps better for science that many of the so-called 'general physiologists' recognize that they are not physiologists at all, and that it would be preferable to style themselves by more accurate descriptive terms such as 'biophysicists,' 'protein chemists,' or 'electrochemists,' as the case may be."[33]

The duplication of programs at Johns Hopkins, Columbia, and Harvard, revolving in large part around competition for grant money, reflected the ongoing tension between academic and medical science programs; it also resulted from intellectual uncertainty. The disciplinary impulse in general physiology was largely the result of the enthusiasm of academic and foundation leaders. Fulton was not the only person unclear about exactly what the subject was. The most obvious uncertainty was whether general physiology was in fact part of physiology, or whether it was a separate subject, equivalent to physicochemical biology. Most scientists accepted the term "general physiology" by 1925 but there was considerable dispute over its content. Each advocate was struggling especially with the nineteenth-century teleomechanist aims implicit in its use. In an area that had been treated mainly in introductory lectures and general reflections, it was not a simple task to set out a reasonably coherent and limited subject matter, a consensual theoretical framework, and a soluble set of research problems.

There were at least three different primary meanings to the word "general." For protozoologists such as Jennings and Mast, general physiology was the physiology of the simplest organisms. Alternatively, it could be the properties common to all organisms, as Loeb, following Bernard, had proposed. The third possibility, as Whitman and others claimed, was that general physiology dealt with the physiology of the organism as a whole. Theoretical frameworks also varied. The explicitly evolutionary theory favored by Jennings had fallen out of favor by 1910; it was replaced by the various forms of physiological holism associated with Henderson and Frank Lillie. Allied to holism was the interest—that William Bayliss of University College, London, had stimulated, and that Ralph Lillie, the University of Chicago chemist Albert P. Mathews, and a number of other lapsed "Loebians" favored—in colloid chemistry. To them, the idea that physiological properties depended upon colloidal structure was appealing in providing a "hard science" justification for their research programs at the same time that it insulated them from direct contact with physical chemistry. Loeb's energeticist framework was a further alternative, intersecting and often opposing both holism and colloidism. The research problems of general physiologists were as varied as the practitioners. They ranged from photosynthesis and bioluminescence to enzyme action, ion transport, protein structure, adsorption, vitamins, irritability, nervous integration, heat production in muscle, the duration of life, and behavior.

Underlying these diverse and often contradictory positions was a deeper tension peculiar to the effort to form general physiology into a scientific discipline. The attempt to define a "discipline" in terms of theory carried with it the rattling skeletons of *Naturphilosophie* and the project of explaining "the nature of life." Mathews, who taught general physiology at the Marine Biological Laboratory from 1903 to 1919, claimed that "psychism" was "as fundamental a property of matter as inertia." Ralph Lillie argued at different times for the necessity of recognizing "Design in Nature" and the existence of "directedness" in all living material. The problem was not so much that self-described "hard working physiologists" such as Lewis V. Heilbrunn disagreed with such claims, but they considered public discussion of doubts about the extension of reductionist methods in bad taste; Heilbrunn was commonplace in caricaturing Ralph Lillie as a typical "elderly biologist" who "leaves his laboratory to publish defeatist or vitalistic views in philosophical journals."[33]

On the other hand, those general physiologists who emphasized the "scientific" through hard-nosed, reductionist research were continually in danger of losing their disciplinary identity and being swallowed up by chemistry. Just as he was becoming an entrepreneur for general physiology, Loeb was shifting from interest in biological questions to a biochemical project that made physical scientists his primary reference group. From 1918 to his death in 1924 he studied the properties of gelatin as a function of pH and the diffusion of ions through artificial membranes; as his interest in chemical problems grew, he deferred biological and physiological questions increasingly to the remote realm of "application."[35] His protégé Northrop followed the same path further. He began as Loeb's assistant at the Rockefeller Institute in 1915 studying nutrition and evolution in *Drosophila*, and then moved to the kinetics of enzyme reactions and the properties of bacterial suspensions. But by 1930, with his success in crystallizing pepsin, he became almost completely caught up in biochemical problems. Hecht gradually molded himself to his identity as a biophysicist, and, as Kohler notes, Cohn also gradually "lost his biological ends in pursuit of more and better chemical means."[36]

Given their differences in outlook and difficulties in gaining recognition it is not surprising that general physiologists were an unusually contentious group with little disciplinary solidarity. They formed no society in the 1920s, and the *Journal of General Physiology* remained a house organ for Loeb and his followers; its exclusivity was in sharp contrast to the inclusive publication policy of the *American Journal of Physiology*.[37] The field stabilized only in the 1940s as a discipline devoted to bioenergetics and the physical chemistry of membrane phenomena, especially as studied in marine organisms. By that time the students of the 1920s generation were mature, and a number were established in academic departments; with the World War II expansion of funding, they believed they no longer needed to depend on medical physiologists, who—according to Heilbrunn—looked down on them and made them feel unwelcome. In 1946 scientists centered around the Marine Biological Laboratory organized the Society of General Physiologists.[38]

The eventual formation of a discipline of general physiology, although important, is historically less interesting than the earlier, more fluid, situation. In those decades the most notable phenomenon was lack of discipline. The major promoters of general physiology came from outside the field—the morphologist Whitman sponsored it at Chicago and the Marine Biological Laboratory as part of his program for an integrated science of biology; the bacteriologist Flexner introduced it at the Rockefeller Institute and promoted it through the Rockefeller philanthropies in his effort to build up the conceptual foundations of medical science. The practitioners of general physiology were indifferent to such aims; they tended to be individualists who had little interest in fixing an academic identity. Loeb moved back and forth between biology, physiology, and eventually chemistry, as he shifted from problem to problem. Lee and Mathews, although less successful, had similarly varied interests. Northrop, a generation younger, was described by Paul De Kruif as "a scientific Lao-Tse"; at the age of thirty he rejected the Columbia physiology professorship because he preferred a life of research and hunting. The "leading" general physiologists were outsiders both intellectually and professionally, glad to find space in enclaves that protected their independence, even if it was defined in increasingly narrow terms.[39]

In the 1920s, however, this lack of discipline began to take a specific direction. The central figure was Crozier, head of Harvard's Laboratory of General Physiology for a decade beginning in 1925. Crozier, in contrast to his predecessors, was passionate about building general physiology; but for him, as for his protégés, the discipline was defined not by subject, theory, or problem, but by actions and "attitudes." He developed general physiology as a model "undiscipline."[40]

Crozier was a poor New Yorker who worked his way through City College examining fish, leaving as his yearbook quotation, "What are literary societies good for anyway?" He studied with Parker and Osterhout at Harvard in the mid-1910s, then rapidly moved up through aggressive research and self-promotion at Illinois, Chicago, and Rutgers, to become head of general physiology at Harvard while still in his early thirties. He immediately antagonized medical school physiologists such as Hallowell Davis with "ungentlemanly" behavior at their seminars; he refused to associate with Henderson, and he also put off most Harvard zoologists. Yet he was the central figure in the university's multimillion dollar Institute of Biology, and between 1927 and 1933 was one of the leading Harvard sponsors of National Research Fellows. He gathered a group of young men that included Gregory Pincus, Hudson Hoagland, Robert Emerson, Edward S. Castle, C. Stacy French, William A. Arnold, T. Cunliffe Barnes, and B. F. Skinner.[41]

The subjects of investigation at the Harvard Laboratory of General Physiology ranged from Crozier's own work on temperature characteristics and enzyme systems to tropisms, photosynthesis, mammalian artificial parthenogenesis, and behavior. What united the department was age (all were under forty), passion for experimentation, and lack of discipline—both professional and, to some extent, personal. They embraced intellectual flux and the cult of scientific genius, and some were personally insufferable. French described the department as a group of "near-amateurs with enthusiasm and interest in a wide range of disciplines." Crozier defended the "intellectual enthusiasm" that "of course makes errors" against "the dull stagnation" that "finds itself in the end doing nothing new." Skinner recalled that he was introduced by a friend as the most conceited person in Cambridge; Hoagland later explicitly compared the arrogance of the laboratory members to that of the molecular biologists of the 1950s.[42]

Crozier was aware from the time of his arrival at Harvard that his institutional position was precarious. He may also have known that his intellectual standing was equally unstable; by the late 1930s he had a reputation as a person who manipulated data, and he did no work of real significance from 1935 to his death twenty years later. The chemist James Bryant Conant, as soon as he became Harvard president in 1933, destroyed Crozier's "empire." He disapproved of Crozier both personally and intellectually, but probably would have eliminated general physiology on organizational grounds, as most other administrators were doing.[43]

Crozier saw general physiology as a way to run an undisciplined, freewheeling scientific operation within the modern framework of academic disciplines. He was able to encourage students—most notably the behaviorist Skinner, and Pincus, developer of the oral contraceptive—who would become some of the most original and provocative life scientists of their generation. General physiology was thus significant as perhaps the first of many transitory structures that have protected undisciplinary scientific activity in this century. Such structures are part of the mythology—and to some extent the public policy—of modern scientific creativity. James D. Watson's *The Double Helix* is the best-known defense of such institutional indefiniteness. While general physiologists did not lead in a direct intellectual line to molecular biologists, they pioneered the form of scientific behavior associated with the double helix.

NOTES

1. Thomas S. Hall, *History of General Physiology*, 2 vols. (Chicago, IL: Univ. of Chicago Press, 1975). Originally published as *Ideas of Life and Matter*, 2 vols. (Chicago, IL: Univ. of Chicago Press, 1969).
2. Robert E. Kohler, *From Medical Chemistry to Biochemistry: The Making of a Biomedical Discipline* (Cambridge, UK: Cambridge Univ. Press, 1982), pp. 108–114, 286–323. Paul F. Cranefield pointed out the peculiar historical position of general physiology in his provocative "The Organic Physics of 1847 and the Biophysics of Today," *J. Hist. Med. Allied Sci.* 12 (1957): 416.
3. Kohler, *Medical Chemistry*, pp. 306, 322–323.
4. Timothy Lenoir, *The Strategy of Life: Teleology and Mechanics in Nineteenth Century German Biology* (Dordrecht, the Netherlands: Reidel, 1982).
5. Ibid.; Rudolph Wagner, *Lehrbuch der speziellen Physiologie*, 3d ed. (Leipzig, Germany: Voss, 1845); Claude Bernard, *Leçons sur les Phénomènes de la Vie Communs aux Animaux et aux Végétaux*, 2 vols. (Paris: Baillière, 1878–1879); Harry Paul, *From Knowledge to Power: The Rise of the Science Empire in France, 1860–1939* (Cambridge, UK: Cambridge Univ. Press, 1986), pp. 93–97.
6. Philip J. Pauly, "The Appearance of Academic Biology in Late Nineteenth-Century America," *J. Hist. Biol.* 17 (1984): 378–382.
7. Ibid., pp. 387–389; on Frederic Schiller Lee's later career at Columbia see chapter III by Alejandra Laszlo in this book. On the delicate questions raised by the interest of women in biology see Margaret Rossiter, *Women Scientists in America: Struggles and Strategies to 1940* (Baltimore, MD: Johns Hopkins Univ. Press, 1982), pp. 73–99.
8. Charles Otis Whitman to William Rainey Harper, n.d. [1892?], Presidents' Papers 18: 6, University of Chicago Archives, Chicago, IL; Whitman, *Report of the Director, Marine Biological Laboratory, 4th Session*, 1891, p. 16; idem, "General Physiology and its Relation to Morphology," *Am. Nat.* 27 (1893): 802–807. On Whitman see chapter VII by Jane Maienschein in this book.
9. On Jacques Loeb see W. J. V. Osterhout, "Jacques Loeb," *Biogr. Mem. Natl. Acad. Sci.* 30 (1930): 318–401; Pauly, *Controlling Life: Jacques Loeb and the Engineering Ideal in Biology* (New York: Oxford Univ. Press, 1987).
10. Pauly, *Controlling Life*, pp. 65–86. See, for example, the announcement in *Anat. Anz.* 7 (1892): 558.
11. Loeb to Nathan Zuntz, 23 September 1894, Dokumentensammlung Darmstädter, Handschriftenabteilung, Staatsbibliothek Preussischer Kulturbesitz, West Berlin, FRG. This letter was written in German, except for the American colloquialism "Pull."
12. Loeb, "The Biological Problems of To-day: Physiology," *Science* 7 (1898): 154–156.
13. Max Verworn, *Allgemeine Physiologie: Ein Grundriss der Lehre vom Leben* (Jena, Germany: Fischer, 1895); Loeb, "Einige Bemerkungen über den Begriff, die Geschichte und Literatur der allgemeinen Physiologie," *Arch. Gesamte Physiol. Mens. Tiere* (*Pfluegers*) 69 (1897): 266–267. On the implications of the cell-state concept see Paul Weindling, "Theories of the Cell State in Imperial Germany," in *Biology, Medicine, and Society, 1840–1940*, ed. Charles Webster (Cambridge, UK: Cambridge Univ. Press, 1981), pp. 99–155.
14. Loeb, "Bemerkungen," 249, 259; idem, "Biological Problems," p. 155.
15. See typescripts of fourteen lectures, dated January through March 1897, in Jacques Loeb Papers, Manuscript Division, Library of Congress, Washington, DC, box 59, file "lecture

material" (hereafter, Loeb Papers); Loeb to Charles Benedict Davenport, 17 February 1897, Charles Benedict Davenport Papers, American Philosophical Society Library, Philadelphia, PA.

16. Approximately one-fourth of Loeb's *The Dynamics of Living Matter* (New York: Macmillan, 1906) discussed the physicochemical properties of protoplasm; the rest summarized the role of electrolytes in tissues, then discussed tropisms, fertilization, heredity, and regeneration. These were the subjects on which Loeb had worked since the late 1880s; while united by his interest in physical chemistry and experimental control, they were not explicitly related to his definitions of general physiology, and they certainly did not provide the basis for a systematic account of a discipline.
17. Loeb to Franklin Paine Mall, 2 February 1898, Franklin Paine Mall Papers, Carnegie Institution of Washington, Department of Embryology Deposit, Allen Mason Chesney Medical Archives, the Johns Hopkins University School of Medicine, Baltimore, MD.
18. Mall to Simon Flexner, 31 October 1902, Simon Flexner Papers, American Philosophical Society; on the Nobel nominations see table in Claire Salomon-Bayet, "Bacteriology and the Nobel Prize Selections, 1901–1920," in *Science, Technology, and Society in the Time of Alfred Nobel* (Oxford, UK: Pergamon, 1982), pp. 392–400.
19. Mall to Loeb, 23 January 1904, Loeb Papers. This action was apparently one of the stimuli for the 1904 revolt, led by Samuel J. Meltzer, against Russell Chittenden's leadership of the American Physiological Society. See Mall to Loeb, 9 April 1905, Loeb Papers. Lee, who had translated Verworn's book, abandoned general physiology around this time to study fatigue. See chapter X by Richard Gillespie in this book.
20. See, for instance, his address at the opening of his laboratory, "The Limitations of Biological Research," *Univ. Calif. Publ. Physiol.* 1 (1903): 33–37. His collection of early papers, *Studies in General Physiology*, 2 vols. (Chicago, IL: Univ. of Chicago Press, 1905), had been in press since 1902.
21. Kohler, *Medical Chemistry*, pp. 149–154.
22. Mall to Loeb, 23 May, 6 November, 23 November 1909, Loeb Papers; Loeb to Anne Loeb, 6 July, 8 July, 11 July 1909, Loeb Papers.
23. Loeb to George Howard Parker, 11 March 1918, Loeb Papers.
24. Loeb to Thomas Hunt Morgan, 29 September 1917, Loeb Papers.
25. The prospectus Loeb originally sent to Flexner fluctuated in reference between a "Journal of Physico-chemical Biology" and "Journal of General Physiology"; he later noted that he adopted "general physiology" at the urging of "friends"—almost certainly Flexner. See Loeb to Flexner, 18 April, 16 May, 24 May, 27 May 1918, Simon Flexner Papers, American Philosophical Society; Loeb to Filippo Bottazzi, 28 December 1920, Loeb Papers.
26. Loeb to Flexner, 7 December 1921, Loeb Papers.
27. General physiology was a good example of the use of a few European precedents—William Bayliss's *Principles of General Physiology* and the status of individual German scholars such as Otto Warburg and Otto Meyerhof—to justify a major American effort.
28. Kohler, *Medical Chemistry*, p. 322.
29. Herbert Spencer Jennings to Raymond Pearl, 23 October 1909, 23 January 1910, Herbert Spencer Jennings Papers, American Philosophical Society; Jennings to Edwin Grant Conklin, 18 July 1909, Edwin Grant Conklin Papers, Firestone Library, Princeton University, Princeton, NJ.
30. Sharon Kingsland, "Raymond Pearl: On the Frontier in the 1920's," *Hum. Biol.* 56 (1984): 15; idem, "A Man Out of Place: Herbert Spencer Jennings at Johns Hopkins, 1906–1938," *Am. Soc. Zool.* (unpublished paper, December 1985).
31. Lee to Nicholas Murray Butler, 14 October 1919, Columbia University Central Files, Low Library, Columbia University, New York, NY; W. J. Crozier to Selig Hecht, 22 March 1926, Selig Hecht Papers, Rare Book and Manuscript Library, Columbia University (hereafter, Hecht Papers); Columbia University College of Physicians and Surgeons, *Announcements*, 1920–1930; H. B. Williams, "Physiology," typescript, 4 p., June 1926, Dean's Office Files, Columbia University College of Physicians and Surgeons, New York, NY.
32. Crozier to Osterhout, "Friday" [June 1928], W. J. V. Osterhout Papers, American Philosophical Society; Lawrence R. Blinks to Pauly, 15 December 1985.
33. John F. Fulton, *Physiology* (Clio Medica V) (New York: Paul Hoeber, 1931), pp. 110–111; see Cranefield, "Organic Physics," 410–411. On human science at Yale in the interwar years see

J. G. Morawski, "Organizing Knowledge and Behavior at Yale's Institute of Human Relations," *Isis* 77 (1986): 219–242.

34. Albert P. Mathews, "The Mechanistic Conception of Life," *Scientia* 5 (1924): 242–252; Ralph S. Lillie, "The Organism as a Whole, From a Physicochemical Viewpoint," (review) *J. Am. Chem. Soc.* 39 (1917): 524; idem, *General Biology and Philosophy of the Organism* (Chicago, IL: Univ. of Chicago Press, 1945); Lewis V. Heilbrunn, *An Outline of General Physiology*, 3d ed. (Philadelphia, PA: Saunders, 1952), p. 7.
35. Loeb to Svante Arrhenius, 8 January 1923, Loeb Papers.
36. Roger M. Herriott, "A Biographical Sketch of John Howard Northrop," *J. Gen. Physiol.* 45 suppl. (1962): 1–15; Kohler, *Medical Chemistry*, p. 315.
37. Pauly's interview with John Howard Northrop, 25 July 1977. Neither Blinks nor C. S. French, students in general physiology in the 1920s, recalled strong attachment to identification as general physiologists at that time; they considered the subject equivalent to biophysics, physicochemical biology, or cell physiology. Blinks to Pauly, 15 December 1985; French to Pauly, 15 May 1986.
38. The "loneliness" of general physiologists in the American Physiological Society resulted apparently from the organization of Society meetings around organ systems. See Heilbrunn to W. O. Fenn, 23 August 1945; Council Minutes, 12 March 1946; Meeting of the Committee on Training of Physiologists, 2 November 1946, all in APS Minute Books, vols. 12 and 13, American Physiological Society, Bethesda, MD. I am grateful to Toby Appel for these references.
39. On Mathews see S. S. Cohen, "Some Struggles of Jacques Loeb, Albert Mathews, and Ernest Just at the Marine Biological Laboratory," *Biol. Bull. Woods Hole* 168, suppl. (1985): 127–136. On Northrop see Paul DeKruif, *The Sweeping Wind* (New York: Harcourt, Brace, & World, 1962), p. 32; Herriott, "Northrop"; Jacques Loeb to Leonard B. Loeb, 19 December 1921, 28 April 1922, Loeb Papers; "Report of the Committee on the Chair of Physiology" (typescript), 3 pp., June 1922, Dean's Office Files, Columbia University College of Physicians and Surgeons, New York, NY.
40. Gregory Pincus to Crozier, 12 November 1929, William John Crozier Papers, Harvard University Archives, Cambridge, MA.
41. Hudson Hoagland and R. T. Mitchell, "William John Crozier: 1892–1955," *Am. J. Psychol.* 69 (1956): 135–138; *Microcosm* (City College of New York Yearbook), 1913, p. 42; Crozier to Osterhout, Friday [June 1928], Parker to Osterhout, 6 June 1928, Osterhout Papers, American Philosophical Society; Pauly's interview with B. F. Skinner, 16 July 1985, Cambridge, MA; Pauly, *Controlling Life*, pp. 183–200.
42. French, "Fifty Years of Photosynthesis," *Annu. Rev. Plant Physiol.* 30 (1979): 1–2; Crozier to W. B. Cannon, 11 April 1940, Walter Bradford Cannon Papers, Harvard Medical Archives, Francis A. Countway Library of Medicine, Harvard Medical School, Boston, MA; Skinner, *The Shaping of a Behaviorist* (New York: Knopf, 1979); Hoagland, *The Road to Yesterday* (Worcester, MA: privately printed, 1974), pp. 60–62.
43. Pincus to Crozier, 12 February 1938, Crozier Papers; Ernest Wolf to Hecht, 27 October 1937, Hecht Papers; see James Reed, *The Birth Control Movement and American Society: From Private Vice to Public Virtue* (Princeton, NJ: Princeton Univ. Press, 1983), pp. 321–326.

IX

Pathologists, Clinicians, and the Role of Pathophysiology

RUSSELL C. MAULITZ

By the last quarter of the nineteenth century, most members of the American medical elite could point with admiration to Rudolf Virchow and his school of cellular pathology. One appealing feature of the Prussian school of pathology, as I have suggested elsewhere, was Virchow's program for the alignment of pathology and physiology. The great Berlin pathologist sought, if not to fuse the two disciplines, then certainly to bring them into closely contiguous relation.[1]

By the 1890s an emerging American stream of voices began to advocate a variant of this same program. In order to advance, went the argument, pathology must cast off its anatomical shackles and begin joining the experimental sciences. Who claimed title to the idea of physiological pathology and the experimental pathology that represented the idea put into practice? At the turn of the century it seemed like pathologists, in concert with physiologists in the American Medical Association (AMA), had laid claim to it. Indeed, a few individuals at elite institutions—men like William MacCallum at Johns Hopkins and Richard Pearce at Pennsylvania—actually began courses in the subject.

However, two sets of forces intervened to divert pathologists from the experimental model. First, the pathologists themselves grew increasingly preoccupied elsewhere: with hospital posts and hence hospital colleagues' demands; with the morphological or anatomical model that correlated cognitively with those demands; and with the problem of what to do with bacteriology, a daughter science that increasingly threatened to dominate its parent. Second, a new group, the clinical investigators, emerged around the time of World War I to mount their own program. That program amounted, in effect, to one of experimental pathology, and served to siphon off the effort and interest that accompanied such activity.

PATHOLOGY AND PHYSIOLOGY AT THE TURN OF THE CENTURY

Interestingly, the early rhetoric of physiological pathology—unlike, say, the program of clinical research that was about to emerge from the Boston-Washington medical elite—was not primarily a creation of the eastern establishment. In 1894, for example, J. Clark Stewart, pathology professor in the medical department of the University in Minneapolis, was among the first to enunciate an expansive program for American pathology. For the proper education of medical students, he declared, pathologists should bring within their grasp related subjects such as clinical microscopy, bacteriology, and "morbid physiology, a most important part of the subject and one much neglected, which treats of the alterations of function caused by disease, with their origin and results."

However, Stewart admitted the wisdom of a certain division of labor. The hospital or faculty pathologist might deal with pathological anatomy, when practicing morbid dissection and hence when teaching, while the "instructor in [clinical] practice. . .can profitably avoid entirely histologic changes and many of the details of the morbid anatomy, concentrating his teaching on what I have included under morbid physiology." [2] If pathologists should do all these things, Stewart continued, the ancient study of *functio laesa* would continue to be the province of the clinician. It was the clinician, after all, who could correlate the symptom with the underlying disturbance. As to the division of labor, Stewart averred, it was often "difficult to draw a line between the teachings of the chairs of practice and pathology, so much so that in some of our best schools the entire didactic work of the two subjects is combined. This avoids repetition, but throws an immense amount of work on the chair [of medicine]." [3]

Stewart's words were both prophetic and naive. His conception of the interlocking roles of clinician and pathologist was soon to be abandoned by other players, as the rules of the game hardened and as pathologists increasingly tried to bring into their fold such subfields as bacteriology and physiological pathology.[4] Instead, what characterized early careers in pathology was what one might call the penultimate career pattern. Basically, although important exceptions can be found, the chair of pathology, or the post of hospital pathologist, was more often than not a way station to the more lucrative and more prestigious clinical posts. Reginald Fitz and Henry Christian of Boston, or William Osler of Montreal and points south, represent exemplars of the pattern.

At the turn of the century, it seemed natural for American pathologists, many aiming for clinical posts, to insist on a physiologically oriented pathology. Such a coherent science of disturbed function could take its rightful position alongside "the other branches of pathology." In 1900 Isaac Levin of New York deemed it the "science which analyzes the observations of the clinic and pathological anatomy and chemistry, and verifies by experiment the conclusions derived from these observations." [5] Like other early proponents of an experimental pathology allied with clinical investigation—many of them trained in Europe—Levin thus seemed to give physiological pathology pride of place as a sort of governor of the other branches, namely clinical pathology, pathological anatomy, and bacteriology.[6]

Physiological pathology, insisted Levin, had its own canons of evidence and analysis, with "methods of research. . .different from those of the other medical sciences," thus bestowing on it "a fundamental importance for medicine as well as for biology." [7] Levin went on to cite a variety of examples supporting his contention, among them the instances of the disruption of circulation in cardiac hypertrophy, the function of inflammatory cells in immunity, the accumulation of carbon dioxide in respiratory disorders, and the effects of toxic substances on digestion.

THE AMERICAN MEDICAL ASSOCIATION

Probably the most important spark for attempts to institutionalize physiological pathology at the turn of the century came, however, from the leaders of the new Section on Pathology and Physiology of the then reform-minded AMA. The section was begun almost by accident through a series of events that culminated in the 1901 merger of two earlier groups, pathology and bacteriology on the one hand, and physiology and dietetics on the other. The new, doubly fortified group thought of itself as a reforming elite and its meeting as the scientific *sanctum sanctorum* of the AMA. It took responsibility for the annual scientific exhibit of the AMA, a traveling "poster session" that moved by rail around the United States displaying rare and unusual pathological material.

As Figure 1 shows, the section's early leadership was drawn from both pathology and physiology. Pathologist Joseph McFarland can be found cheek by jowl with physiologists Walter Cannon and Yandell Henderson. Equally

SECTION ON PATHOLOGY AND PHYSIOLOGY

1901	Chairman, F. B. Wynn, Indianapolis. Secretary, J. McFarland, Philadelphia. Delegate, W. A. Evans, Chicago.
1902	Chairman, Victor C. Vaughan, Ann Arbor, Mich Secretary, J. McFarland, Philadelphia. Delegate, E. R. LeCount, Chicago.
1903	Chairman, J. McFarland, Philadelphia. Secretary, H. A. Christian, Boston. Delegate, A. P. Ohlmacher, Detroit.
1904	Chairman, Winfield S. Hall, Chicago. Secretary, H. A. Christian, Boston. Delegate, J. McFarland, Philadelphia.
1905	Chairman, H. A. Christian, Boston. Secretary, Walter L. Bierring, Iowa City. Delegate, J. McFarland, Philadelphia.
1906	Chairman, Walter L. Bierring, Iowa City. Secretary, Walter B. Cannon, Boston. Delegate, Winfield S. Hall, Chicago.
1907	Chairman, Walter B. Cannon, Boston. Secretary, M. J. Rosenau, Washington, D. C. Delegate, W. M. L. Coplin, Philadelphia.
1908	Chairman, M. J. Rosenau, Washington, D. C. Secretary, H. Gideon Wells, Chicago. Delegate, H. A. Christian, Boston.
1909	Chairman, H. Gideon Wells, Chicago. Secretary, Yandell Henderson, New Haven, Conn. Delegate, Walter B. Cannon, Boston.
1910	Chairman, Yandell Henderson, New Haven, Conn. Secretary, Leo. Loeb, St. Louis. Delegate, Walter L. Bierring, Iowa City
1911	Chairman, Leo Loeb, St. Louis. Vice Chairman, William Ophüls, San Francisco. Secretary, A. W. Hewlett, Ann Arbor, Mich. Delegate, E. R. LeCount, Chicago.
1912	Chairman, A. W. Hewlett, Ann Arbor, Mich. Vice Chairman, A. J. Carlson, Chicago. Secretary, William Ophüls, San Francisco. Delegate, Yandell Henderson, New Haven, Conn.
1913	Chairman, William Ophüls, San Francisco. Vice Chairman, L. B. Wilson, Rochester, Minn. Secretary, A. J. Carlson, Chicago. Delegate, Richard M. Pearce, Philadelphia.
1914	Chairman, A. J. Carlson, Chicago. Vice, Chairman, L. B. Wilson, Rochester, Minn. Secretary, Frederick P. Gay, Berkeley, Calif. Delegate, A W. Hewlett, Ann Arbor, Mich.

LIST OF OFFICERS—Continued

1915	Chairman, Frederick P. Gay, Berkeley, Calif. Vice Chairman, James Ewing, New York. Secretary, Isabella C. Herb, Chicago. Delegate, A. W. Hewlett, Ann Arbor, Mich.
1916	Chairman, James Ewing, New York. Vice Chairman, C. W. Green, Columbia, Mo. Secretary, Isabella C. Herb, Chicago. Delegate, H. Gideon Wells, Chicago.
1917	Chairman, Louis B. Wilson, Rochester, Minn. Vice Chairman, Francis Carter Wood, New York. Secretary, Isabella C. Herb, Chicago.
1918	Chairman, Francis Carter Wood, New York. Vice Chairman, Isabella C. Herb, Chicago. Secretary, J. J. Moore, Chicago. Delegate, E. R. LeCount, Chicago.
1919	Chairman, Howard T. Karsner, Cleveland. Secretary, J. J. Moore, Chicago. Delegate, E. R. LeCount, Chicago.
1920	Chairman, E. R. LeCount, Chicago. Vice Chairman, William S. Carter, Galveston, Texas. Secretary, J. J. Moore, Chicago. Delegate, James Ewing, New York.
1921	Chairman, John A. Kolmer, Philadelphia. Secretary, J. J. Moore, Chicago. Delegate, D. J. Davis, Chicago.
1922	Chairman, Arno B. Luckhardt, Chicago. Vice Chairman, Kenneth M. Lynch, Dallas, Texas. Secretary, J. J. Moore, Chicago. Delegate, D. J. Davis, Chicago.
1923	Chairman, Kenneth M. Lynch, Dallas, Texas. Vice Chairman, C. W. Greene, Columbia, Mo Secretary, J. J. Moore, Chicago. Delegate, D. J. Davis, Chicago.
1924	Chairman, Charles W. Greene, Columbia, Mo. Vice Chairman, Harry J. Corper, Denver. Secretary, J. J. Moore, Chicago. Delegate, D. J. Davis, Chicago.
1925	Chairman, H. J. Corper, Denver. Vice Chairman, R. G. Hoskins, Columbus, Ohio. Secretary, J. J. Moore, Chicago. Delegate, D. J. Davis, Chicago.

Figure 1: *Early leaders in the Section on Pathology and Physiology.*

clear is the fact that the leadership was an amalgam of eastern establishment gentlemen—individuals who would have felt at home with the swells of the Interurban Clinical Club—and earnest midwesterners like Frank B. Wynn. Indeed, the traveling exhibit was Wynn's pet project, an activity viewed by many as the linchpin of the section in its earliest years.[8]

In June of 1904 the secretary of the AMA Pathology and Physiology Section, McFarland of Philadelphia, discussed its goals at an annual AMA meeting in Atlantic City. He saw it primarily as an opportunity to create a common meeting ground between separate disciplines. A year later, however, the Chicago physiologist Winfield Hall articulated a much grander vision, lamenting the fact that "one searches the chapters of general pathology in vain for any sections that deal adequately with the problem of the physiology of disease." Hall defined the problem of physiological pathology—i.e., who should do it?—as well as anyone before or since:

> The question might be asked, "Why should not the physiologist cover this field and present to the students normal and pathologic physiology?" We will say in answer to that question: The physiologist presents normal physiology during the first two years of the medical course. The students have not progressed far enough in the study of pathologic conditions to enter understandingly into a consideration of the relation between symptomatology and pathology. Thus we find that important as this field is, neither the physiologist nor the pathologist—one for the lack of preparation of his students, the other for the lack of time—can take up this most important subject. The student therefore enters on the study of clinical medicine and surgery with a knowledge of normal structure and function, and also a knowledge of diseased structure, but no knowledge of diseased function.[9]

Who, therefore, asked Hall, should take responsibility for this "very important phase of pathology and physiology [that] has as yet received scant attention from medical teachers"? It mattered not, he answered, whether such a person hailed from pathology, physiology, or medicine, as long as he was versed in all three. He cited Ludolph Krehl's (1904) textbook, *Pathologische Physiologie*, just then being translated by Albion Hewlett of San Francisco, another section stalwart who later moved on to Ann Arbor. The Krehl textbook soon reached a wide American audience in Hewlett's English version.[10]

The Section on Pathology and Physiology lasted until the mid-1930s, publishing its final *Transactions* volume ca. 1931. After World War I, its role was in large measure assumed by the more explicitly structured Federation of American Societies for Experimental Biology. As a temporary alliance of disciplines, the AMA Section on Pathology and Physiology had achieved considerable success. What may seem particularly remarkable was the lack of conflict between the physiologists and the pathologists, although one should recall that careers still possessed a high degree of plasticity. Thus both Christian of Boston, speaking in 1905–1906 on the "phases of recent work in pathology and physiology," and Walter Bierring of Iowa City, speaking a year later on "the need [for] medical cooperation in laboratory work," could sound the same note of sweet reason. There was much work for everyone, they argued, with new demands daily placed on section members by government commissions and, increasingly, by privately endowed laboratories in the major cities. Both Christian and Bierring were in fact already leaning toward a sort of tacit separate-but-equal arrangement between pathology and physiology. Both also proceeded from their pathology chairs to the professorships of clinical medicine, suggesting that each may have seen the laboratory, whether pathological or physiological, as a way station to higher positions.

Christian and Bierring stressed above all the importance of physical and intellectual proximity between hospital and laboratory.[11] Based on the German model, this idea of proximity needed (according to Bierring) to be reflected in the general orientation of the clinical chief, with "his broad grasp of the bearing of pathology, physiology and chemistry on clinical medicine [and who] has associated about him a corps of laboratory workers." Bierring claimed in 1907 that

> it is but a matter of recent memory when our entire conception of pathology consisted of the study of morbid anatomy or histologic lesions. As the scientific viewpoint of pathology has enlarged[,] physiology and physiologic [sic] chemistry have been accorded their due recognition as most important factors in advancing the knowledge of the nature of disease processes, the latter, in fact, having largely overshadowed the former. It is a further significant fact that in a process like diabetes with an unknown morbid anatomy our knowledge of the pathologic physiology. . .is more complete than in many disease conditions with definite lesions.[12]

When Bierring spoke of clinicians with the "broad grasp," and of morbid anatomy as newly "overshadowed" by physiological chemistry, he was undoubtedly pursuing another, parallel agenda. Something new was clearly afoot—not, I would contend, a new subdiscipline of pathology or physiology, but rather the premonition of a new impulse, that of clinical investigation. Adumbrated by the inveterate and indefatigable society-maker Samuel Meltzer, and clearly defined in the late 1910s and 1920s, clinical investigation was more than just a new activity. It was a new twist to an old ideology: science at the bedside.[13]

In a 1904 address delivered at the dedication of a new laboratories building at the University of Pennsylvania, George Dock, writing from Ann Arbor but soon to become the avatar of medical educational reform in St. Louis, tackled the issue of the proper relations between the disciplines. For Dock, the advent of experimental pathology in American institutions had been a proper corrective to "some of the errors and excesses of pathological anatomy." Traditional morbid anatomy had neglected the functional study now being pursued by the Meltzerians, leading to "such erroneous conceptions as that which made pneumonia 'lung fever,' and typhoid an 'enteric fever.'" Yet morbid anatomy had also led to much that was good in "the critical examination of alleged results," and deserved to be fostered with an adequate supply of its own teachers ("nothing but participation in knowledge will prevent the rusting of the faculties").[14]

For Dock, indeed, the fundamental source of new knowledge lay not at the crossroads of physiology and pathology but in morbid anatomy. His own program represented a research version of the "Cabot exercises" soon to be initiated by Richard Cabot and Cannon as the clinicopathological exercises of the Massachusetts General Hospital. For the clinician there was the "great need. . .of giving hospitals. . .means of research—laboratories, apparatus, and trained workers—to have a sufficiently large corps of clinicians to carry on modern methods of investigating and treating the sick." Pathology was beset not so much by a lack of physiological inquiry as by the inattention to postmortem dissection, "the neglect of pathological anatomy." [15]

In Dock's view, pathological anatomy would continue to be the "bridge spanning the gap between the fundamental sciences and clinical medicine." [16] And indeed he had put his finger on the feature that most clearly distinguished pathology from the other laboratory sciences of medicine: the desire to be both "basic" and the bridge to the "clinical." (Contrast any statement about the relation between pathology and the clinical with Henderson's view of the isolation of physiology: while pathology and bacteriology had successfully forged links with the clinical, physiology had "led a life of mere self-development—like a proud virgin, beautiful, cold and sterile.")[17]

MAY 1913: THE AMERICAN SOCIETY FOR EXPERIMENTAL PATHOLOGY

Some pathologists disagreed with Dock's emphasis on anatomical pathology. They saw functional studies as a key link between their field and the other basic sciences. Hence by World War I, physiological pathology was already in an ambiguous position. Who was to conduct its research and teaching? Pathologists who wished to dabble in physiology? Experimentalists trained in pathology? Physiologists interested in derangements of function? Physiologically minded clinicians? In 1913 a small coterie of east coast reformers, mainly pathologists, made a conscious attempt to line up with the physiologists, pharmacologists, and biological chemists in order to give physiological pathology greater visibility. Fifteen men met in Washington in May of 1913 to consider a new constituent group, the American Society for Experimental Pathology of the larger Federation of American Societies for Experimental Biology.[18] Virtually the same group had been toiling for a decade and more to insure that physiological pathology maintained a presence in the new "Flexnerian" medical curricula.

As early as 1906, noting that "in a general way the lectures of Julius Cohnheim formed a model" for his course on pathological physiology at Johns Hopkins, MacCallum described the manner in which the study of disturbed function might be approached:

> In this [course of] study. . .it was planned to produce experimentally such lesions of the various organs as are commonly met with in the wards, and any available methods were used to this end. It may be remarked in passing that ideal uncomplicated lesions may be produced in this way which will illustrate the principle involved[,] often in a way seldom seen in clinical cases.[19]

At the University of Pennsylvania, Pearce undertook a similar course on experimental pathology, vowing to "fill the gap which exists between the autopsy room and the laboratory of pathological histology on the one hand and the clinic on the other." [20] As precedents he cited MacCallum's course, Cohnheim's "classical experiments" in Berlin and Breslau, Peyton Rous's course at Michigan (where the Baltimorean Rous served briefly as a pathology instructor), and the postgraduate physicians' demonstration courses at Columbia. Pearce held forth to ninety students while experimentally demonstrating processes like inflammation ("ear of rabbit exposed to water at 56° for three minutes—ear of rabbit 24 hours after rubbing with croton oil"); hydropericardium; and "'deglutition' [aspiration] pneumonia, due to foreign particles in bronchi." Pearce concluded that the "possibilities of enlarging [the course's] scope and of improving its presentation are numerous and will be evident to any one familiar with the methods of physiology and chemistry as applied to experimental pathology." [21]

However, courses like these were few and far between. What evidence I have found suggests that research followed suit: the surge of interest in physiological pathology that was apparent during the first few years of the century soon leveled off and actually lost ground to pathologists' several other pursuits.

PATHOLOGY AS REAL ESTATE

Figures 2 and 3 illustrate this relative loss of ground in terms of laboratory real estate. William Councilman's letter to Eugene Opie, dated 21 May 1910, provides a rough sketch of the pathology laboratories at Harvard. Two wings were separated by an amphitheater-library. One wing was devoted to teaching, the other to investigation. Of the major early departments of pathology, Harvard's was probably the least physiologically oriented, partly because of the interests of its early faculty and perhaps also because of the proximity of an extraordinarily strong physiology department. Within the pathology department, anatomy and bacteriology were the strong suits. (Of the department's early leaders only Burt Wolbach, in view of his work on vitamin deficiency states, is a partial exception.)[22]

The Harvard pathology laboratory was relatively compact; assistants and students labored in minuscule quarters in the research wing. There is not enough detail to establish or develop the suspicion that other fields began to outstrip physiological pathology.

What one needs to find, therefore, is an early example of the pathology lab "exploded out" into its subdisciplines. That happened, almost literally, at Johns Hopkins ten years later. Near the end of 1920 a fire destroyed much of the old, modest space in which Councilman had begun his academic pathology career back in the 1880s, shown in Figure 4. This had been the laboratory in which William H. Welch and MacCallum had built up the Hopkins department. After the fire, the remains of the building were razed and in mid-1922 a new one was begun thanks partly to a grant of $400,000 from the General Education Board of the Rockefeller Foundation. The plan of the structure finally opened in November 1923 is shown in Figure 3.[23]

The third and fourth floor plans of the new building are particularly instructive. There was space for both research and (in one large third floor classroom above the entrance hall) teaching in physiological pathology. Even here, however, the main business of pathology was anatomical and bacteriological.[24] As pathological physiology lost ground, bacteriology gained ever more space within the broad domain of pathology.

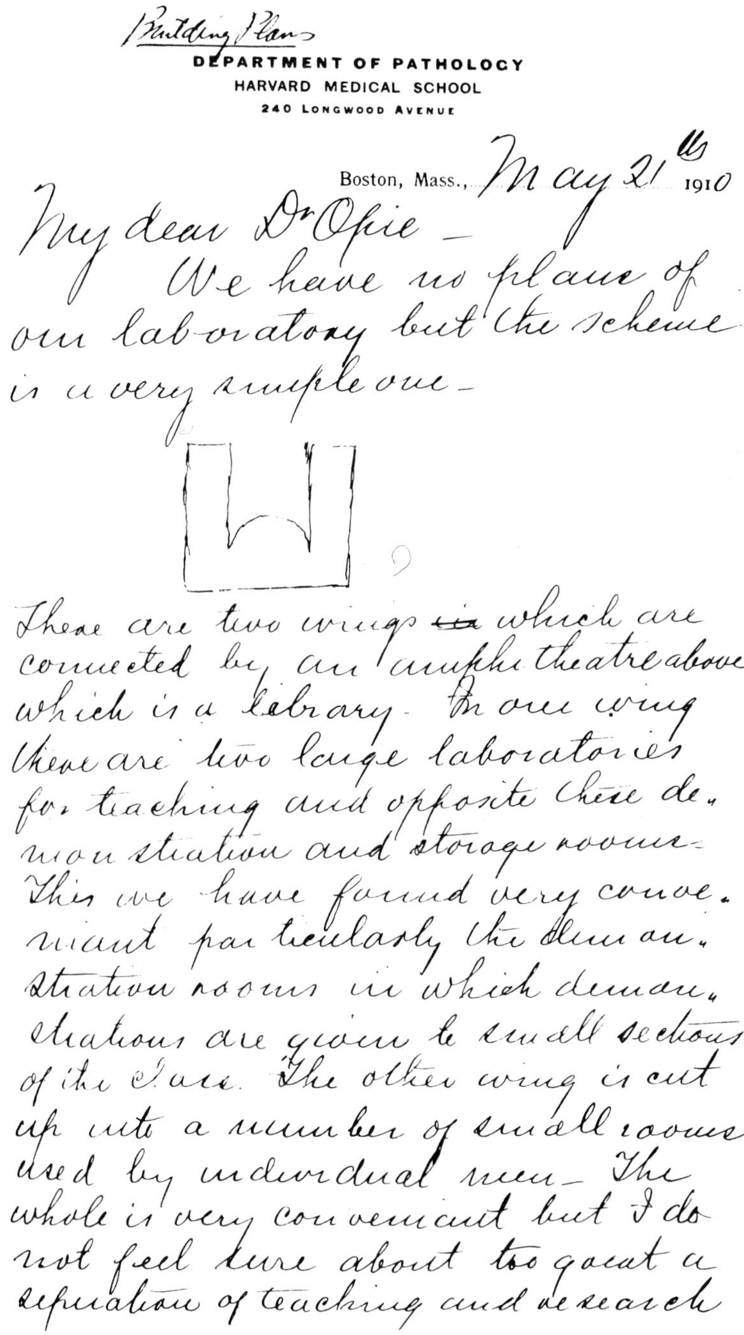

Building Plans

DEPARTMENT OF PATHOLOGY
HARVARD MEDICAL SCHOOL
240 LONGWOOD AVENUE

Boston, Mass., May 21st 1910

My dear Dr Opie —

We have no plans of our laboratory but the scheme is a very simple one —

There are two wings which are connected by an amphitheatre above which is a library. In one wing there are two large laboratories for teaching and opposite these demonstration and storage rooms. This we have found very convenient particularly the demonstration rooms in which demonstrations are given to small sections of the class. The other wing is cut up into a number of small rooms used by individual men — The whole is very convenient but I do not feel sure about too great a separation of teaching and research

Figure 2: *Letter, with figure, Councilman to Opie.*

THE TRIUMPH OF THE CLINICIANS

Between World Wars I and II, clinical investigators intensified their efforts to develop a science of disease that relied heavily on the methods and models

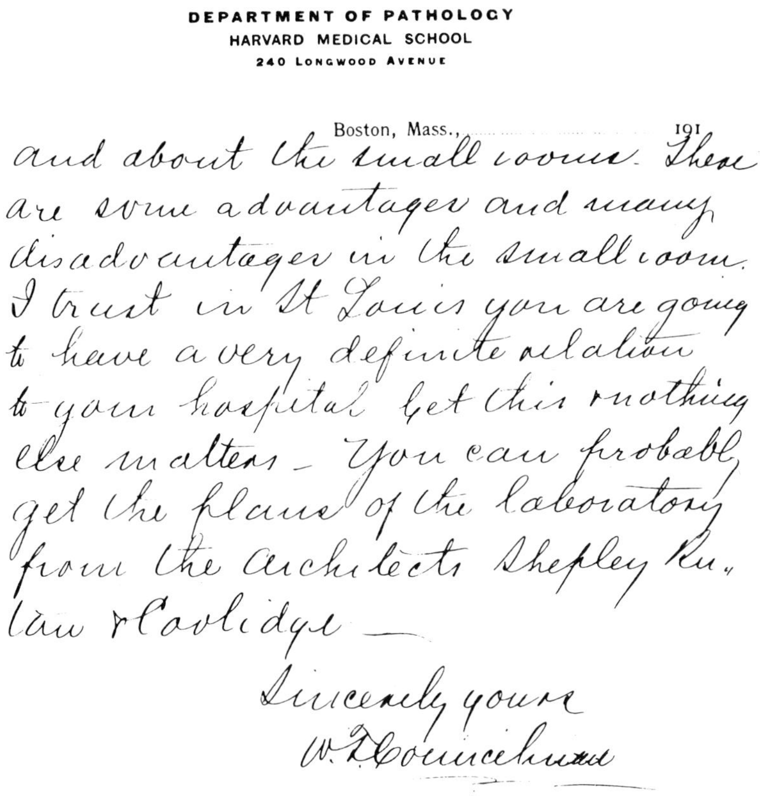

DEPARTMENT OF PATHOLOGY
HARVARD MEDICAL SCHOOL
240 LONGWOOD AVENUE

Boston, Mass., 191

and about the small rooms. There are some advantages and many disadvantages in the small room. I trust in St Louis you are going to have a very definite relation to your hospital. Get this nothing else matters. You can probably get the plans of the laboratory from the architects Shepley Rutan & Coolidge —

Sincerely yours
W. T. Councilman

Figure 2—*Continued*

of physiological pathology. These efforts paralleled those of the experimental pathologists and others perceived as doing "pure medical science." While morbid anatomists went in one direction (their activities fostered by the growing importance of general hospital settings) and the experimental pathologists went in another (fostered by the growing importance of specialized research centers), the practitioners of the new clinical science developed their own version of pathophysiological investigation.

Individuals at some remove from a Hopkins or a Rockefeller were fully aware of such developments. Their observations give meaning to the peculiar role of the pathologist as "bridge figure." Jacques Loeb's brother Leo, for example, writing in 1912 from the Skin and Cancer Hospital in St. Louis, told of how clinicians had become "dissatisfied with the attitude of the [pure laboratory] investigator. In the first place the practitioner feels that he does not receive sufficiently direct assistance in his work from the laboratory worker," giving rise to the "criticism that the [latter's] work is too theoretic [and] has not sufficient bearing on those aspects of disease [that] bring practical results." [25]

Who was to blame for this situation? According to Leo Loeb, "it is especially the pathologist against whom the criticism is directed." Get out of the research laboratory, he admonished the pathologist, and back to the microtome. "There is no doubt that the clinician needs assistance and that he himself is usually

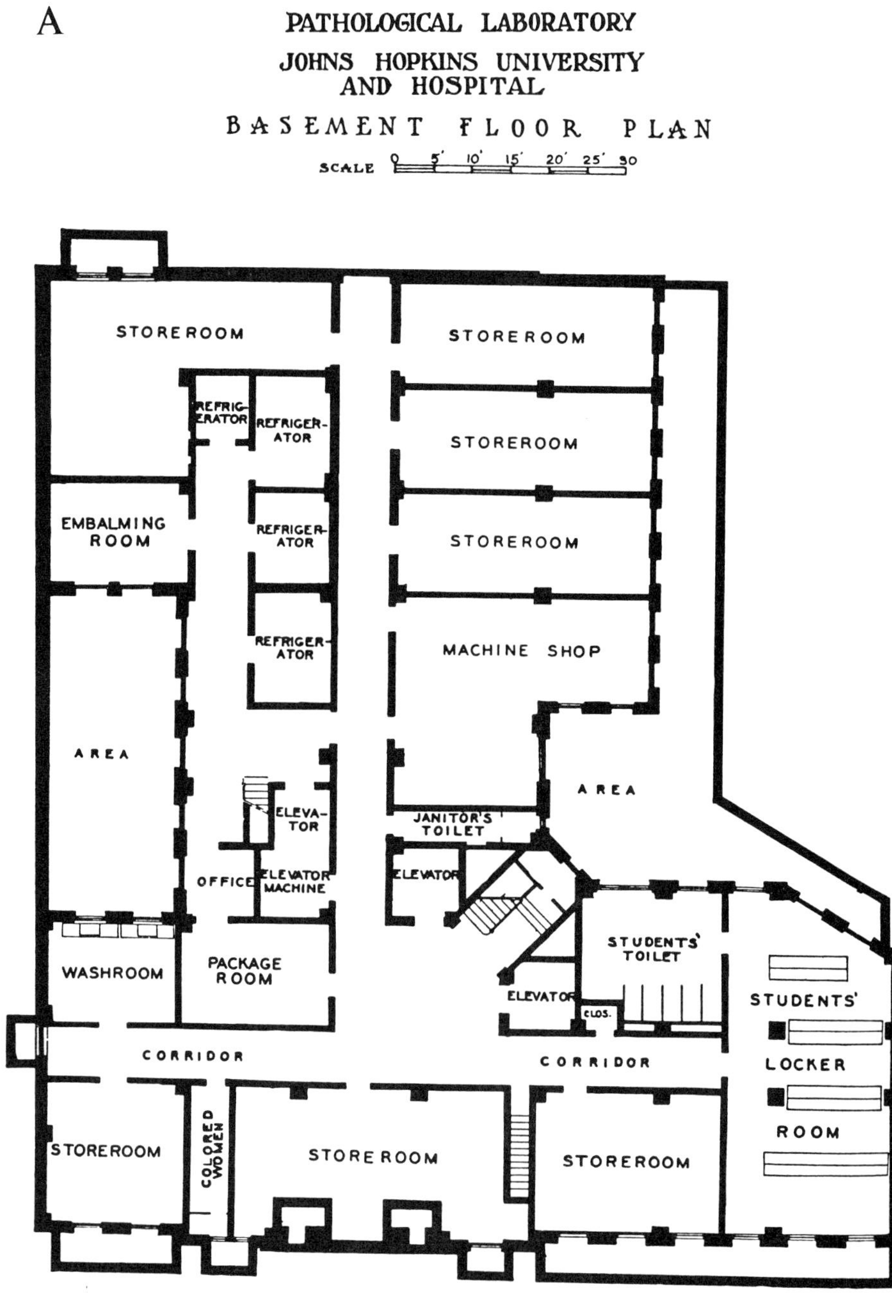

Figure 3A: *New pathological laboratory space at Johns Hopkins, 1920 [From MacCallum, 1925 (see n. 23).]*

not in a position to carry out certain pathologic and biochemic examinations which would aid him." [26] What of the work a few pathologists had undertaken in the experimental laboratory? "Under these conditions it seems, in general, preferable that this [work] be entrusted to special workers." [27] Who were these

B

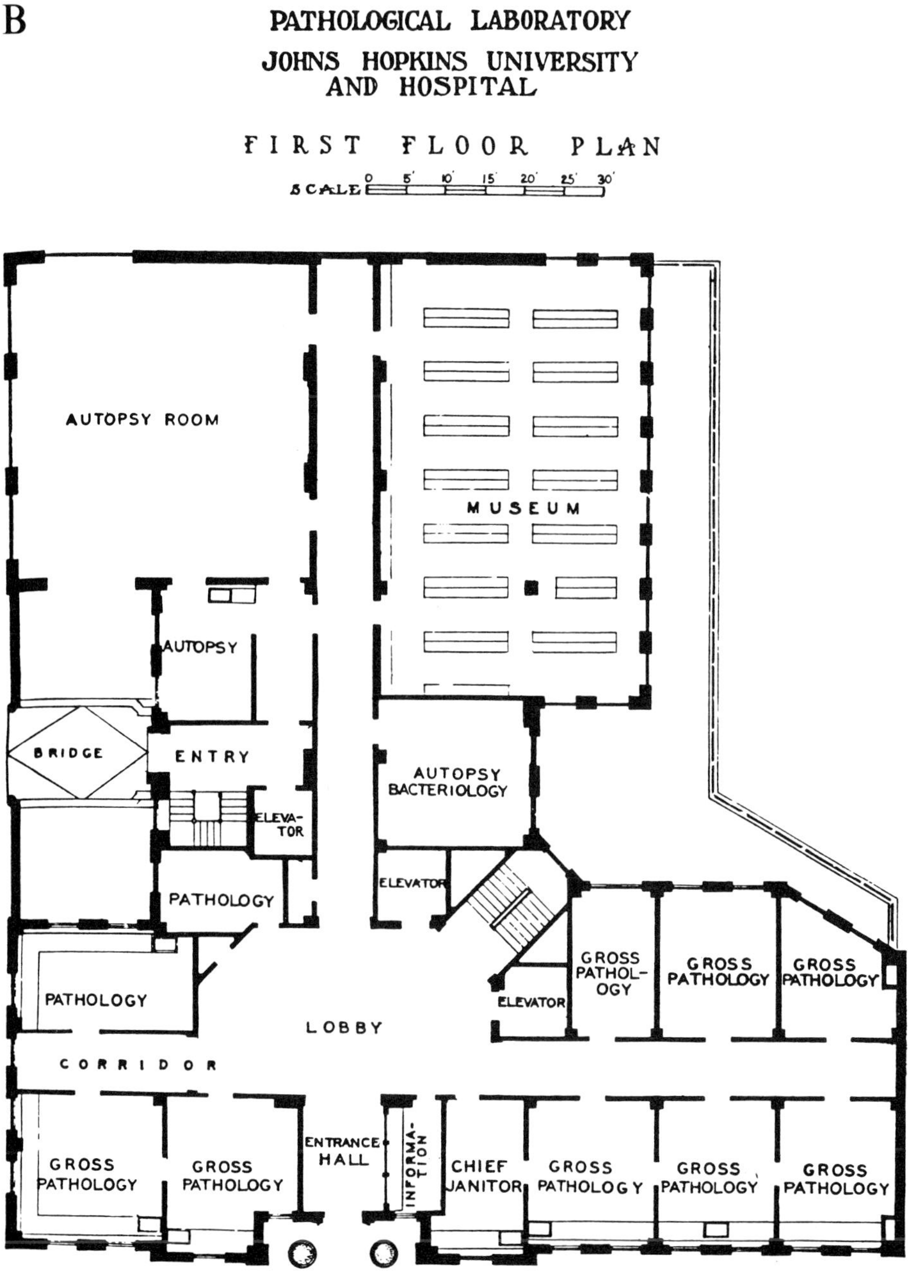

Figure 3B

special workers? While Leo Loeb may have been calling in part for technicians, he also clearly had in mind something like Meltzer's "standing army of regulars"—clinicians able to cross the great divide between ward and laboratory, and thereby fill the gap left by scientists now receding into ever more abstruse investigations.

C

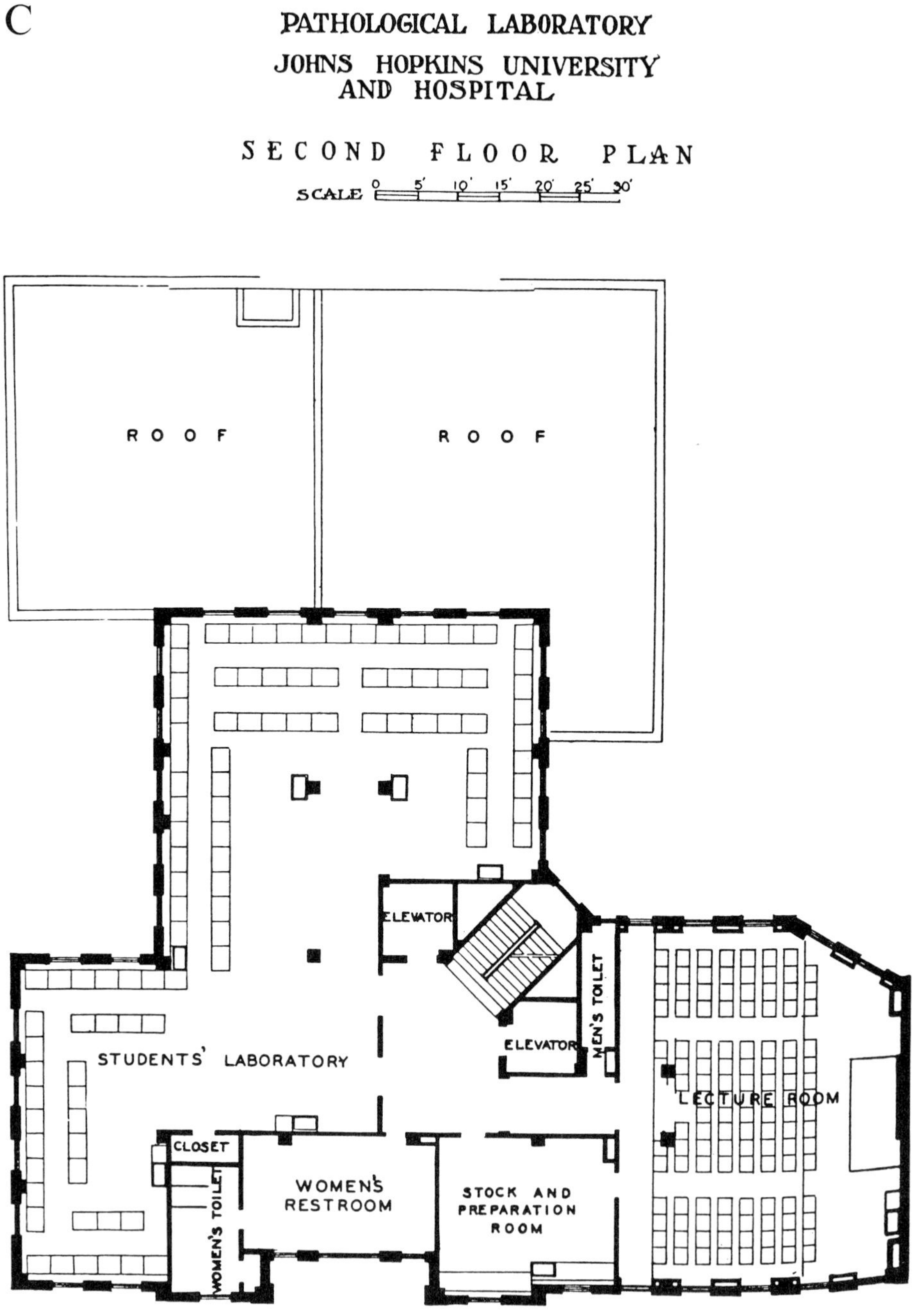

Figure 3C

A year later, in 1913, shortly before the founding of the Society for Experimental Pathology, Hewlett of Ann Arbor echoed Leo Loeb's views in a discussion of the relations between physiological pathology and the burgeoning specialty of internal medicine.[28] Hewlett's argument was ingenious and

D

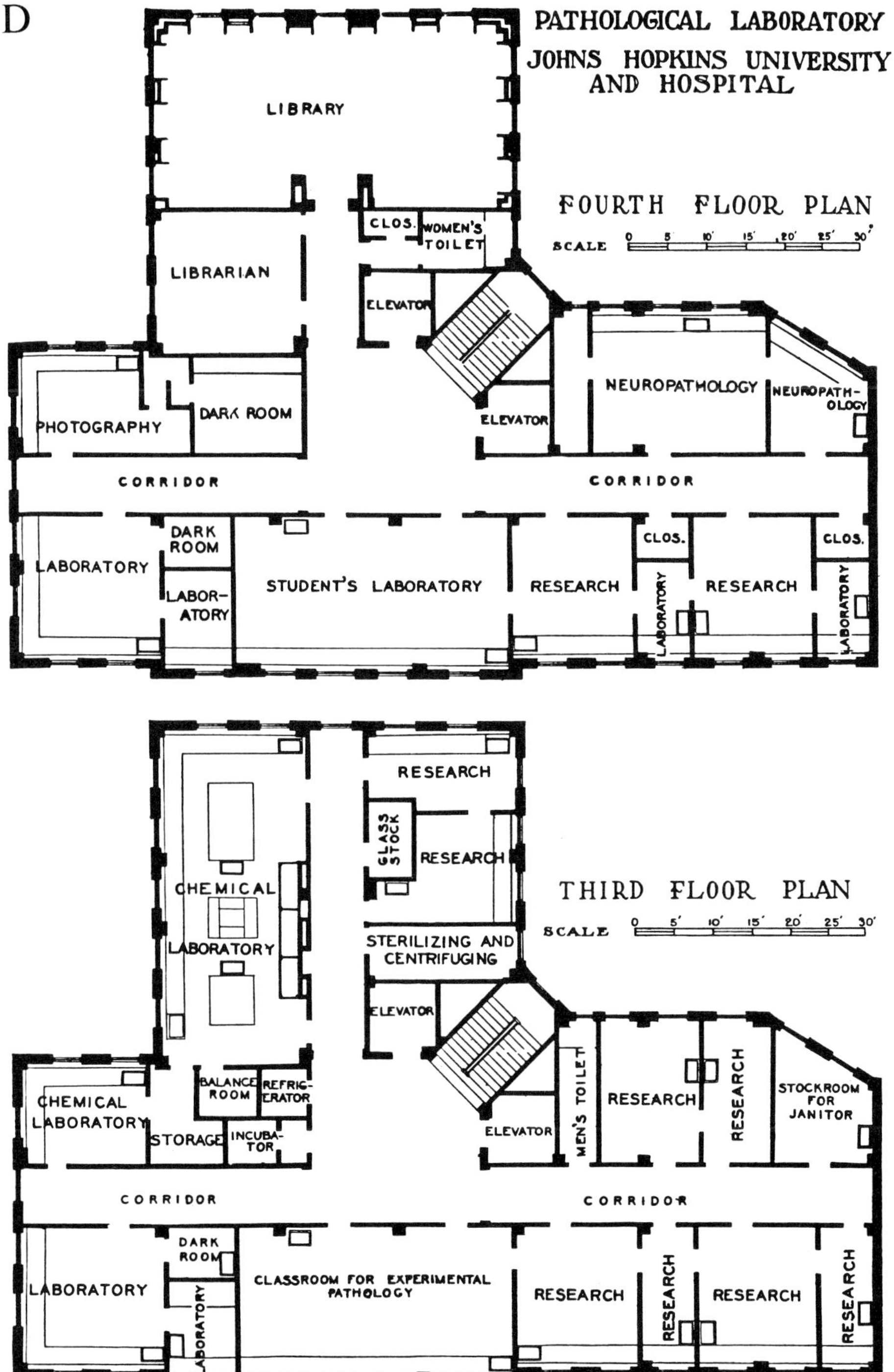
PATHOLOGICAL LABORATORY
JOHNS HOPKINS UNIVERSITY AND HOSPITAL
FOURTH FLOOR PLAN
SCALE 0 5' 10' 15' 20' 25' 30'
LIBRARY
CLOS.
WOMEN'S TOILET
LIBRARIAN
ELEVATOR
PHOTOGRAPHY
DARK ROOM
NEUROPATHOLOGY
NEUROPATH-OLOGY
ELEVATOR
CORRIDOR
CORRIDOR
DARK ROOM
LABORATORY
LABOR-ATORY
STUDENT'S LABORATORY
RESEARCH
CLOS.
LABORATORY
RESEARCH
CLOS.
LABORATORY
RESEARCH
GLASS STOCK
RESEARCH
CHEMICAL LABORATORY
THIRD FLOOR PLAN
SCALE 0 5' 10' 15' 20' 25' 30'
STERILIZING AND CENTRIFUGING
ELEVATOR
BALANCE ROOM
REFRIG-ERATOR
CHEMICAL LABORATORY
STORAGE
INCUBA-TOR
ELEVATOR
MEN'S TOILET
RESEARCH
RESEARCH
STOCKROOM FOR JANITOR
CORRIDOR
CORRIDOR
DARK ROOM
LABORATORY
LABORATORY
CLASSROOM FOR EXPERIMENTAL PATHOLOGY
RESEARCH
RESEARCH
RESEARCH
RESEARCH

Figure 3D

E

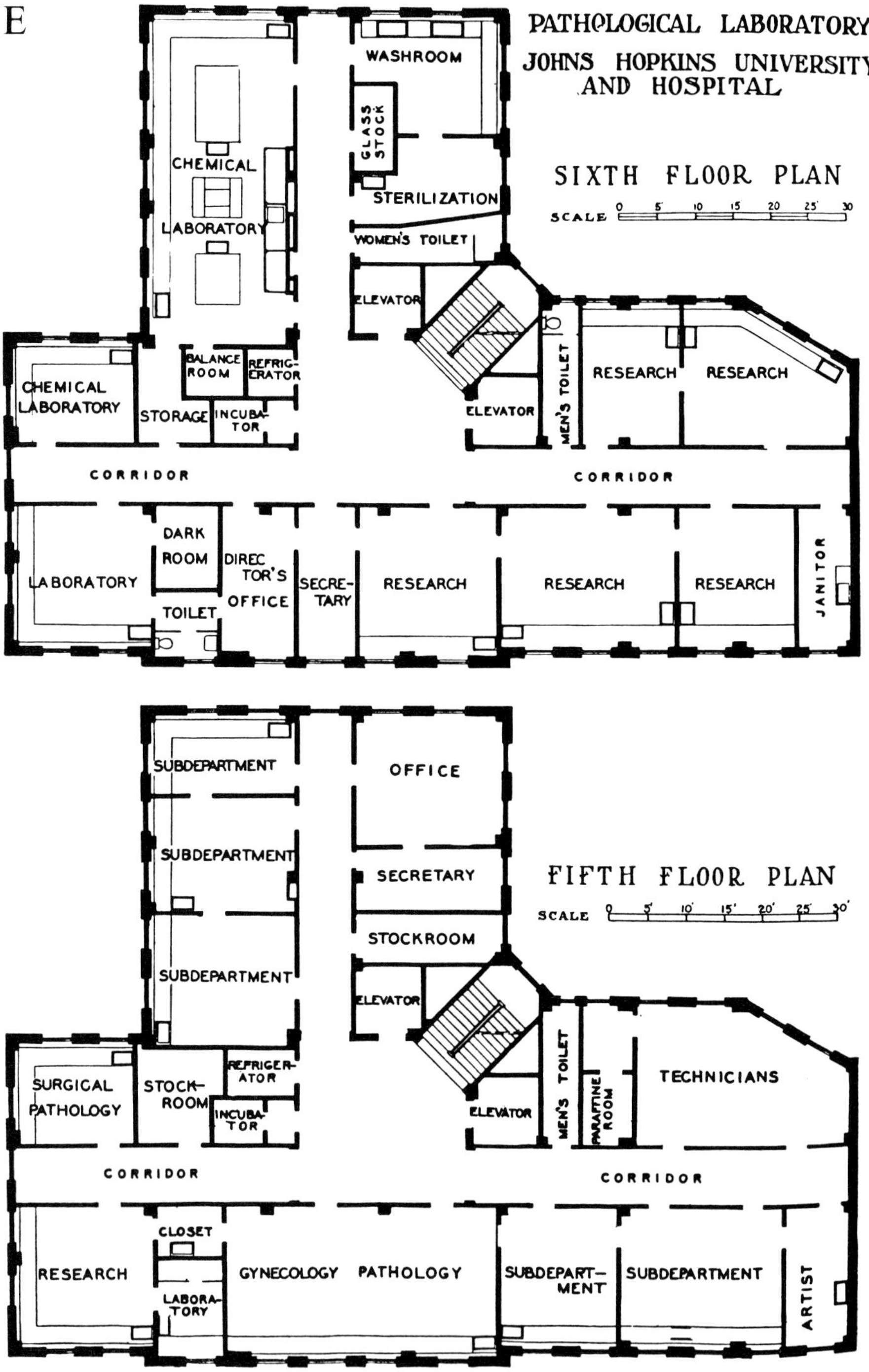
PATHOLOGICAL LABORATORY
JOHNS HOPKINS UNIVERSITY
AND HOSPITAL
SIXTH FLOOR PLAN
SCALE 0 5 10 15 20 25 30
WASHROOM
GLASS STOCK
CHEMICAL LABORATORY
STERILIZATION
WOMEN'S TOILET
ELEVATOR
BALANCE ROOM
REFRIGERATOR
CHEMICAL LABORATORY
STORAGE
INCUBATOR
ELEVATOR
MEN'S TOILET
RESEARCH
RESEARCH
CORRIDOR
CORRIDOR
DARK ROOM
LABORATORY
DIRECTOR'S OFFICE
TOILET
SECRETARY
RESEARCH
RESEARCH
RESEARCH
JANITOR
SUBDEPARTMENT
OFFICE
SUBDEPARTMENT
SECRETARY
FIFTH FLOOR PLAN
SCALE 0 5' 10' 15' 20' 25' 30'
STOCKROOM
SUBDEPARTMENT
ELEVATOR
SURGICAL PATHOLOGY
STOCKROOM
REFRIGERATOR
INCUBATOR
ELEVATOR
MEN'S TOILET
PARAFFINE ROOM
TECHNICIANS
CORRIDOR
CORRIDOR
CLOSET
RESEARCH
LABORATORY
GYNECOLOGY PATHOLOGY
SUBDEPARTMENT
SUBDEPARTMENT
ARTIST

Figure 3E

F

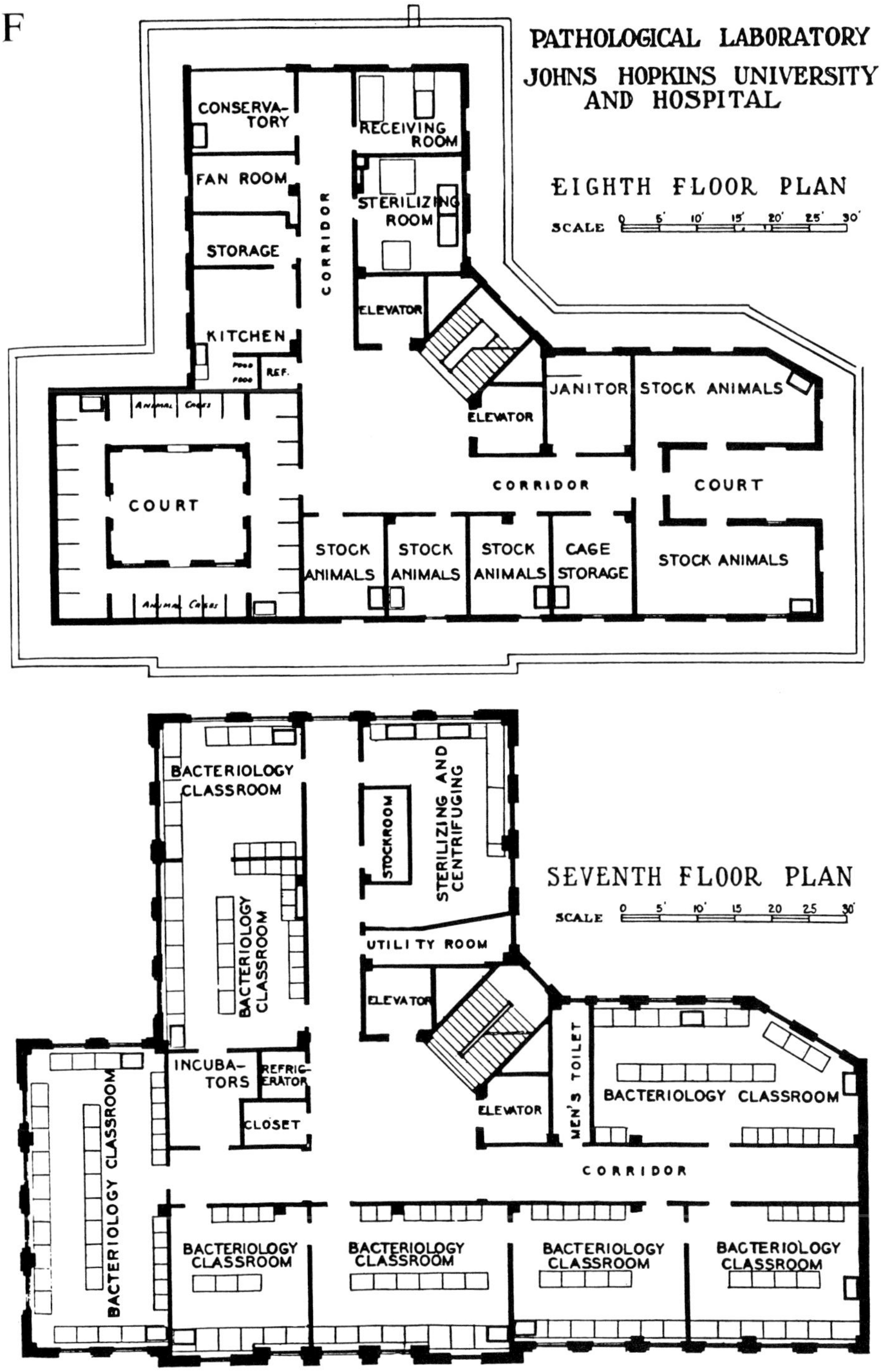
PATHOLOGICAL LABORATORY
JOHNS HOPKINS UNIVERSITY AND HOSPITAL
EIGHTH FLOOR PLAN
SCALE 0 5' 10' 15' 20' 25' 30'
CONSERVATORY
FAN ROOM
STORAGE
KITCHEN
REF.
CORRIDOR
RECEIVING ROOM
STERILIZING ROOM
ELEVATOR
ELEVATOR
JANITOR
STOCK ANIMALS
COURT
CORRIDOR
COURT
STOCK ANIMALS
STOCK ANIMALS
STOCK ANIMALS
CAGE STORAGE
STOCK ANIMALS
SEVENTH FLOOR PLAN
SCALE 0 5' 10' 15 20 25 30'
BACTERIOLOGY CLASSROOM
BACTERIOLOGY CLASSROOM
STOCKROOM
STERILIZING AND CENTRIFUGING
UTILITY ROOM
ELEVATOR
INCUBATORS
REFRIGERATOR
CLOSET
ELEVATOR
MEN'S TOILET
BACTERIOLOGY CLASSROOM
BACTERIOLOGY CLASSROOM
CORRIDOR
BACTERIOLOGY CLASSROOM
BACTERIOLOGY CLASSROOM
BACTERIOLOGY CLASSROOM
BACTERIOLOGY CLASSROOM

Figure 3F

provides keen insight into how institutional and cognitive shifts go hand in hand. Morbid anatomy, he argued, had benefited surgery greatly, providing "a basis for both diagnosis and treatment." Recently, however, internists had "learned to recognize clearly that even when the underlying anatomic changes are unknown or inaccessible to treatment, nevertheless much may be done for the patient by correcting so far as possible the essential functional disturbances. . . ." Physiological pathology, one of the critical bases for medicine, with an "importance. . .now generally recognized," had still "acquired no definite place in the medical curriculum, in the clinic, or in the laboratory." [29]

Hewlett had thus defined the need for work in physiological pathology with an important new twist. The surgeons, he said, still have their (anatomical) pathologists but the internists' (experimental) pathologists are few and far between. In this same paper in 1913, in a prescient section entitled "workers in pathologic physiology," Hewlett ruminated on ways to rectify the problem.

> [T]o a lesser extent the problems of pathologic physiology also, so far as these can be studied on animals, are now passing into the hands of workers in experimental medicine or experimental pathology. We cannot but welcome the advent of such departments [e.g., research medicine at Pennsylvania], which furnish a vigorous stimulus to the study of clinical problems, and teach younger men the scientific aspects of internal medicine; but here again we should not lose sight of the fact that the problems of experimental medicine are derived in the main from the clinic and that the closest interrelation must exist between the two fields.[30]

Hewlett's remarks on the training and deployment of these "younger men" echoed Meltzer:

> internal medicine is not simply an application of the healing art; it is also a rapidly growing science. . . . [I]n order to have that continuous stream of scientific productiveness which brings rapid and stable advances in our knowledge, the conditions of work must be favorable both to the exceptional person and to the larger number of earnest and capable men who make up the rank and file of scientific investigators.[31]

In 1920 pathologist Howard Karsner, a Philadelphian transplanted since 1914 to Cleveland, took the argument a step further.[32] Karsner might well have enjoyed a discussion of the history of disciplines, for he discussed the teaching of "the pathology of function" in clear and candid terms. Pathology, he said, "is not a science [translate 'discipline'] in the same sense that medicine itself is not a science." It is, rather, a constellation of methods, including those of the premedical sciences as well as the traditional pure sciences. Given the "widespread application of physiology and chemistry in internal medicine, [and] the important utilization of immunologic methods in the diagnosis and treatment of disease," physicians must be well versed in these new aspects of physiological pathology. With respect to the best site for such training, Karsner professed no ardent designs on it, conceding that the "place in which pathology of function is to be taught cannot be single; it must be multiple." [33]

Karsner himself was neither a clinician nor a pathologist attempting the metamorphosis to clinician. He was simply a pathologist (albeit a wide-ranging one with the sort of combined anatomical and physiological interests that placed him in classic Virchovian mold), thus underscoring the basic point: the new community of which he was a part after World War I was not especially concerned with the place of physiological pathology. Karsner was willing to cede it to other disciplines: a decentralization of knowledge.

Who were the beneficiaries of this decentralized knowledge about deranged function? Increasingly, they were clinicians like Theodore Janeway and Christian in the United States, or Humphry Rolleston and T. Clifford Allbutt in Britain. In the 1910s Janeway, for example, mounted an assault on one of the quintessential problems of pathophysiology: hypertension.[34] For clinicians, hypertension was indeed one of those problems central to medicine at the bedside, but one that was thought to be susceptible to elucidation by means of laboratory investigation. Yet—a critical point—this investigation was to be conducted in a variety of ways. Physiology and chemistry for such clinicians were at least as important, if not more important, than anatomy.

This was the eclectic viewpoint shared by the authors assembled by the Josiah Macy, Jr., Foundation in 1932–1933 to survey the past generation's labors on the problem of arteriosclerosis. Virchow's successor at Berlin, Ludwig Aschoff, came together with men like Esmond Long, MacCallum, Soma Weiss, Gideon Wells, Karsner, and the ubiquitous Rockefeller man Alfred Cohn, to discuss a clinical dilemma—atherogenesis—that clearly outstripped all the participants' individual laboratory paradigms.[35] In earlier decades, inflammation and neoplasia had served similarly as clinical problems into which investigators could inject their own favored laboratory styles, for example Cohnheim on inflammation, or Levin on cancer metastases.

In the 1910s and 1920s the attention of clinical investigators turned increasingly to disorders of the renal and cardiovascular systems. In the Macy Foundation volume on arteriosclerosis, Cohn and the volume's sponsor, Ludwig Kast of the Macy Foundation, deftly summarized the promise and the functional thrust of the new work on relations between various types of vascular disease. "Care must be taken," asserted Cohn,

> to keep distinct different arterial affections. Arterial hypertension obviously is a malady which must be sharply differentiated, first from Bright's disease, and second from arteriosclerosis. . . . [T]he issue has become a subject of investigation. Great stress need, therefore, no longer be laid on these distinctions; that these conditions exist and that they require separation is generally understood and acknowledged.[36]

Kast displayed a keen sense of how this might be accomplished and why it needed to be. Departing from the point made by several authors that medical successes in the acute infectious diseases had unmasked the full impact of degenerative disorders, Kast complained that while these

> diseases [of aging] constitute a very important problem in medicine and in human economy, they have attracted less interest on the part of investigators than their importance seems to invite or require. Concerning arteriosclerosis the reason may be found in the impasse to which purely morphological studies have come. Though morphology will no doubt continue in future studies to document the processes which lead to the deterioration of blood vessels, it will not be their main approach. . . .As in other fields of medicine, the shift of interest from the study of structure to the study of functions suggests new viewpoints.[37]

A

PLATE 51. PATHOLOGICAL BUILDING.

MAIN AND SECOND FLOOR PLANS.

FIGURE 1. Main floor plan.

B	Bacteriological rooms, 17′ 0″ × 32′ 0″, 16′ 9″ × 16′ 7″.	M	Morgue, 17′ 0″ × 29′ 8″.
P	Cremating furnace.	V	Ventilating chimneys, 3′ 0″ × 3′ 6″.
P R	Private research, 16′ 6″ × 19′ 9″.	E	Main entrance.
A	Autopsy theatre, 29′ 8″ × 38′ 2″.	S	Students' entrance.
W R	Waiting room and library, 16′ 6″ × 24′ 4″.	D	Lift, 2′ 9″ × 2′ 10″.
		W C	Water closet.

FIGURE 2. Second floor plan.

P H	Pathological Histology, 17′ 0″ × 37′ 0″.	B Y	Balcony.
D L	Director's laboratory, 16′ 7″ × 16′ 9″.	V	Ventilating chimneys, 3′ 0″ × 3′ 6″.
M M	Museum, 16′ 6″ × 26′ 3″.	W C	Water closets.
E P	Experimental Pathology, 17′ 0″ × 29′ 8″.	E	Main entrance.
P	Photographic rooms, 16′ 6″ × 18′ 0″ and 6′ 0″ × 12′ 0″.	S	Students' stairs.
		S L	Stairs to attic.

Another investigator cast in the same mold as Cohn was Christian. Confronting what he called the "nephritis problem," Christian described the difficulty of establishing the missing functional link, discernible in studies dating all the way back to 1827 and Richard Bright, between the symptoms of renal failure and the anatomical view "of kidneys from men dead of nephritis." Ever since Bright, he lamented, pathological anatomy had reigned:

> [S]ometimes [anatomical diagnosis] has been the acknowledged goal, at other times, though not so stated, it is evidently the aim of the investigator. Little by little it has become recognized that such an attainment has not been approached with any closeness and I think. . .it is now regarded as improbable that we will ever be able to correlate closely postmortem anatomical appearances with the functional disturbances of the kidney during life, at least so long as present pathologic technic and classification continue in use.[38]

It would serve little purpose to discuss here the details of Christian's proposed methods of functional determination in nephritis. What is more

PLATE-51

B

PATHOLOGICAL BUILDING
MAIN AND SECOND FLOOR PLANS

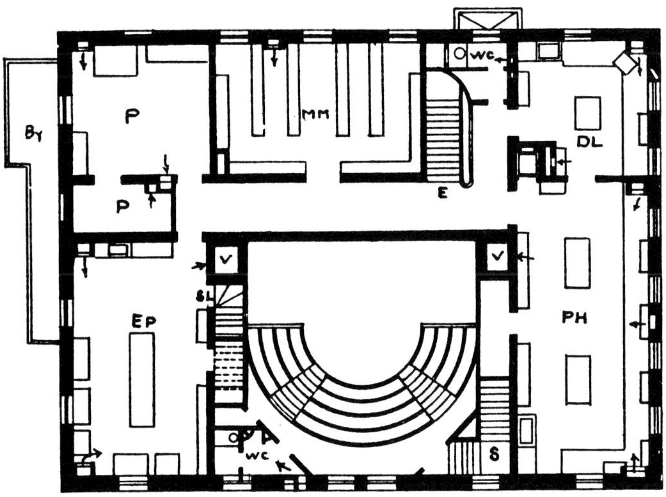

-FIG. 2-

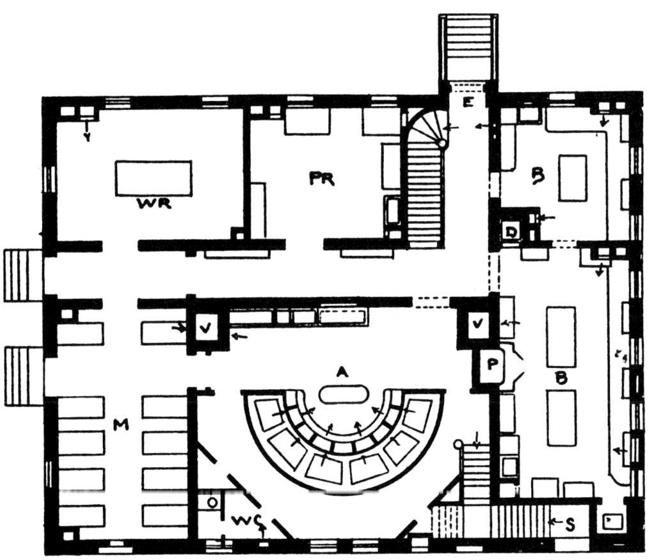

-FIG. 1-

important is a sense of the spirit of that approach. That spirit can be stated simply: as much as he wished to throw off the yoke of anatomical domination, Christian also warned against the physiologist—or the clinical investigator of

C

PLATE 52. PATHOLOGICAL BUILDING.

LONGITUDINAL AND TRANSVERSE SECTIONS.

FIGURE 1. Longitudinal section north and south.

B	Cellar or basement floor.	V	Ventilating chimneys.
D	Main floor.	A C	Accelerating steam coils.
E	Second floor.	F	Cremating furnace.
A	Autopsy operating table.	S	Skylight.

FIGURE 2. Transverse section east and west.

B	Cellar or basement floor.	V	Ventilating chimney.
D	Main floor.	H	Main stairs.
E	Second floor.	W C V	Water closet vent, 16″×16″
A	Autopsy operating table.	S	Skylight.

CREMATING FURNACE.

FIGURE 3. Elevation of furnace.

FIGURE 4. Plan of furnace at *a* of Fig. 3.

FIGURE 5. Plan of furnace at *b* of Fig. 3.

FIGURE 6. Plan of furnace at *c* of Fig. 3.

FIGURE 7. Transverse section at *d* of Fig. 3.

FIGURE 8. Transverse section at *e* of Fig. 3.

FIGURE 9. Longitudinal section at *h k* of Fig. 3.

---

function—getting too far out ahead of the clinician. He thus reserved his fullest measure of scorn for one particular genre of researcher: "that type of clinician who, with but a scant glance at his patient and no conception of the patient's symptoms, rushes to his laboratory tests with the idea that scientific accuracy increases by the square of the distance from the bedside and by the cube of the time spent in carrying out a test out of touch with the sick man."[39]

As the foregoing examples demonstrate, clinical investigation, although performed in the name of understanding physiological pathology, took that endeavor further afield than ever, professionally and institutionally, from the traditional core of pathology—from what pathologists actually did.

D

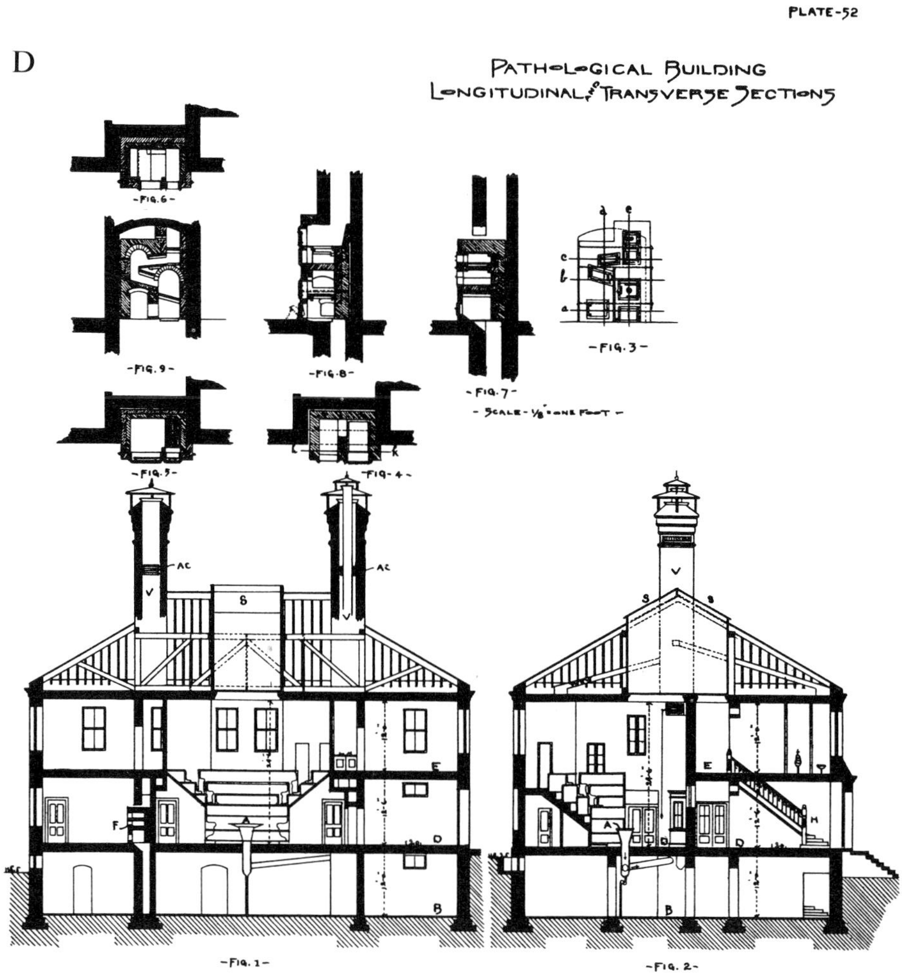

Figure 4: *Building in which Councilman began his academic pathology career in the 1880s.* **A:** *key to main and second floor plans shown in* **Figure 4B; C:** *key to longitudinal and transverse sections shown in* **Figure 4D.**

THE FATE OF THE PATHOLOGISTS: HOSPITALS

And what exactly did they do? Perhaps the pathologists could have regrouped, but they were preoccupied with matters other than physiological pathology. A series of distinct, if overlapping, concerns can be identified. The first, the lure of the hospital and its service demands on the pathologist, continued to exert a constant pressure. As early as 1898 Councilman had identified the special influence of the hospital on the pathologist as opposed to other, purely laboratory-based medical scientists when he wrote that "hospitals have been erected everywhere; and [in them] there are men whose work is only in the [anatomical and clinical pathology] laboratory and who do not come at all, or only rarely, in connection with the sick."[40] At the same time Councilman noted that "pathology must explain the lesion." A simple, elegant statement: experiment is needed but in the end the lesion is paramount.

The hospital also remained the context for the continued tug exerted by the autopsy table and microtome bench. Indeed, the continued importance of pathological anatomy is a *leitmotif* that may be traced in the programmatic writings of most pathologists in the first half of this century. Thus in 1905, Dock, in phrases that resonate with those of Bierring, could entreat an audience to credit his

> plea for the further cultivation of pathological anatomy for the medical student and the physician. It is natural that the immense strides of bacteriology and cellular diagnosis should overshadow the older topic and even make it seem unnecessary. Pathological anatomy was one of the most important factors in the development of medical knowledge, but [sic] it is still essential. . . .[41]

THE FATE OF THE PATHOLOGISTS: BACTERIOLOGY

George Dock (d. 1951) lived nearly half a century after these remarks, and hence would learn just how prescient they were. The chief preoccupation of most pathologists between World Wars I and II, except for those already captured by the morphological model and the hospital context that fostered it, was their relationship with bacteriology. Just as more time and energy was spent working out physiology's relationship with biochemistry than it did with pathology or the other basic science disciplines, so too did pathology have to deal with the problem of its daughter science bacteriology. In fact, many investigators in the 1910s and 1920s considered bacteriology the experimental wellspring of pathology. Typical of this group were Simon Flexner and his student Frederick Gay of Berkeley, who declared in 1916 that the "study of the infectious diseases is still, in spite of thirty years of increasingly successful endeavor, the greater field of promise for the investigating pathologist."[42]

It helped that pathology and bacteriology remained institutionally bound together in many places until the late 1930s. Together they presented a formidable "draw" at both the theoretical and applied levels, in the laboratory and the classroom.[43] That relationship was reflected in the deliberations of the special Committee on Research in Pathology convened in 1916 by Pearce of Pennsylvania. At the behest of the committee of 100 on scientific research of the American Association for the Advancement of Science, Pearce was to report in forty-five days on the sense of a group comprising himself, Wells, Rous, and Janeway. Here is how they saw the "nature of work to be aided":

> In the awarding of grants that preference be given to problems in etiology, pathological physiology and chemistry, as representing the most profitable lines of investigation. Under this arrangement bacteriology, protozoology, and immunology are to be construed as branches of pathology. Clinical studies may be included if the problem is susceptible of the use of methods of physiology, chemistry or immunology, in solving definite problems of pathological physiology. . . .Purely morphological studies, on account of the ease with which they can be conducted in any well equipped laboratory, without extra outlay for apparatus or supplies, will be assisted only when some new principle or some unusual material is involved.[44]

The committee of 100 had $100,000 to spend across many scientific disciplines. Pearce's committee was given $4,000 to spend, the equivalent perhaps of one academic salary at the middle or junior rank. That was not going to buy them a great deal of physiological pathology.

The question also remained as to who should perform microbiological diagnosis and investigation. From the beginning there had been issues of parenthood, in the United States as in Germany, with preventive medicine and public health vying with pathology for title to the adoption papers for microbiology. However, during the first three decades of the century, pathology departments by and large succeeded in laying claim to bacteriology, only to find the young sprout threatening to outgrow its parent discipline. As early as 1904 Dock appears to have seen that coming. By the 1930s it was clear, especially with the explosion of viral research, that microbiology was too large to be contained within pathology. In that decade, an increasing number of medical school deans engineered the two disciplines' fission, so that bacteriology might stand on its own, with everything that implied in terms of careers and patrons.[45]

Typical, once again, was the University of Pennsylvania, where, by the turn of the century, bacteriological teaching and research had moved—literally as well as metaphorically—across the street: out of the atrophied Institute of Hygiene and into the Department of Pathology. Alexander C. Abbott, successor to John Shaw Billings, had carried the burden of most hygiene-related subjects until 1928, when he proposed to retire. (One suspects that during this period many Flexner-era academic bacteriologists and pathologists did, in fact, reach retirement). William Pepper, the dean at Pennsylvania, appointed a committee on bacteriology to determine the feasibility of a new, separate department. At its initial meeting in January, the committee moved quickly to urge the establishment of a preventive medicine department, from which they recommended "that Bacteriology as one of the fundamental sciences be separated." They suggested "that bacteriology for students of the second year be transferred to the Department of Pathology. . . .This arrangement will bring about an intimate relation with the fundamental sciences upon which the development of medical bacteriology and immunology in large part depends and will not prevent the establishment of an independent department of bacteriology should the need for it become evident."[46]

The need was soon abundantly evident. The committee met four times in the spring of 1928, taking careful note of changes afoot in other schools in the process of dividing bacteriology from pathology. By summer's end they reported the following:

> Consideration of the intrinsic importance of the subjects of Bacteriology, Immunology, Filterable Viruses, together with the position that these subjects occupy in other leading medical schools in this country, made it obvious that an adequate department of these subjects would require an annual budget of at least $35,000 to $45,000 administered by a full Professor as the head of the Department aided by an adequate staff. . . .It was equally apparent that proposals on such a scale would be entirely inappropriate at the present moment and it was therefore determined to proceed on a much smaller basis that was recognized as inadequate and temporary. It was therefore determined to seek for a younger man of the grade of an Assistant or Associate Professor at a salary of about

> $4,500 to $5,000, and to this end over a score of letters were addressed to pathologists and bacteriologists throughout the country. . . .
> It is further suggested that the department be known as the Department of Medical Microbiology, and that it should include activities in medical parasitology as well as in bacteriology, immunology, filterable viruses, etc.[47]

By the late 1930s this scene had been replicated in most institutions. As bacteriology moved toward independent status, asserting its independence from the parent discipline of pathology, representatives of that discipline saw one of their most important bulwarks of experimentalism fade as well.

WORLD WAR II AND BEYOND

By the beginning of World War II, then, a new configuration had emerged, characterized by separate departments of pathology and microbiology, and by a continued anatomical thrust in the pathology departments of both academic and nonacademic (hospital-based) institutions, sometimes with a rather thin veneer of experimentalism. (Arguably that veneer was applied only in the most elite of the academic centers.) Experimental pathology, even at this point, was not confined to such a thin layer of one disciplinary elite. Rather, it had already diffused out, much as Karsner had foreseen twenty years before, into a wide range of departments. In the elite teaching institutions just before and just after World War II, physiological pathology, as a concept and as a domain of work, occupied much the same niche that molecular biology would enjoy a generation later.[48]

The establishment in 1935 of board-certifying examinations in pathology applied the fixative, as it were, to the role of the pathologist. Pathologists' activities were reflected in their textbooks. Those textbooks were anatomical and, to a lesser extent, chemical ("clinical chemistry"). Yet after World War II, the publication of major texts on physiological pathology proceeded apace. Almost all the authors came, as they had before World War II, from the corps of clinical investigators. In 1950, for example, William Sodeman of New Orleans published the first edition of his magisterial *Pathophysiology: Mechanisms of Disease.* Sodeman, a professor of tropical medicine at the University of Louisiana, enlisted twenty-four other contributors. Their collective ancestry could be traced to the work of Cohnheim and his disciples in Germany. Almost all of the contributors were key clinical investigators of the late interwar and early post–World War II period—men such as William Bean, Irvine Page, and Joseph Kirsner. It seems that only one contributor (William Castle of Harvard) was a member of the Society for Experimental Pathology.[49]

In his text, Sodeman described his intent to approach "problems of disease in. . .[internal medicine] from the standpoint of disturbed physiology. [It is] not intended to take the place of standard texts on physiology or textbooks of medicine. . .but does try to bridge the gap between them by presenting a clinical picture of disease seen as physiologic dysfunction." [50] Sodeman produced a second edition some six years later, and other authors in the 1960s and 1970s followed suit with similar efforts, both in content and in the complexion of their joint authorship.[51]

Hence the period after World War II mainly accelerated and solidified the

X

Industrial Fatigue and the Discipline of Physiology

RICHARD GILLESPIE

In the summer of 1915 a worker inspecting shrapnel cases at a munitions factory in Birmingham, England described the conditions under which he and his fellow workers produced materiel for the British army. Men and women, skilled workers and inexperienced recruits such as himself, labored twelve hours a day, for as long as seventy-nine hours over a seven-day week. Observing the lathe operators toward the end of their shift, he noted in his diary:

> It were interesting to know how far their last hour's output falls short of the high-water mark attained sometime in the forenoon. Their first hour is well-known to be less effective than a later when they have warmed to work. But the degree to which persistence fritters out despite all incentives as the day proceeds and drags on into overtime deceives enquiry. The individual differences must be great. Overtime must be uneconomical and unhygienic plan of heightening output.

Describing his own sensations in the last two hours of work, he recorded that "fatigue and ennui press heavier and more insistently." Such an assessment could have been made by any observant and literate worker, but it is significant that in this case the worker was Charles S. Sherrington, professor of physiology at Oxford, future president of the Royal Society, and from 1918 the first chairman of the British government's Industrial Fatigue Research Board. Sherrington, fifty-eight years old, had been determined to experience first-hand the fatigue that resulted from long and arduous work.[1]

Industrial fatigue became the central concept in the physiology of work and the leading area of applied physiology in the 1910s and 1920s. Sherrington was probably the only physiologist who attempted to experience fatigue so directly, but physiologists throughout Europe and North America sought ways to apply physiological principles to the organization, conditions, and hours of work. Provoked by production shortfalls during the First World War, several governments established committees of inquiry and research programs, all designed to reduce fatigue and increase productive efficiency. Physiologists in turn believed industrial fatigue could define a new social role for physiology. No longer would physiologists be confined to the pursuit of academic knowledge or the service role of teaching preclinical sciences to medical students; now the laws and discoveries of physiology could be applied to social problems of national significance.

The term "fatigue" was used as loosely as it was broadly; it could be defined in a variety of ways in physiology (muscular fatigue, changes in the chemical composition of the muscles and blood, or reduction in physiological performance) and in psychology (mental lassitude or slow reaction times); at the same time it referred to a serious workplace disorder, resulting in lost production, accidents, ill health, and even industrial unrest; and it was perceived to be a symptom of modern industrial society, brought about by the precipitous application of scientific and technical discoveries and disregard for basic human and social needs. The diversity of meanings gave industrial fatigue its power; the concept brought together scientific knowledge in physiology with broader industrial and social problems, while pointing to a vision of a society ordered by objective physiological laws.

Yet industrial fatigue only remained a viable scientific and social concept for at most twenty years; thereafter its meanings fragmented and dissolved. Formulated in the years prior to the First World War, the concept flourished during the war and reconstruction, only to wither gradually during the 1920s. Physiologists found themselves in competition with other scientific experts, notably psychologists and physicians, over who would gain a niche in industry. Industrial fatigue placed too much emphasis on altering the working environment to prove attractive to managers, who preferred psychologists' promise of selecting workers for the existing environment. Government support, which had been crucial to the development of the concept of industrial fatigue, declined in the 1920s as the state withdrew from direct intervention in the organization of industry. At the same time, the scientific meaning of industrial fatigue became more problematic, as physiologists encountered difficulties in applying physiological principles to work. The study of industrial fatigue survived in an attentuated form, but research became increasingly academic and its findings were no longer considered applicable to normal industrial work.

In this chapter I use the construction, development, and decline of the concept of industrial fatigue as a probe into how physiologists in the early twentieth century defined their discipline and perceived their social role. The relationship between the knowledge and power of a profession is a complex one. It can be tempting to assume that a profession's power is derived from its control of an objective body of knowledge, and that therefore power and authority flow naturally from expertise. This simple formulation is undermined by the repeated demonstration that social interests are embedded in knowledge claims—that knowledge is as much dependent on social power as the reverse. In this case, the development of the concept of industrial fatigue required an extraordinary extension of laboratory research on frog muscle so that it now applied to the daily variation in output of industrial workers. Physiologists were only able to transform what was little more than a metaphor into scientific knowledge by obtaining social power from reformers, industrialists, and the capitalist state, which they did through promises of improving industrial and national efficiency.

Once knowledge claims have attained scientific legitimacy, they can exercise power over society. Scientific concepts are ways of ordering the social as well as the natural world. They help define the relationship between the scientific discipline and the rest of society, drawing the boundaries within which scientific experts can control social and political action. As we shall see, the concept of industrial fatigue brought the organization of industrial work within the realm of physiology. What once had been the subjective feeling of the individual worker was now objectified into a scientific entity and social problem, the solution to which depended on the intervention of physiologists—Sherrington's feelings of fatigue and ennui were irrelevant to the science of industrial physiology. I argue that physiologists attempted to use industrial fatigue as a tactic to extend their expertise and authority from the laboratory into society, thereby opening a new social role for their profession.[2]

THE SCIENCE OF MUSCULAR FATIGUE

The study of muscular fatigue occupied a central place in the physiology of the late nineteenth and early twentieth centuries. Indeed, muscular fatigue characterized the new experimental physiology of the second half of the nineteenth century, for it consisted of the accurate recording of the movements of prepared muscles in the laboratory, using newly developed physiological instruments. American physiology followed closely in the wake of European developments; articles on muscular contraction and fatigue were common in the early volumes of the *American Journal of Physiology*, first published in 1898, while William H. Howell's *American Textbook of Physiology* (1896, 1901) devoted a lengthy chapter to the topic. The ability to conduct experimental, laboratory-based research attracted physiologists to the subject; Frederic Schiller Lee, professor of physiology at Columbia University and one of the leading American researchers of the subject, observed that so much was known about muscular fatigue "owing to the unequaled opportunity of applying to the study of muscular activity the exact methods of the physicists and the chemists."[3]

The standard experiment involved the measurement of contractions of the isolated gastrocnemius muscle of a frog—the favored animal because its nerves and muscles remained irritable for several hours after it had been killed. By irritating the nerve with electricity, chemicals, or extreme temperature, the physiologist could measure the muscular contractions with a myogram, which amplified the contractions and recorded them on a plate or cylinder covered with soot. Elaborate versions of the experiment included weights attached to the muscle to increase the amount of work, and a clock to make and break the electrical impulse at set intervals. The resulting myogram recorded the height of each contraction. The typical myogram showed an increase in the strength of the early contractions (known as the staircase or treppe), followed, as fatigue set in, by a gradual decline in contractions. A period of rest followed by further irritation of the muscle would result in a new staircase, then the inevitable onset of fatigue.[4]

Investigation of fatigue in humans was necessarily more indirect. Here physiologists used an ergograph, developed by Italian physiologist Angelo

Mosso around 1884, to measure the voluntary contractions of a group of muscles. Typically the subject would be required to repeatedly lift a finger to which a weight had been attached; the instrument measured the amount of total work by multiplying the weight by the height the finger was lifted. Later researchers introduced other instruments, such as the dynanometer, which required the subject to squeeze a spring.[5] Whatever their instruments or techniques, however, physiologists would have agreed with the definition of fatigue in Howell's textbook:

> The condition of muscular fatigue is characterised by lessened irritability, decrease in the rate and vigor with which the muscle contracts and liberates energy, and a still greater decrease in the rate with which it relaxes and recovers its normal form. In a sense, whatever induces such a state can be said to cause fatigue, but it is perhaps best to restrict the term to the form of fatigue which is produced by excessive functional activity.[6]

Knowledge of fatigue was still open to a great deal of uncertainty at the start of the century. Studies of the metabolic changes during fatigue had revealed a number of physiological changes, including the loss of carbohydrate and the presence of pathological acids. Lee had determined that the presence of lactic acid, carbon dioxide, and monopotassium phosphate in frog's muscle was associated with fatigue. Physiologists debated whether fatigue occurred first in muscles or the central nervous system. Sherrington's research indicated that fatigue occurred at the synapse, and that the nervous system thereby protected the muscle from irreversible damage. A few psychologists, on the other hand, cautioned physiologists that fatigue was inextricably bound up with the subject's psychological state, which would affect any experimental results.[7]

The study of muscular fatigue was considered basic research, however, and American and British physiologists did not dwell on, or even perceive, the potential applications of their knowledge. An occasional reference was made to Mosso's experiments with soldiers on forced marches, in which he found that rests allowed recovery from fatigue and speculated that this was because the body was given time to expel fatigue products.[8] No physiologists seem to have responded immediately to Mosso's book on fatigue, translated into English in 1904, in which he called for physiological research on the effects of industrial work, especially on women and children.[9] Gerald Geison has shown that the fruits of physiological research in this period rarely found their way into clinical practice. Muscular fatigue was no exception.[10]

FATIGUE AND EFFICIENCY

Within the space of a few years, however, physiological research on muscular fatigue had become a central element of debates over industrial efficiency, scientific management, working conditions, and the hours of work. Fatigue, a scientific concept whose range of meanings was restricted to academic physiological knowledge, suddenly resonated with new meanings, as it became associated with overwork, ill health, racial degeneration, and most importantly, inefficiency. This transformation cannot be attributed to a new scientific discovery opening up the possibility for the application of physiology to a major social issue; rather, it was the reverse. Social and political changes created an environment in which new meanings were constructed for the concept of fatigue. Industrial fatigue emerged out of the interactions of several factors: the development of scientific management in industry; increasing concern in an imperialist age with national efficiency; and a new confidence—voiced by the state, social reformers, and scientists alike—that social problems and conflict could be resolved by the application of scientific knowledge.

Josephine Goldmark's *Fatigue and Efficiency: A Study in Industry* (1912) was the first American work on the subject and set the tone for future research and writing. Her study of fatigue stemmed from a request in 1907 by lawyer and social reformer Louis Brandeis for a brief on the study of women's working hours, to be used by the defense in a Supreme Court challenge to the constitutionality of state laws that restricted women's hours of labor. As chairman of the Committee on the Legal Defense of Labor Laws of the National Consumer's League, a leading social reformer, and sister-in-law of Brandeis, Goldmark (1877–1950) was an obvious choice for the task. The extensive brief prepared by Goldmark and Brandeis for Muller versus Oregon was successful in persuading the Supreme Court to uphold the states' power to legislate in this area, the court declaring that the working woman's "physical nature and the evil effects of overwork upon her and her future children justify legislation to protect her from the greed as well as the passion of men." Goldmark, in her book, gathered the material from the briefs into an extended argument for the protection of women workers, drawing extensively on the experience of European countries and citing the evidence from factory inspectors, commissions, and scientists.[11]

The first paragraph of Goldmark's book made clear that she considered physiological evidence the most persuasive authority in finding a solution to what others might regard as a political and economic, or perhaps moral issue; the book, she declared, "draws upon the scientific study of fatigue—one of the most modern inquiries of physiological, chemical, and psychological science—for aid in the practical problem of reducing the long working day in industry." Goldmark echoed the faith of many progressive era reformers in declaring that only scientific knowledge could provide the proper basis and authority for rational progress. Not that economic or moral arguments disappeared from her rhetoric, or that of other reformers; rather, science provided the base to which these other appeals could be attached. The structure of the book reflected this ordering: the introduction was followed by a chapter summarizing the physiological research on fatigue, drawing especially on the writings of Mosso, Lee, and British physiologist Michael Foster; only then did she turn to the more traditional testimony of factory inspectors and social investigators.

The central argument of *Fatigue and Efficiency* was that daily output was subject to physiological laws, which were detectable beneath the peculiarities of individual workers and work practices:

> Just as we have seen diagramatically in the laboratory, the sequence of treppe, maximum effort, fatigue, and exhaustion, so in these experiments we see how working capacity increases during the second and third hours of work, falling as

> fatigue gains towards the noon hour, rising again slightly after food and rest at noon, to decline more rapidly to a minimum in the afternoon.

Goldmark rejected the employers' widespread belief that profits were made in the last hour of the day, citing three earlier industrial "experiments" with shorter working hours. In 1893 the British firm of Mather and Platt reduced the working day from nine to eight hours at its Salford Iron Works; production in the ensuing year was actually slightly higher than for the average of the preceding six years, and the company switched permanently to an eight-hour day and a forty-eight-hour week. Similar experiments at the Engis Chemical Works in Belgium in the 1890s concluded that the workers were capable of producing as much in eight hours as in ten, and that total production costs fell 20 percent. Finally, German physicist, industrialist, and social reformer Ernst Abbe introduced the eight-hour day at the Zeiss Optical Works in 1900, and provided detailed statistical evidence of increased efficiency, which he ascribed to a reduction in fatigue. Hereafter these tests were repeatedly cited in the literature of industrial fatigue.[12]

Goldmark was of course not the only person advocating the application of scientific laws to the organization of industry. In 1911, the year before *Fatigue and Efficiency*, Frederick Winslow Taylor had published *Principles of Scientific Management*, a work that summarized his experience in the restructuring and rationalization of American industry over the previous two decades. Goldmark applauded scientific management for its application of scientific principles to work, but she was a sufficiently astute and committed social reformer to warn against possible abuses of scientific management by unscrupulous bosses concerned only to increase the intensity of work. She felt ambivalent about the working conditions being produced by scientific management—what she saw as the subordination of the worker to the machine and the suppression of all individuality and spontaneity—but resigned herself to the fact that this was an inexorable process. All that one could do was "seek to imbue the purely economic view of the workers, as units of production, by a broader physiological spirit."[13]

The effect of *Fatigue and Efficiency* was to draw laboratory research in physiology into close proximity with social reform, industrial management, and the politics of work. It was not simply a case of physiological research being appropriated by social reformers and efficiency experts for their own ends; physiologists were quick to claim industrial fatigue as an important part of their discipline. Indeed, the introduction to Goldmark's book was written by Columbia University physiologist Lee, who had guided Goldmark through the literature on fatigue. Lee suggested that much human misery was due to the failure to live in accordance with physiological laws, generally because of a widespread ignorance of the findings of physiologists; industrial fatigue was clear evidence of this. Part of the responsibility lay with physiologists who, like other scientists, spent their time in research and rarely tried to instruct the outside world. Lee was to do much to change that situation in the following decade, becoming the leading American physiologist to conduct research in and popularize industrial fatigue.[14]

Frederic Schiller Lee (1859–1939) was a major figure in American physiology. He had been one of the first graduate students in physiology at Johns Hopkins University, completing his Ph.D. in 1885. After a year studying with Carl Ludwig and Max von Frey at Leipzig, he spent four years as an instructor at Bryn Mawr, then moved to Columbia in 1891, becoming a professor in 1904 and chairman of the department in 1909. His professional accomplishments included the presidencies of the Society for Experimental Biology and Medicine (1908–1910), the Harvey Society (1912–1914), and at the peak of his activities on industrial fatigue, the American Physiological Society (APS) (1917–1919). It is probable that the environment at Columbia sensitized Lee to the need to apply his scientific expertise to social problems. In the 1890s Columbia University president Seth Low had worked unceasingly to turn the university into a source of experts that would provide a scientific rationale for social reform. Lee was thus working in an institution that exemplified the progressive era's fascination with expertise and efficiency. Thus in 1911 Lee delivered a series of popular lectures, subsequently published as *The Scientific Features of Modern Medicine* (1911), in which he argued that the public should show more trust in modern medicine, for it was now based on scientific experimentation.[15]

In 1916 Lee fired the first shot in what proved to be an intensive campaign on industrial fatigue and for the establishment of a new science of industrial physiology. In a paper read before the Industrial Hygiene Section of the American Public Health Association and published as the lead article in an issue of *Science*, Lee answered the rhetorical question he had posed in the title, "Is the Eight-Hour Working-Day Rational?":

> In order to enable the wisest decision of the question to be made there is needed not mere opinions—not the opinions of employers, however broad-minded or narrow-minded; or of laborers, however industrious or indolent; or of labor leaders, however generous or selfish their ambitions; or of the laity, however philanthropic their motives; or of statesmen, whether they are impelled by high idealism or by practical politics; but a rigidly scientific study of the question, through the medium of laboratory tests, of the physiological effects of different occupations and the physiological capacities of different laborers and a resultant classification, on a physiological basis, of work and workers.

Lee claimed that laboratory studies of muscular fatigue and industrial experiments with shorter hours proved the existence of a physiological law whereby a reduction in hours would result in an increase in output. Much more research needed to be conducted before the optimum hours of labor could be determined, and clearly these would vary, depending on the work and the worker, but in the interim he indicated that science supported the universal introduction of the eight-hour day.[16]

Lee was doing more than giving the imprimatur of a leading physiologist to a campaign for shorter hours—he was proposing the study of industrial fatigue as an important new area of physiological research, drawing on the physiologist's experience in laboratory studies of fatigue. The research was already under way: Lee and his colleague Ernest L. Scott had published a paper in the *American Journal of Physiology* in which they subjected cats to controlled levels of humidity and temperature, killed them, and tested the working capacity of the muscles. Work decreased as atmospheric conditions worsened, thereby providing scientific confirmation of the findings of government inquiries on the occurrence of fatigue in hot and humid factories. A modest piece of research, but suggestive of more elaborate laboratory studies of industrial fatigue.[17] Lee was proposing that the physiologist leave the laboratory to study industrial conditions, the varying demands of work, and the physiological capacity of workers. This was not a vague, unrealizable dream, for he was well aware that such a research program was under way in Great Britain.

British interest in industrial fatigue roughly paralleled American developments, and reflected a similar response to perceived problems of industrial and national efficiency. Awareness of Britain's lagging economic and political power and low birthrate at the turn of the century, of which Britain's loss in the Boer War seemed symptomatic, sparked an extended campaign for political and social reform. An emphasis on efficiency and management across the political spectrum was translated into reforms in education, the civil service, the military, and social welfare, and as in America, the middle and upper classes placed their faith in the application of scientific knowledge by trained experts. Academics joined with politicians, administrators, and industrialists to revitalize the nation.[18]

Industrial fatigue was no exception. The committee structure of the British Association for the Advancement of Science (BAAS) provided the initial focus. A Committee on Mental and Muscular Fatigue, established in 1909 and chaired by Sherrington, surveyed the literature and conducted modest experiments; its function appears to have been to provide information to the Committee on Mental and Physical Characters Involved in Education.[19] The application of the concept of fatigue to industrial problems did not occur until 1913 when, in recognition of American and European developments, the BAAS formed the Committee on the Question of Fatigue from the Economic Standpoint. Much of the research was conducted by Philip Sargant Florence (1890–1982), a young Cambridge graduate who subsequently became professor of economics at Birmingham. Florence developed a bibliography on fatigue of over 1,000 items, consulted with Lee, and prepared lengthy reports summarizing the current state of knowledge. Unlike Goldmark and Lee's work, however, the British committee had semiofficial status, a fact underscored when one of its members, A. F. Stanley Kent, professor of physiology at Bristol, was appointed by the British government to develop measures of industrial fatigue that could be used by the government Factory Inspectorate.[20] The First World War soon interfered with these activities—but rather than bringing them to a halt it intensified the effort to identify and reduce industrial fatigue, with new official organizations being established in both Britain and the United States.

WAR AND INDUSTRIAL FATIGUE

The First World War transformed industrial fatigue from an industrial problem into an issue on which the survival of the nation depended, for industrial production was more critical than military prowess. Industrial fatigue was now a problem for the state, not only individual factories and companies. Physiological research on industrial fatigue flourished under state patronage during the war, conducted in government agencies and coordinated by official committees. Physiologists hoped too that the power of the state and its institutions could be mobilized to ensure that the recommendations of physiologists would be implemented in industry, although here they had mixed success. The concept of industrial fatigue peaked during the war and reconstruction because the physiologists' expertise was ratified by the social power of the state.

"War is the greatest of all agents of change," George Orwell once observed, "It speeds up all processes, wipes out minor distinctions, brings reality to the surface." The First World War played such a role in cementing relationships between science, industry, and the state. Where government support of scientific research for the benefit of industry had previously occurred in a piecemeal fashion, by the end of the war it was organized into national bodies that continued to function in peacetime. In Britain the Department of Scientific and Industrial Research (DSIR) and the Medical Research Committee (MRC) and in the United States the National Research Council (NRC) provided institutional mechanisms for the national coordination of science. These organizations provided a focus for the activities of university scientists whose wartime experiences had demonstrated to them the value of basic research in industrial, military, and social applications, and alerted them to the fact that government and industry could be persuaded to become benefactors of scientific research.[21]

British physiologists were successful in demonstrating the value of their knowledge for the war effort, especially on the home front, where long hours and hazardous conditions threatened the health of industrial workers. As soldiers dug into their trenches for a long campaign, the government took measures to direct the production of munitions, a massive operation in its own right, employing 3,400,000 workers by the end of the war. The pressure to produce combined with rapid expansion brought about appalling conditions, many of them permitted by the government. During 1915 it suspended normal union workshop practices and lifted limitations on the hours of work, which were extended to sixty-seven hours per week, although in practice they could be longer; these were the conditions that Sherrington observed during his two months in a Birmingham munitions factory. At the same time, however, the government established the Health of Munition Workers Committee, which was empowered to conduct research and make recommendations to the Ministry of Munitions. The committee, consisting primarily of factory inspectors, government medical officers, and medical researchers, included Sir Walter Fletcher, formerly a Cambridge physiologist and coauthor of the classic paper on lactic acid in muscle metabolism, and now secretary of the MRC, and Leonard E. Hill, professor of physiology at London Hospital Medical School and director of the MRC's Department of Applied Physiology and Hygiene. For the next three years the committee directed research on working conditions (TNT poisoning, ventilation, poison gases), accidents, sickness and lost time, housing, the dietary needs of workers, and of course, industrial fatigue. The studies on fatigue were conducted by, among others, Florence from the BAAS committee, and Oxford physiologist H. M. Vernon. Demands by unions for the reintroduction of shorter hours found support in the work of the committee, which argued that a reduction of hours would improve the health of the workers without impeding production, and in 1916 the government made some moves to limit hours of work. Medical research was now forming the basis of government control of working conditions in the largest sector of British industry; although as in the United States, concern with the welfare of workers was always mediated by overriding commitments to national efficiency and scientific management.[22]

With the end of the war in sight the government determined to continue the work of the Health of Munition Workers Committee into peacetime. In July 1918 the Industrial Fatigue Research Board (IFRB) was established as a joint board of the MRC and DSIR. Sherrington was appointed chairman of the new board; among its members were Fletcher; Winifred Cullis, professor of physiology at London; E. L. Collis, professor of preventive medicine at Cardiff; T. M. Legge, medical inspector of factories; Charles S. Myers, a physician and industrial psychologist; and three industrialists.[23] The board was a substantial undertaking; in the first year it was employing eight full-time and two part-time investigators, thirteen assistant investigators, and fifteen clerks. In the twenty years between the wars it produced eighty-four reports on fatigue, accidents, hours of work, industrial psychology, working conditions, time and motion study, telegraphists' cramp, and many related subjects.[24]

The study of industrial fatigue in the United States took a similar course during the war, although American physiologists failed in their attempts to establish a postwar equivalent of the IFRB. In May 1917, a month after America's entry into the war, the government's Council of National Defense (CND), an advisory organization to ensure the cooperation of industry and labor in the war effort, established a Subcommittee on Industrial Fatigue as part of its Committee on Labor; Lee was the force behind the committee. At its first meeting, held in his laboratory, he was appointed its executive secretary, he chose many of the committee members, and he directed its research. Chairman of the committee was Thomas Darlington, a physician and secretary of the health and welfare committee of the American Iron and Steel Institute; other members included Florence, fresh from his work with the BAAS committee on fatigue, and now studying for a Ph.D. at Columbia; Josephine Goldmark; Ernest G. Martin, professor of physiology at Stanford; David L. Edsall, professor of clinical medicine at Harvard; J. W. Schereschewsky, surgeon in charge of industrial hygiene studies at the Public Health Service; Ernest L. Scott, a physiologist at Columbia; A. H. Ryan, professor of physiology at Tufts Medical College; Raymond Dodge, professor of psychology at Wesleyan; and Robert E. Chaddock, a statistician at Columbia.[25]

Lee's committee never enjoyed the status or access to funds of its British counterpart. The CND was concerned with coordination rather than the direct regulation of production in private industry, and Lee and Darlington were unsuccessful in obtaining funds for research from it. However, Schereschewsky was able to obtain the financial support of the Public Health Service

(PHS), and an arrangement was made whereby the committee's researchers became assistants of the PHS; Lee acquired the title of consulting physiologist, a position he held until 1925. In November 1917 Lee and Schereschewsky approached the War Industries Board with the suggestion that it find a way to implement any recommendations that the committee might make, and as a result President Wilson and Secretary of War Baker considered moving the committee's work into the Department of Labor, where it would be able to influence more directly the labor practices of industry. However, nothing came of these proposals, and the committee was left on the perimeter of government wartime organizations.[26]

Lacking the power to recommend or implement industrial practices that would reduce fatigue, Lee's committee concentrated on research. In addition the limited funds meant that the committee was obliged to focus its research efforts carefully; the result was a study that in many respects was superior to those of the more prolific British researchers. The major investigation entailed a study of the output, hours of work, accidents, and working conditions in two factories that worked on an eight-hour shift and a ten-hour shift—respectively the Highland Park plant of the Ford Motor Company and the metals manufacturing plant of Scovill Manufacturing Company in Waterbury, Connecticut. The majority of the field research was conducted by Florence, with the assistance of Lee and Goldmark; at the same time Martin, Ryan, and Scott worked on developing physiological measures of industrial fatigue, first developing them in the laboratory and then testing them in the factories; and Albert B. Hastings, a graduate student in biochemistry at Columbia, conducted his Ph.D. research on fatigue products in the blood.[27]

As the war drew to a close Lee worked hard to convert the wartime activities into peacetime organizations that would further research into industrial fatigue. Visiting Britain at the end of the war and again in 1920, he was on both occasions impressed by the work of the IFRB and the amount of interest in the subject, and sought to establish similar acceptance of the field in the United States. Lee and Martin hoped that in due course they would be able to establish a department or institute of industrial physiology, but at the end of the war concentrated on more achievable initiatives.[28]

The most important institutional niche appeared to be the NRC, established to organize the scientific effort for the war, but now reorganized as the prime organization for the national coordination of science, and with special emphasis on the application of scientific research to industrial problems. In May 1918 the Medical Sciences Division of the NRC appointed the Subcommittee on Fatigue in Industrial Pursuits, with Lee as chairman and the members of the wartime committee as its members; in December 1919 this committee appears to have been replaced by the Committee on Industrial Physiology, again chaired by Lee. Lee was also an active member of the Committee on Medical Problems in Industry, which included fatigue in its brief.[29] Institutions for the dissemination and application of studies of industrial fatigue were also emerging: the American Association of Industrial Physicians and Surgeons had been founded in 1916 (Schereschewsky was its president from 1916–1918); while the *Journal of Industrial Hygiene* was established in 1919 as the first major academic journal in the field. Edited by two medical scientists who had been directly involved with fatigue studies, Edsall and Kent, and with Lee and Schereschewsky among its associate editors, it is not surprising that the first two volumes of the journal each contained five articles on industrial fatigue.[30]

Emerging from their government committees at the end of the war, the fatigue researchers in both Britain and the United States sought ways to ensure that their findings would be implemented in peacetime. Lee published a popular summary of fatigue research, *The Human Machine and Industrial Efficiency* (1918) and gave lectures around the country to physicians, industrial leaders, and the public; Florence published *Economics of Fatigue and Unrest* (1924), in which he eschewed Lee's technocratic vision in favor of a critique of the capitalist organization of work and called for an unspecified degree of worker participation. In Britain, one of the IFRB's researchers, Vernon, summarized the IFRB's work in his *Industrial Fatigue and Efficiency* (1921), and board member Charles S. Myers did much to publicize industrial fatigue and the work of the board.[31]

By the beginning of the 1920s, the concept of industrial fatigue had a solid existence, its reality constructed out of a plethora of industrial and laboratory experiments. Fatigue was defined in two distinct ways: either by a measure of industrial performance, such as output, or by one of several physiological measures of the state of individual workers. Output was by far the most popular means of measuring fatigue; production figures were relatively easy to obtain, and output linked fatigue firmly with productive efficiency. Moreover, as we have already seen, output curves could be compared with laboratory studies of fatigue in frog muscle. Physiologists followed Goldmark in regarding the similarities in the curves as evidence of the applicability of physiology to the study of work—the comparison was particularly useful in popular works—but aware that other factors intervened in the workplace, they steered away from claiming that the curves reflected identical processes. Lee, for example, wrote that the curves were "not unlike" one another and that this "suggests, although it does not make certain, a similar fundamental interpretation."[32] Vernon noted in particular the way a "practice effect" (improved neuromuscular coordination as the day progressed) and an "end spurt" (increased concentration toward the end of the day in order to reach production quotas) would tend to level out the output curve and conceal the degree of fatigue; on the other hand, increasing monotony or conscious restriction of output would also reduce output and thus overestimate fatigue. These other factors had to be controlled if a true measure of physiological fatigue was to be obtained, but as Florence observed, unwillingness and incapacity shaded off into one another, and in the final analysis it was impossible to separate physiological and psychological factors.[33] Yet despite these difficulties and qualifications, investigators felt fairly secure in using output as a measure of fatigue, so long as care was taken to control for other variables. Other measures of productive efficiency, such as accidents, lost time, or turnover, were also used, primarily to show that a reduction of fatigue would reduce their incidence.[34]

Physiological measures constituted the second way of measuring fatigue. Much of the research for these fatigue tests was conducted by the American physiologists: Martin developed a spring-muscle test that measured the strength of the worker; Ryan experimented with a vascular skin-reaction test, in which the skin was marked with a blunt instrument and measurement taken of the time it took for the white streak to begin to disappear; and Hastings analyzed the concentration of hydrogen ions and other chemical changes in the urine. In Britain, Kent experimented with blood pressure and variations in the acuity of sight and hearing. In each case there was some relationship between the test and fatigue at the end of the day, although the correlations were by no means exact. Nevertheless, physiologists argued not unreasonably that the research program was still young; industry could look forward to the day when a simple and practical physiological measure of industrial fatigue was available.[35]

Fatigue researchers were willing to give practical advice to industry, but their recommendations were broad and inexact. However, the researchers argued that it was still too early to set precise hours of work and rest periods based on physiological research. Florence had observed during a Committee of Industrial Fatigue (CIF) meeting that "all of the practical suggestions of the English Committee have nothing to do with the research side of the work," and the British researchers could have returned the compliment. The comparative study by Florence and the CIF at the Ford and Scovill plants showed that production at Ford, with eight-hour shifts, was more efficient than in the ten-hour shifts at Scovill, where efficiency was measured by the degree to which hourly output remained stable. This, however, was based on a statistical analysis of output data rather than physiological research per se, although of course the similarity to laboratory studies along with the incidence of accidents and lost time was taken as evidence that physiological fatigue was the major variable.[36] Moreover, as Florence admitted to a CIF meeting, fatigue and production varied between factories, manufacturing processes, and individuals:

> I cannot imagine myself recommending a definite length of working day. Research has shown that such work varies more than you would expect. You would naturally think, beforehand, that all the operatives of a factory would be pretty much the same in their fatiguing effect; really there is an enormous difference.[37]

Here again, this did not strike the researchers as an overwhelming obstacle; it only increased the complexity of the problem and underscored that physiological research on industrial fatigue was a long-term research program.

The concept of industrial fatigue reached its zenith in the years during and immediately after the war, pushed along by the impetus of research programs and institutions created during the war and reconstruction, and by the scientists' conviction that government, industry, and the public supported the research and desired its implementation. Where two decades previously "fatigue" had referred simply to physiological changes in muscle, its meaning had now expanded to include the hours and conditions of industrial work. Laboratory experiments by physiologists were now accepted by government and at least some industrialists as relevant to the organization of work, and through their fatigue studies physiologists had stepped into a world in which their expertise was applicable to social problems.

INDUSTRIAL PHYSIOLOGY AND THE DISCIPLINE OF PHYSIOLOGY

Once established as a legitimate scientific concept, "industrial fatigue" was used by physiologists to redraw the boundaries of the discipline of physiology. Physiologists came to associate the study of industrial fatigue with basic professional and institutional changes in their discipline, to the extent that their discussion of industrial fatigue resonated with their dreams of professional independence and social utility. The science of industrial fatigue proved a powerful resource, opening up a new applied science of industrial physiology, contributing to the creation of a new professional ideology that stressed the direct application of physiology to social problems, and redefining traditional relationships between physiology and medical education.

Lee coined the term "industrial physiology" to refer to the expansion of industrial fatigue to include the whole field of industrial hygiene and even the organization of industrial work. Industrial physiologists would bring the scientific method to the organization of work, integrate the findings of other disciplines, and ensure that engineers and managers proceeded in a scientific and efficient way. In the years following the war Lee spent his time proselytizing for the establishment of industrial physiology, so much so that his name became intimately connected with the subject. The study of industrial fatigue had established a scientific and to some extent practical basis for the field. It was another matter, however, to persuade owners and managers, workers, or scientists from other disciplines that industrial physiology provided the solution to a host of industrial problems; in short, that physiology should acquire authority over the workplace. Lee's writings in this period constitute an extended argument for being given that power.[38]

Science itself lent authority to the enterprise. The movements for national and social efficiency, brought to their peak during the war, had intensified faith in scientific expertise and expanded the institutional mechanisms through which experts could exercise their power. The title page of Lee's *Human Machine* encapsulated these developments and established his authority by listing his scientific and social credentials: Ph.D., LL.D., Dalton professor of physiology, Columbia University, president of the APS, consulting physiologist to the United States PHS, chairman of the Subcommittee on Fatigue in Industrial Pursuits of the NRC, and executive secretary of the Divisional Committee on Industrial Fatigue under the Advisory Commission of the Council of National Defense. In the text itself he drew a parallel between the development of scientific medicine and industrialization. In both cases there was a conflict between empiricism and tradition on one side and science on the other; scientific medicine had succeeded through the introduction of the experimental method, and industry must do the same by supporting experimentation in both laboratories and factories.[39]

Industrial leaders and managers required more tangible and immediate benefits, however, and Lee sought to persuade them that industrial physiology would solve some central problems of industrial capitalism, among them:

> How fatigue can be kept down to its lowest reasonable limit, how the working power of the individual can be maintained from day to day and from week to

> week and be made to yield a maximum output without detriment to itself and others—in other words, how the human machine can be used so as to obtain from it the most profit.[40]

Lee's conception of industrial fatigue was thus moving away from that developed by Goldmark or even the pioneering industrialists at the turn of the century. Industrial fatigue for Goldmark stemmed directly from her concern to defend protective labor legislation; while she stressed that shorter hours were also more productive, her concern with labor reform was paramount. In his 1916 article on the eight-hour day, Lee had rejected the partisanship of both labor reformers and industrialists, choosing instead the higher ground of science, and arguing that profits should be subordinated to physiological principles. By 1918, Lee was more interested in gaining the backing of industry—for permission to conduct research in factories, the adoption of industrial physiologists' recommendations, and the employment of industrial hygienists. Goldmark's priorities were reversed, and the health of workers appeared rather as an outcome of efficient and profitable production.[41]

Establishing industrial physiology meant also that workers must subject themselves to the dictates of science. Lee, along with many of the fatigue researchers, insisted that workers must give up all of their traditional techniques of restricting output: "The worker should not willfully limit his production, but should work according to his capacity." The researchers were aware that restriction of output was primarily a defensive response to the constant lowering of piece rates, but claimed that they would be able to establish proper rates based on sound physiological principles, bringing science to time and motion studies and scientific management. Indeed Lee had no time for independent unions, privately expressing the view that union strength was lamentable and irrational. Scientific experts, especially industrial physiologists, could work together with industrial leaders to produce an efficient and rational system of production.[42] The role of scientific knowledge in reducing the power of workers was clear in the concept of industrial fatigue itself; all the researchers agreed that the workers' feelings of fatigue were no guide to the existence of objective, physiological fatigue, which could only be detected by the industrial physiologist.[43]

For industrial physiology to become the primary science of industry it also had to assert its authority over the other scientific experts already established or seeking a niche in the factory—notably scientific management experts, industrial psychologists, and physicians. Lee tackled each of these groups, arguing in each case that only physiology could provide the true scientific basis for their work. He noted that although scientific management experts loudly condemned fatigue, they had no scientific basis on which they could set piece rates, other than basing rates on the fastest worker (time and motion experts often adjusted for fatigue by simply adding five or ten percent to their carefully measured times). As for psychology, Lee argued that the tests developed by industrial psychologists were best incorporated into industrial physiology, which gathered knowledge about all aspects of the human machine. Industrial medicine was an easier case, for here Lee could assume that industrial physicians would as a matter of course rely on the findings of physiology. In Lee's view industrial physiology was poised to become the preeminent science of work.[44]

Industrial fatigue was also perceived as defining a new social role for physiology as a profession and scientific discipline, for it exemplified the application of physiological research to areas of broad social concern outside its traditional influence on clinical medicine. It pointed to the conceptual and institutional means by which physiological knowledge of human abilities, limitations, and requirements would be used to guide social policy in education, nutrition, health, and work. Long before he cycled off to work in a Birmingham munitions factory or became chairman of the IFRB, Sherrington expressed to Lee his belief that physiology "should be part now of the equipment of positive knowledge for every citizen—man and woman—because of his responsibilities and her responsibilities toward family and community in regard to health, that basis of all well-regulated activity."[45] The study of industrial fatigue during the First World War appeared to be an excellent example of the use of physiology to establish "well-regulated activity" in the workplace, but it was not the only one. Walter Fletcher, reflecting on "The Aims and Boundaries of Physiology" in his 1921 presidential address to section I of the BAAS, observed that the stresses put on the healthy body during this war had made "the physiologist and his methods indispensable," whether it be studying shock and exposure among soldiers, industrial fatigue, or food rationing on the home front:

> The nation as a whole, and those especially who have the government, public or private, of large groups of men in their hands, have learned that obedience to physiological laws is a first necessity for the maintenance of the body machinery in health and for its effective and harmonious use. . . . Physiologists have a new vista before them of immense public usefulness.[46]

Lee was in complete agreement with these sentiments, arguing that the university physiologists should no longer be regarded as idealistic and impractical, for industrial fatigue had shown itself to be "preeminently practical."[47]

British and American physiologists also suggested that industrial fatigue could assist in restructuring the traditional relationships between physiology and medical education, although in different ways in each country. British physiology had long been an independent scientific discipline, firmly established in the schools and universities, and with a secondary role as a preclinical science in medical education. As secretary of the MRC, Fletcher supported this academic independence from medicine, but saw too the need to bring physiology down to earth, ensuring that its knowledge filtered into medical education as well as social policy. By establishing its practical value during the war, through industrial fatigue and other studies, the divorce between physiology and medical practice was lessening.[48]

In the United States, on the other hand, physiology had been established as, and remained, a preclinical science in the medical school; the European style of general physiology had only a toehold in American universities. The role of physiology, and to some extent its intellectual styles, were shaped by its institutional location, which in turn was determined by the reform of medical education.[49] Lee's interest in industrial fatigue, and the meanings he invested in it, become clearer when we consider his position at Columbia University and his attitude to the reorganization of medical education. When Lee arrived at Columbia in 1891, he was appointed to a position at the College of Physicians and Surgeons, which had only recently been absorbed as the university's medical school. When Columbia moved uptown to Morningside Heights, Lee and the physiological laboratory were forced to remain in the medical school at the midtown site. Thereafter, Lee was always searching for ways to shift physiology onto the main campus, where he believed that it belonged with the sciences of physics, chemistry, biology, and psychology; but he failed in his attempt to have physiology made part of the pure science and philosophy faculties, and his protracted campaign to shift the medical school and a university hospital to the Columbia site came to nought. This is not to say that he did not accept the value of physiology in medical education; on the contrary he worked tirelessly to expand physiological training for physicians, especially into undergraduate and postgraduate courses, although with only modest success. Industrial fatigue seems to have provided an avenue for Lee to escape the frustrations of trying to improve the physiological component of medical education and the stultifying effects of being confined to the service role of teaching a preclinical science to physicians, who would all too quickly forget what they had learned once they entered the clinical years. In one fell swoop, industrial fatigue sidestepped the whole problem of the relationship between physiology and medicine; physiologists could now intervene directly and improve the health and efficiency of their society. In addition, it was hoped that the demand for industrial physiologists by industry and government agencies would open up a market for physiology graduates that was independent from medicine.[50]

Industrial fatigue thus acquired multiple layers of meanings, stretching far beyond the simple identification of fatigue in frog muscle with industrial output. For physiologists, industrial fatigue meant an expanded social role for their discipline, in which physiological knowledge of human abilities, limitations, and requirements would be used to guide social policy. And it implied a redefinition of the relationship between physiology and medicine, in which physiologists would gain independence from and a degree of control over the medical profession.

THE DECLINE OF INDUSTRIAL FATIGUE

If meanings can crystallize around a concept such as industrial fatigue, they can also dissolve, and by the early 1920s several factors were weakening the combination of meanings that gave industrial fatigue its power. No single solvent was sufficient or perhaps even necessary, but in combination they resulted by the 1930s in a severe restriction of the concept of industrial fatigue to a narrow range of physiological research, increasing doubt that physiological studies of fatigue were meaningful when applied to industry, a decline in research in and the application of industrial physiology, and the stripping of industrial fatigue's original reform impulses.

Part of the difficulty arose from the fact that industrial fatigue had multiple meanings, depending on social interests or professional and disciplinary perceptions—paradoxically, it had been this diversity of meanings that had made industrial fatigue such an instant success. As the political and industrial environment changed, however, the forces that had held fatigue together shifted and realigned, leaving the concept to disintegrate. The tension had always been there, well concealed in the literature but evident, for example, in the meetings of Lee's wartime committee. We have already seen how Lee increasingly envisaged industrial fatigue as the technocratic application of physiological principles to an unnecessarily politicized social problem, whereas Goldmark retained her commitment to feminism and social reform, seeing the scientific findings as supporting rather than replacing political action. At least they agreed on the need to combine social action and science, whereas David Edsall, professor of clinical medicine at Harvard and coeditor of the *Journal of Industrial Hygiene*, saw industrial fatigue as strictly a scientific problem, complaining of Lee's book that "there is too much tendency to hitch up the medical work and welfare work together. The medical work is vastly more important and this connection is damaging to it."[51] Concern with national efficiency during the war could hold these divergent views together, but proved insufficient in peacetime.

The cessation of direct state intervention in industrial production at the end of the war meant that government agencies became less interested in funding or implementing the findings of industrial fatigue researchers. This was clearest in Britain, where the IFRB was almost disbanded during the postwar recession and survived with a reduced budget. Moreover, the board's terms of reference restricted it to research and prevented it from setting guidelines for industry. In America the PHS continued to fund a physiologist in Lee's laboratory, but it lacked the finances and political power to implement Lee's scheme for a corps of industrial physiologists. In both countries the attention of employers and government shifted from working conditions to the problem of containing worker militancy, and industrial fatigue was overshadowed by endless discussions of industrial relations."[52]

A second factor was that research seemed to increase the complexity of industrial fatigue and early hopes for a quick and practical solution gave way to recognition of the need for long-term research programs. As we have seen, the literature published during and after the war had shown that the relationship between fatigue and output was highly complicated, and the practical value of physiological fatigue tests doubtful. The problems with industrial fatigue were analyzed at length by Reynold Spaeth, an associate in physiological hygiene at the Johns Hopkins School of Hygiene and Public Health, and published prominently in the first issue of the *Journal of Industrial Hygiene*. Of laboratory studies of muscular fatigue, he observed that from the point of view of industrial fatigue, what one needed to know was how muscle "may operate so as to produce the greatest amount of work, in a given period of time with the least fatigue. And, strange as it may seem, this precise question has never, as far was we know, been experimentally answered by the physiologists." Of the industrial research conducted by physiologists he was equally skeptical: "How tired a worker may become without endangering his health, that is, where fatigue ceases to be normal and becomes cumulative, neither the physiologists nor anyone else has thus far succeeded in demonstrating."

Psychologists became more vocal in their criticism of physiologists' attempts to develop a practical fatigue test, insisting that it would be impossible to control for psychological factors; IFRB psychologist Bernard Muscio wrote such a persuasive critique that the board gave up all attempts to develop a test.[53]

It would be a mistake, however, to interpret the decline of industrial fatigue as simply the scientific failure of a research program, or as the empirical refutation of overblown hopes. The difficulties cited by postwar critics had all been addressed during the meetings of the CIF, but the researchers had nevertheless retained confidence in industrial fatigue. When Harvard physiologist Cecil K. Drinker, in a critical review of Goldmark and Hopkins' report on the CIF research, noted that "fatigue is undefined as a physiological entity," Goldmark and Hopkins retorted that science proceeded by the experimental study of such postulates: "No more than medical science in the laboratory, can industrial science in the factory or, indeed, any applied science wait always upon ultimate proof." Problems of evidence and proof could not by themselves undermine the concept.[54]

After the First World War, physiologists suddenly found themselves in serious competition with other groups that claimed expertise in areas that overlapped with industrial fatigue. Psychologists were especially effective in challenging physiology, both on the intellectual and organizational fronts. British psychologists were the most vocal in attacking industrial fatigue, in part because they were competing with physiologists within a single organization, the IFRB; but their criticisms added to the difficulties of physiologists in America as well.[55] Equally important, psychologists engaged in a concerted campaign after the war to establish industrial psychology as a field of academic inquiry and practical techniques. Selection tests developed for the army during the war quickly found their way into industry; this process was aided by the establishment of consulting organizations such as the Scott Company and Psychological Corporation, both of which were controlled by leading academic psychologists. The psychologists proved more adept at using existing institutions such as the NRC for their purposes. Lee looked on in amazement and dismay in 1919 when the psychologists took control of an Engineering Foundation proposal to study the social and medical aspects of industrial work; the initial proposal had placed industrial fatigue first in a list of relevant studies, but the organization that emerged, the Personnel Research Federation, and its *Journal of Personnel Research* were controlled by psychologists and other social scientists, and industrial physiology received little attention. There is little doubt that managers found industrial psychology, which concentrated on selecting the correct worker for the job, as ideologically and practically preferable to industrial physiology, which emphasized altering the conditions of work to suit the worker.[56]

The other significant group standing between physiologists and industry was industrial physicians, who had begun to enter industry in large numbers with the introduction of workmens' compensation laws in the 1910s. Few industrial physicians managed to gain much independence of judgment and action from their employers, however, and industrial medicine consisted for the most part of physical examinations when hiring workers and treatment of accident cases. An industrial physician might be able to influence management on such matters as temperature and ventilation, but hours of work or the organization of production were areas in which managers and engineers would broach no interference. There was therefore little incentive for physicians to concern themselves overly with physiologists' work on industrial fatigue; guides to industrial medicine and hygiene would discuss fatigue, but as just one part of a broader field. Indeed, many of the practical suggestions made by physiologists—the importance of rest periods, proper seating, and selection of workers—were proposals that physicians could put into practice without the assistance of physiologists.[57] In addition, managers' obsession with immediate results to current problems conflicted with physiologists' hopes for long-term research into industrial fatigue. In 1919 Ryan had taken a full-time position with Scoville to direct a company-funded research program in industrial physiology, effectively the continuation of the CIF research, but in 1922 he resigned and returned to Tufts, frustrated by the way in which trivial daily requests had overwhelmed his research plans.[58]

The meanings of industrial fatigue thus narrowed substantially in the 1920s and 1930s: the reform impulses disappeared, the industrial applications shrank, physiological research became increasingly academic and limited in its industrial relevance, and researchers came to doubt the value of the concept of industrial fatigue. By 1927, for example, Lee and Sherrington were placing fatigue in quotation marks, even in their correspondence to one another. Two years later the IFRB changed its name to the Industrial Health Research Board, for the concept of industrial fatigue had lost much of its social applicability.[59]

The shifting meanings of industrial fatigue can best be captured by looking at the leading American research program on industrial fatigue between the wars. In 1927 the Laura Spelman Rockefeller Memorial agreed to fund a program of "Research in Industrial Hazards" at the Harvard Business School, led jointly by psychologist and social theorist Elton Mayo and Harvard physiologist L. J. Henderson. The two men's research was only loosely related, but they shared an interest in industrial fatigue; Mayo participated in and coordinated the famous Hawthorne experiments on worker motivation and management techniques, while Henderson directed the Harvard Fatigue Laboratory, where physiologists and biochemists conducted basic research on fatigue. Industrial fatigue provided the rationale for their joint program; it was expected that the laboratory would provide basic knowledge and techniques that could be applied by Mayo in industry.[60]

However, the subsequent research and theorizing at Harvard served to restrict further the power of industrial fatigue. Before coming to Harvard, Mayo had argued that industrial fatigue led to psychiatric disorders among workers and eventually to industrial unrest; fatigue was a recognizable subject on which he could base his less-familiar study of industrial psychiatry, and he hoped to show a connection between workers' physiological condition and their psychiatric and social adaptation to work. As the Hawthorne experiments developed, however, Mayo had less need to rely on industrial fatigue, and the first chapter of his *Human Problems of an Industrial Civilization* (1933) consisted of a detailed critique. Surveying the complexity of the fatigue research, he concluded, "One begins to wonder at this point whether the word 'fatigue' is not itself in serious danger of being overworked." He asserted that the type of fatigue shown in the laboratory was rare in industry, because of increased mechanization of arduous jobs, and a process of "natural selection" of workers for the remaining physical tasks. Physiological tests on the Hawthorne workers had shown that they were in a steady state, or equilibrium, and this was true of most factory workers. Industrial fatigue ceased to be a major industrial problem; workers' psychiatric states, not working conditions, were the major factor influencing productive efficiency.[61]

In the fatigue laboratory, research by Henderson's colleagues focused increasingly on basic physiological problems and was concerned with only the most extreme industrial conditions. David Bruce Dill and Arlie V. Bock followed Henderson's work on the biochemistry of the blood by conducting detailed laboratory experiments on the interaction of metabolic and respiratory functions. They concluded that industrial work did not upset the metabolic equilibrium: "Fatigue, if it is experienced, cannot be charged to an upset in physiological equilibrium." Industrial fatigue only occurred among a few workers in the most strenuous conditions, and even then was not due to any single factor; steel workers laboring in high temperatures suffered heat cramps due to salt depletion, while aviators suffered from the lack of oxygen. In each case specific measures could be taken to prevent the problem. Under the influence of the fatigue laboratory, industrial fatigue shrank to a fraction of its former size; where before, fatigue had been defined by output or accidents, and thus applied in all workplaces, now it was defined as a physiological change that occurred in only the most unusual situations. Where before, the failure of fatigue tests had been taken to indicate a problem with the tests, now it was interpreted as undermining the very concept of fatigue.[62]

This narrow construction of industrial fatigue reflected Mayo and Henderson's conservative political outlook. Mayo's rejection of fatigue was part of his elaboration of a management ideology that stressed the importance of creating a social organization that took into account the human needs of the workers; an understanding supervisor would do more for production than rests or shorter hours. This view became the new orthodoxy of the human relations movement, disseminated in the writings on the Hawthorne experiments. In a development symbolic of the decline of industrial fatigue between the wars, Henderson chaired the NRC Committee on Work in Industry from 1937–1941; the published report was titled *Fatigue of Workers: Its Relation to Industrial Production* but it was devoted primarily to discussion of the Hawthorne experiments, and was written by George Homans, a sociologist.[63]

CONCLUSION

By the time the Second World War had again made industrial productivity an issue of national survival, the concept of industrial fatigue had lost its potency. The industrial problems remained and physiologists retained their disciplinary aspirations, but "industrial fatigue" had lost its intellectual coherence and social power, and no longer seemed the basis for a comprehensive solution to the organization of work. Physiologists found a niche in the war effort, but this time with the military, and research on fatigue was incorporated into the field of aviation medicine.[64]

The history of the concept of industrial fatigue reveals the degree to which scientific concepts are constructed from their social context as much as from laboratory results. The process is particularly clear in this case, where the scientific knowledge had immediate practical applications. Industrial fatigue became a viable scientific concept because of the convergence of a widespread faith in the social utility of science, a political concern that industrial inefficiency would threaten national survival, and a scientific discipline eager to apply its physiological knowledge to social problems. The existence of physiological knowledge about muscular fatigue was perhaps the least important of the factors that contributed to the construction of the concept of industrial fatigue. Once the social conditions changed, the concept of industrial fatigue lost its political and cognitive power.

In this paper I have explored the shifting meanings of industrial fatigue from the point of view of physiologists, trying to show how the study of a scientific concept can reveal much about the social context of a scientific discipline. I have not attempted to determine just what effect the concept of industrial fatigue and the techniques of industrial physiology had on the practice and organization of work or on the workers' experience of fatigue—an enormous task. Indeed, in one sense workers only experienced industrial fatigue as long as physiologists, governments, and employers recognized its existence, for many workers could be persuaded that their symptoms were instead due to laziness, poor working habits, an unhealthy lifestyle, or psychiatric problems. I have done no more than give an impressionistic account of the responses of the state, industrial managers and engineers, workers, and physicians. Yet these are important questions demanding further research, not least because physiologists are once more entering the workplace, responding in particular to workers' demands for the study of repetitive strain injury, stress, and the physiological effects of shiftwork.

NOTES

1. H. M. Sinclair, "Sherrington and Industrial Fatigue," *Notes Rec. R. Soc.* 39 (1984): 91–104, quotes on 102–104.
2. For two discussions of the social construction of scientific knowledge and its application to the history of science, see Steven Shapin, "History of Science and its Sociological Reconstructions," *Hist. Sci.* 20 (1982): 157–211; Henrika Kuklick, "The Sociology of Knowledge: Retrospect and Prospect," *Annu. Rev. Sociol.* 9 (1983): 287–310. On the interrelationship of scientific knowledge and disciplinary and professional power, see Pierre Bourdieu, "The Specificity of the Scientific Field and the Social Conditions of the Progress of Reason," *Soc. Sci. Inform.* 14 (1975): 19–47; Charles Rosenberg, "Toward an Ecology of Knowledge," in *The Organization of Knowledge in Modern America, 1860–1920*, ed. Alexandra Oleson and John Voss (Baltimore, MD: Johns Hopkins Univ. Press, 1979), pp. 440–455; Terence Johnson, *Professions and Powers* (London: Macmillan, 1972). On the social meanings of scientific concepts in the biomedical sciences and medical practice, see Ludwik Fleck, *Genesis and Development of a Scientific Fact* (Chicago, IL: Univ. of Chicago Press, 1979); Karl Figlio, "How Does Illness Mediate Social Relations? Workmen's Compensation and Medico-Legal Practices, 1890–1940," in *The Problem of Medical Knowledge*, ed. Peter Wright and Andrew Treacher (Edinburgh: Edinburgh Univ. Press, 1982), pp. 174–224; Bruno Latour, "Give Me

a Laboratory and I Will Raise the World," in *Science Observed*, ed. Karin Knorr-Cetina and Michael Mulkay (Beverly Hills, CA: Sage, 1983), pp. 141–170. Warner has recently called for more attention to be paid in the history of medicine to the social meanings of scientific knowledge; John Harley Warner, "Science in Medicine," *Osiris* 1 (1985): 37–58.

3. William Henry Howell, ed., *An American Textbook of Physiology*, vol. 2 (Philadelphia, PA: Saunders, 1901), pp. 17–170 (the chapter, "General Physiology of Muscle and Nerve," was written by Warren P. Lombard, professor of physiology at the University of Michigan); Frederic S. Lee, "Fatigue," in *Harvey Lectures 1905–6* (Philadelphia, PA: Lippincott, 1906), pp. 169–194, quote on 170.
4. Howell, *American Textbook of Physiology*, pp. 17–170.
5. Ibid.; Angelo Mosso, *Fatigue*, trans. M. Drummond and W. B. Drummond (London: 1904); Shepherd Ivory Franz, "On the Methods of Estimating the Force of Voluntary Contractions and on Fatigue," *Am. J. Physiol.* 4 (1900/1901): 348–372.
6. Howell, *American Textbook of Physiology*, p. 70.
7. Lee, "Fatigue," 169–194; Lee, "The Nature of Muscle Fatigue," *Proc. Am. Physiol. Soc.* (1898/1899): xi; William McDougall, "The Conditions of Fatigue in the Nervous System," *Brain* 32 (1909): 256–268. British physiologist, psychologist, and anthropologist W. H. R. Rivers used a placebo to counteract the effects of the subject's interest in his research on the influence of alcohol and caffeine on fatigue; see Rivers, *The Influence of Alcohol and Other Drugs on Fatigue* (London: Arnold, 1908).
8. See, for example, Howell, *American Textbook of Physiology*, p. 72.
9. Mosso, *Fatigue*, pp. 171–174.
10. Gerald L. Geison, "Divided We Stand: Physiologists and Clinicians in the American Context," in *The Therapeutic Revolution*, ed. Morris J. Vogel and Charles E. Rosenberg (Philadelphia, PA: Univ. of Pennsylvania Press, 1979), pp. 67–90.
11. Josephine Goldmark, *Fatigue and Efficiency: A Study in Industry* (New York: Charities Publication Committee, 1912), quote on p. 252; the material used in this and subsequent briefs is contained in pt. 2. On the history of protective labor legislation for working women and the significance of Muller versus Oregon, see Alice Kessler-Harris, *Out to Work: A History of Wage-Earning Women in the United States* (New York: Oxford Univ. Press, 1982), pp. 180–214, esp. 186–187.
12. Goldmark, *Fatigue and Efficiency*, pp. 121–174, quotes on 34, 121.
13. Goldmark, *Fatigue and Efficiency*, pp. 191–209, quote on 286. On scientific management, see Frederick Winslow Taylor, *The Principles of Scientific Management* (New York: Harper, 1911); Daniel Nelson, *Managers and Workers* (Madison: Univ. of Wisconsin Press, 1975); David Montgomery, *Workers' Control in America* (Cambridge, UK: Cambridge Univ. Press, 1979). On the spread of scientific management and the efficiency movement through American society, see Samuel Haber, *Efficiency and Uplift: Scientific Management in the Progressive Era, 1890–1920* (Chicago, IL: Univ. of Chicago Press, 1964).
14. Ibid., p. v–vi. At least two other researchers, Francis Benedict and Edward Cathcart at the Nutrition Laboratory of the Carnegie Institution of Washington were aware of the possibility of applying physiological research to a study of industrial efficiency; Benedict and Cathcart, *Muscular Work* (Washington, DC: Carnegie Institution, 1913), esp. pp. 3–4.
15. H. B. Williams, "Frederic Schiller Lee," *Science* 91 (1940): 133; *Nat. Cyclopaedia Am. Biogr.* 29: 24–25; Philip J. Pauly, "The Appearance of Academic Biology in Late Nineteenth-Century America," *J. Hist. Biol.* 17 (1984): 369–397, esp. 387–389; Douglas Sloan, "Science in New York City, 1867–1907," *Isis* 71 (1980): 35–76; Frederic S. Lee, *The Scientific Features of Modern Medicine* (New York: Columbia Univ. Press, 1911), esp. chapt. 8.
16. Lee, "Is the Eight-Hour Working Day Rational?" *Science* 44 (1916): 727–735, quote on 734.
17. Lee and Ernest L. Scott, "The Action of Temperature and Humidity on the Working Power of Muscles and the Sugar of the Blood," *Am. J. Physiol.* 40 (1916): 486–501.
18. G. R. Searle, *The Quest for National Efficiency: A Study in British Politics and Political Thought, 1899–1914* (Berkeley: Univ. of California Press, 1971).
19. Reports of the Committee on Mental and Muscular Fatigue are in *Report BAAS, 1910* (London: John Murray, 1911), pp. 292–297; *1911*, pp. 174–175. The committee included psychologist McDougall and J. S. MacDonald, professor of physiology at Sheffield.
20. Reports of the Committee on the Question of Fatigue from an Economic Standpoint are in

Report BAAS, 1914 (London: John Murray, 1915), pp. 175–176; *1915*, pp. 283–349; *1916*, pp. 251–274; A. F. Stanley Kent, *Interim Report on an Investigation of Industrial Fatigue by Physiological Methods* (London, 1915), Cd 8056; Kent, *Second Interim Report* (London, 1916), Cd 8335. The Committee on the Question of Fatigue was chaired by J. H. Muirhead, professor of philosophy at Birmingham, UK and its membership was dominated by economists and physicians from Birmingham.

21. A. C. Seward, ed., *Science and the Nation* (Cambridge, UK: Cambridge Univ. Press, 1917); Robert M. Yerkes, ed., *The New World of Science* (New York: Century, 1920).
22. Health of Munition Workers Committee, *Industrial Fatigue and its Causes* (London: HMSO, 1916), Cd 8213; Health of Munition Workers Committee, *Final Report: Industrial Health and Efficiency* (London: HMSO, 1918), Cd 9065; Noelle Whiteside, "Industrial Welfare and Labour Regulation in Britain at the Time of the First World War," *Int. Rev. Soc. Hist.* 25 (1980): 307–331; *Medical Research Council, Annual Report, 1916–1917* (London: HMSO, 1917), pp. 17–19, 83. On Fletcher see Robert E. Kohler, "Walter Fletcher, F. G. Hopkins, and the Dunn Institute of Biochemistry," *Isis* 69 (1978): 331–355, esp. 333–336; the classic paper was W. M. Fletcher and F. G. Hopkins, "Lactic Acid in Amphibian Muscle," *J. Physiol. Lond.* 35 (1907): 247–309.
23. Industrial Fatigue Research Board (IFRB), *Report* (1920), pp. 2–6, quote on 5. Sherrington resigned the chairmanship in 1921 upon becoming president of the Royal Society, but remained a member of the board.
24. Ibid. (1920), p. 22; for a list of IFRB publications see Industrial Health Research Board, *Industrial Health in War* (London: HMSO, 1940), pp. 29–36.
25. Minutes of the Committee of Industrial Fatigue (CIF) are held in the Frederic S. Lee Papers (hereafter, FSL), Butler Library, Columbia University, New York, box 6; see also "Memorandum on Studies of Industrial Fatigue," *Public Health Rep.* 34 (1919): 731. Phillip Sargant Florence's dissertation on industrial fatigue was published as *Use of Factory Statistics in the Investigation of Industrial Fatigue* (New York: Columbia Univ. Press, 1918).
26. CIF Minutes, 7 June 1917, 7 March 1918. On the Council of National Defense and other government wartime agencies, see Ellis Hawley, *The Great War and the Search for a Modern Order* (New York: St. Martin's, 1979), pp. 20–37.
27. The major study of the CIF was Goldmark and Mary D. Hopkins, *Comparison of an Eight-Hour Plant and a Ten-Hour Plant*, Public Health Service Bull. no. 106 (Washington, DC: GPO, 1920); see also Albert Baird Hastings, *The Physiology of Fatigue*, Public Health Service Bull. no. 117 (Washington, DC: GPO, 1921); Scott, "The Present Status of our Knowledge of Fatigue Products," *Public Health Rep.* 33 (1918): 605–610; A. H. Ryan, "The Quantitative Measurement of General Fatigue," *Am. J. Physiol.* 45 (1918): 537–538. Hastings went on to become professor of biochemistry at Chicago (1926–1935) and at Harvard (1935–1958).
28. E. G. Martin to Lee, 26 March 1918, FSL, box 2; Lee to W. H. Howell, 6 January 1919, FSL, box 1; Lee to J. R. Murlin, 1 September 1920, FSL, box 2.
29. See National Research Council, "Physiology Committee. Sub-Committee on Fatigue in Industrial Pursuits," in National Research Council Archives (hereafter, NRC), National Academy of Sciences, Washington, DC, Division of Medicine, Committee on Physiology; Henry A. Christian to Lee, 11 December 1919, NRC, Committee on Industrial Physiology, 1919–1920; "Minutes of Sub-Committee on Medical Problems in Industry," 17 September 1919, NRC, Sub-Committee on Medical Problems in Industry, 1919.
30. Ralph Chester Williams, *The United States Public Health Service, 1798–1950* (Washington, DC: USPHS, 1951), pp. 279–286; Henry B. Selleck, *Occupational Health in America* (Detroit, MI: Wayne State Univ. Press, 1962).
31. A partial list of their publications includes Lee, *The Human Machine and Industrial Efficiency* (New York: Longman, Green, 1918); Lee, "The Human Machine in the Factory," *Harvey Lectures, 1917–19* (Philadelphia, PA: Lippincott, 1920), pp. 216–233; Florence, *Economics of Fatigue and Unrest, and the Efficiency of Labour in English and American Industry* (New York: Henry Holt, 1924); H. M. Vernon, *Industrial Fatigue and Efficiency* (London: Routledge, 1921); Charles S. Myers, "Industrial Overstrain and Unrest," in *Lectures in Industrial Administration*, ed. Bernard Muscio (London: Pitman, 1920), pp. 172–184; Myers, *Mind and Work* (New York: Putnam's, 1921); Myers, "Industrial Fatigue," *Lancet* 1 (1925): 905–908.
32. Lee, *Human Machine*, p. 13; see Goldmark and Hopkins, *Comparison*, p. 29 for a similarly qualified analogy.

33. Vernon, *Industrial Fatigue*, p. 34; Lee, *Human Machine*, p. 13; Florence, *Economics of Fatigue and Unrest*, pp. 383–384.
34. Vernon, however, held that accidents did not measure fatigue, and that they were due rather to the workers' carelessness (the employers' traditional explanation) and the "psychical factor"; see Vernon, *Industrial Fatigue*, pp. 179–226.
35. Lee, *Human Machine*, pp. 14–18; Florence, *Economics of Fatigue and Unrest*, pp. 359–371. Vernon, however, chose not to treat fatigue tests in his book, arguing that all were impractical for use in industry; see Vernon, *Industrial Fatigue*, pp. 3–4.
36. Lee, *Human Machine*, pp. 24–78; Vernon, *Industrial Fatigue*, passim; Florence in CIF, 26 June 1918; see summary of the conclusions and recommendations of the CIF study in Goldmark and Hopkins, *Comparison*, p. 26.
37. Florence in CIF, 26 June 1918.
38. Even British researchers associated industrial physiology with Lee; see Sir Thomas Oliver to Lee, 30 May 1930, FSL, box 2.
39. Lee, *Human Machine*, p. 103; Lee, "The New Science of Industrial Physiology," *Public Health Rep.* 34 (1919): 723–728, esp. 724.
40. Lee, "Industrial Efficiency," *Public Health Rep.* 33 (1918): 29–35, quote on 30.
41. For an extended discussion of the emphasis on productivity by members of the British IFRB and Factory Inspectorate, see Whiteside, "Industrial Welfare and Labour Regulation."
42. Lee, *Human Machine*, pp. 39–42, quote on 100; see also the discussion of restriction of output or stereotyping in Goldmark and Hopkins, *Comparison*, pp. 85–94; Lee to Newton D. Baker, 5 October 1923, FSL, box 1.
43. See, for example, Lee, "Is the Eight-Hour Working Day Rational?" p. 729; Myers, *Mind and Work*, p. 41.
44. Lee, *Human Machine*, pp. 90–95; Lee, "Human Machine in the Factory," p. 233; on the use of fatigue allowances in time and motion studies, see William Gomberg, *A Trade Union Analysis of Time Study* (Chicago, IL: Science Research Associates, 1948), pp. 71–82.
45. Sherrington to Lee, 20 December 1908, FSL, box 2.
46. Fletcher, "The Aims and Boundaries of Physiology," *Report BAAS, 1921*, pp. 125–142, quotes on 131, 132.
47. Lee, "Human Machine in the Factory," p. 233.
48. On British physiology, see Geison, *Michael Foster and the Cambridge School of Physiology* (Princeton, NJ: Princeton Univ. Press, 1978).
49. On the relationship between physiology and medical education, see Geison, "Divided We Stand"; Kohler, *From Medical Chemistry to Biochemistry* (Cambridge, UK: Cambridge Univ. Press, 1982); Pauly, "Appearance of Academic Biology"; and Alejandra Laszlo, chapter III, this book.
50. On Lee's attempt to have physiology made part of the science faculty at Columbia, see Pauly, "Appearance of Academic Biology," pp. 287–289; on his campaign to shift the College of Physicians and Surgeons to Morningside Heights, see Lee to Sherrington, 21 October 1912, FSL, box 2; for his views and proposals on medical education, see Lee, "A Proposed Undergraduate Course in Clinical Physiology," *J. Am. Med. Assoc.* 64 (1916): 639–641; Lee, Richard M. Pearce, and W. B. Cannon, "Medical Research in its Relation to Medical Schools," *J. Am. Med. Assoc.* 68 (1917): 1075–1079.
51. CIF, 26 June 1918.
52. IFRB, *Report*, 1921, pp. 5–8; 1922, pp. 4–7; on the PHS research, see Frederick Bonner Flinn, "Industrial Aspects of Human Fatigue" *J. Personnel Res.* 4 (1923–1924): 285–293
53. Reynold A. Spaeth, "The Problem of Fatigue," *J. Ind. Hyg.* 1 (1919/1920): 22–53, quote on 25; Spaeth, "The Prevention of Fatigue in Manufacturing Industries," *J. Ind. Hyg.* 1 (1919–1920): 435–447, quote on 435; Muscio, "Is a Fatigue Test Possible?" *Br. J. Psychol.* 12 (1921/1922): 31–46. For other criticisms of fatigue, see Emery R. Hayhurst, "Points in the Detection of Industrial Fatigue and Measures for its Possible Complete Elimination," *J. Ind. Hyg.* 2 (1920/1921): 256–258; Henry C. Link, "A Practical Study in Industrial Fatigue," *J. Ind. Hyg.* 1 (1919/1920): 233–237.
54. C. K. Drinker, *J. Ind. Hyg. Abstr.* 2 (1920/1921): 109–115, quote on 115; Goldmark and Hopkins, *Comparison*, quote on p. 350. Lee and Edsall had forced refinements in Florence's definition of fatigue, while psychologist Raymond Dodge had made criticisms similar to those later made by Muscio; see CIF 22 October 1917, 26 November 1917.

55. Muscio, "Is a Fatigue Test Possible?"; Myers, *Mind and Work*, esp. pp. v–vi; Myers, "The Study of Fatigue," *J. Personnel Res.* 3 (1924/1925): 321–334. Lee and the American physiologists took Muscio's criticisms seriously; see Martin to Lee, 15 June 1922, FSL, box 2.
56. Alfred D. Flinn, "Problems of Industrial Personnel: Tentative Scheme for Research," 20 August 1919, NRC, Executive Board, Committee on Industrial Personnel Research; *NRC Annual Report*, 1920, pp. 14–15; 1921/1922, pp. 13–14; Lee to R. G. Hussey (Division of Medical Sciences, NRC), 16 June 1919, FSL, box 1; Robert M. Yerkes, "What is Personnel Research?" *J. Personnel Res.* 1 (1922/1923): 56–63. On the history of industrial psychology, see Loren Baritz, *The Servants of Power* (Middletown, CT: Wesleyan Univ. Press, 1960).
57. Spaeth, "Prevention of Fatigue," pp. 435–447. Surveys of industrial medicine and hygiene include W. Irving Clark, Jr., *Health Service in Industry* (New York: Macmillan, 1922); National Industrial Conference Board, *The Physician in Industry* (New York: NICB, 1922); J. D. Hackett, *Health Maintenance in Industry* (Chicago, IL: Shaw, 1925).
58. Lee to Surgeon General, PHS, 26 September 1919; Ryan to J. W. Schereschewsky, 13 April 1922, National Archives, Washington, DC, Record Group 90 (Public Health Service), 2048M/197.
59. Sherrington to Lee, 31 October 1927, FSL, box 2; A. Landsborough Thomson, *Half a Century of Medical Research*, vol. 2 (London: HMSO, 1973–1975), pp. 167–168.
60. Richard Gillespie, "Manufacturing Knowledge: A History of the Hawthorne Experiments," (Ph.D. diss., Univ. of Pennsylvania, 1985); S. M. Horvath and E. D. Horvath, *The Harvard Fatigue Laboratory: Its History and Contribution* (Englewood Cliffs, NJ: Prentice-Hall, 1973).
61. Elton Mayo, "Revery and Industrial Fatigue," *J. Personnel Res.* 3 (1924/1925): 273–281; Mayo, *The Human Problems of an Industrial Civilization* (New York: Macmillan, 1933), chapt. 1, quote on p. 8.
62. D. B. Dill et al., "Industrial Fatigue," *J. Ind. Hyg.* 18 (1936): 417–431, quote on 418; Dill, "Applied Physiology," *Annu. Rev. Physiol.* 1 (1939): 551–576.
63. See, for example, F. J. Roethlisberger and W. J. Dickson, *Management and the Worker* (Cambridge, MA: Harvard Univ. Press, 1939), chapt. 5; George C. Homans, *Fatigue of Workers: Its Relation to Industrial Production* (New York: Reinhold, 1941); Gillespie, "Manufacturing Knowledge."
64. See, for example, the first volume of the *Journal of Applied Physiology*, established by the American Physiological Society in 1948.

XI

Physiological Identity of American Sex Researchers Between the Two World Wars

DIANA E. LONG

Sexology as a field of medicine and social science was not named until after World War II, but an organized field of sex research was promoted by the Committee for Research in Problems of Sex (CRPS) in the 1920s and 1930s with funding from the Rockefeller Foundation and support from the National Research Council (NRC). The field benefitted from the Rockefeller Foundation's desire to reform both sex and medical education during this period. Originally funded by the John D. Rockefeller Jr. Bureau of Social Hygiene and a subunit of the Medical Sciences Division of the NRC, the committee became, after 1933, a joint subunit of the Medical and Natural Sciences divisions of the Rockefeller Foundation.[1] The major achievement of this sex research project during the interwar period was a new understanding of sex function and its regulation in higher animals and humans. This new understanding was summarized for the first time in *Sex and Internal Secretions*, published by the CRPS in 1932 and in an enlarged edition in 1939.[2]

Endocrinology, the study of the hormones (internal secretions) of the ductless glands, was the field in which the funding of the CRPS had the greatest impact in the 1930s. This is evident if one looks at the new textbook or the publication of articles by those whom the CRPS funded in its first two decades. By 1938, these grantees were publishing fully one quarter of their articles in *Endocrinology*, the journal of the Association for the Study of Internal Secretions that had been founded in 1916.[3]

During the interwar period, the new field of endocrinology was closely associated with the older discipline of physiology that is the subject of this book. Physiology was the discipline that described organic functions and their regulation in higher animals and humans. The question I want to address here is whether or not the sex research supported by the CRPS was seen by its authors as part of American physiology. More specifically, I will discuss the sex research published in the official journal of the American Physiological Society (APS), the *American Journal of Physiology* (*AJP*). I will analyze the sex researchers in an attempt to enlarge our understanding of the opportunities and the relationships of physiologists to other disciplinary communities during this period. Throughout this chapter, I use the term "sex research/researcher" in quotes to indicate research sponsored by the CRPS. Twenty-nine grantees (thirty-one individuals) published 175 articles between 1923 and 1947 in the *AJP* that acknowledged the support of the committee.

The CRPS was a committee of the NRC funded by the Rockefeller Foundation. The biomedical members of the committee in the early years were Robert M. Yerkes, a psychobiologist who served as chair of the CRPS, Walter B. Cannon, doyen of American medical science, and Frank R. Lillie, the "captain" of a growing biological enterprise at the University of Chicago.[4] The two other original members of the committee were Katherine B. Davis, director of the Bureau of Social Hygiene, which handled the Rockefeller funds from 1921 until 1931, and Thomas W. Salmon, psychiatrist and leader of the mental hygiene movement. By 1947, these innovative and well-established statesmen of American science had awarded grants to 104 researchers comprising 101 granting units (grantees). By then, CRPS grantees had published about 1,800 articles and books in a wide variety of journals of medical, natural, and social science that explicitly acknowledged support of the CRPS.

A glance at the *AJP* before the CRPS came into existence makes it abundantly clear that the physiologists had discovered both sex and endocrine physiology before 1922. Thus volume 41, for July–November 1916, has seven articles on endocrinology, including one by Cannon, and six on sexual reproduction, including one by endocrinologist Oscar Riddle. The field of sex research did not become completely dependent on the CRPS after it was created in 1921. In 1928, the *AJP* published 257 articles, nineteen on sex and thirty-eight on endocrinology. Twelve of the nineteen articles on sex were listed as supported by the CRPS, yielding a percentage of 63 that was fairly constant in the late 1920s and 1930s. Even after 1923, CRPS grantees continued to publish articles that were not supported by that committee. Between 1916 and 1937, twenty-three CRPS grantees published articles in the *AJP* that were independent of CRPS support. They had other resources.

Yet the 175 articles published in the *AJP* by CRPS grantees between 1923 and 1941 suggest that a community of interest[5] in sex physiology had formed in the United States with the encouragement and support of the committee. The community of "sex researchers" that chose to publish in the *AJP* was small, consisting of thirty-one individuals. Table 1 shows the results if the corpus of sex physiology is broken down into five-year periods.

During the 1920s and 1930s, when the *AJP* was edited by Donald Hooker, the criteria for publication were twofold: (1) the inherent quality of the research and (2) the appropriateness of the article for publication in a journal of physiology. Hooker was himself a grantee of the CRPS and a longtime contributor to the *AJP* who chose to publish the results of his sex research elsewhere. His very definite views of the definition of physiology were articulated in the annual reports he wrote as editor of the *AJP* for the APS in which he justified his acceptance or rejection of articles for publication. The reports are especially informative between 1929 and 1933, when the APS went through an organizational crisis that produced a "new statement of policy" in the 1929 annual report and an editorial board of trustees in 1933. The editorial board was divided up by fields of which endocrinology, but not sex research, was one. The endocrinologist on the board was Roy Hoskins,

Table 1
Breakdown of Published Articles

1923–1927	29 articles from	6 grantees
1928–1932	59 articles from	14 grantees
1933–1937	49 articles from	17 grantees
1938–1942	32 articles from	13 grantees
1943–1947	6 articles from	5 grantees

American Journal of Physiology articles: 175 by 31 people as 29 grantees

the editor of *Endocrinology*, who had once applied for but not received a grant from the CRPS.

Under Hooker, the *AJP* was a journal of laboratory research, focusing on the study of function and control in vertebrate animals.[6] Hooker and his editorial board agreed that the 175 articles by CRPS researchers were appropriate for the *AJP*.

However, if these 175 articles belonged to physiology, as Hooker and the board defined it, can we also say that the authors and supervisors of those publications were physiologists? The answer to this question will tell us a good deal about the meaning of that term in the United States in the 1920s and 1930s when the field of sex research was developing in its American context.

In an effort to point toward an answer to the question, I want to offer a group portrait of the thirty men and one woman who published sex research in the *AJP*. I will explore when they chose to publish the results of their research and what professional societies they chose to join. I will try to find several ways to ask about their disciplinary self-definition before turning to their official history by Sophie Aberle and George Corner for further information about disciplinary affiliations.

WHERE THE "SEX RESEARCHERS" PUBLISHED

The "sex researchers" published in the *AJP* when they wanted to reach physiologists. These thirty-one scholars, on twenty-nine grants, (who produced a total of about 1,100 publications for the CRPS) published from one to twenty-seven articles each in the *AJP*. (See Table 6 for the articles they wrote for the *AJP*.)

Obviously the thirty-one scientists in question here published only a portion of their sex work in the *AJP*, which for them ranked second in popularity to the *Proceedings of the Society for Experimental Biology and Medicine*. They published even more frequently in *Anatomical Reviews*, but a majority of these were abstracts rather than full-fledged articles. The *AJP*, while not the journal of first choice for the "sex researchers," was clearly very important to them during the early years of the funding of sex research. Table 2 summarizes the publishing history of this group of "sex physiologists," that is, those funded by the CRPS who sometimes published in the *AJP*.[7]

Table 2
Journals in Which the Thirty-One CRPS Grantees Published

Journal	Number of Articles	Sex Physiologists	
		Number	Percent
Am. J. Physiol.	175	31	100
Proc. Soc. Exp. Biol. Med.	191	15	49
Endocrinology	115	14	46
Anat. Rev.	261	12	39
Articles	103	12	
Abstracts	158	10	
Am. J. Anat.	18	8	26
J. Exp. Zool.	28	4	13
Physiol. Zool.	37	3	10

DISCIPLINARY SELF-DEFINITIONS BY THE "SEX RESEARCHERS"

If thirty-one of the "sex researchers" published articles in the *AJP*, did they do so as physiologists? Were they sufficiently committed to the discipline of physiology to belong to the APS and thus to receive the *AJP* regularly and have the opportunity to discuss their sex research at Society meetings? Of the thirty-one "sex researchers," twenty belonged to the APS, more than belonged to any other scientific society except the American Association for the Advancement of Science. By comparison, only fifteen of them belonged to the American Association of Anatomists. As a matter of fact, Cannon, Philip Bard, and Henry Bazett were elected presidents of the APS in 1914, 1941, and 1950 respectively. Fifteen other CRPS grantees who did not publish a sex physiology article in the *AJP* were also members of the APS.

This data on society memberships is culled from *American Men of Science*,[8] from which we can learn still more about the status of this group of thirty-one "sex researchers." By 1938, only five of them—George Parker, Cannon, Lillie, Herbert M. Evans, and Carl Hartman had been elected to the National Academy of Sciences. By 1955, however, this honored group included five others: Bard, Frederick Hisaw, Charles Huggins, P. E. Smith, and Calvin Stone.[9] In 1938, sixteen of the sex researchers won "stars" in *American Men of Science* that indicated their prestige among their peers in science. Twenty-four of the thirty-one had already authored articles in the *AJP* before or while they received grants from the CRPS. They had already been recognized as producers of worthwhile physiological research.

How did these thirty men and one woman define themselves in the sixth edition (1938) of *American Men of Science*? They had three opportunities to indicate their disciplinary identity: in naming their department, in naming their research field, and in the self-identifying label given at the beginning of each entry. The last was perhaps the most important index of disciplinary affiliation and the answers given by our thirty-one "sex physiologists" are revealing, if confusing.

They fall into four groups: physiologists, zoologists, anatomists, and clinicians. "Clinician" refers to those who published on human sexuality as members of clinical medical departments, or as psychologists; it also includes one chemist, Reuben Gustavson, who was a joint CRPS grantee with the gynecologist Robert Frank (Table 3).

Only eleven of the thirty-one CRPS grantees who published as physiologists in the *AJP* identified themselves as physiologists. The departmental affiliations of these eleven generally accord with their self-identification but not uniformly. The departments listed for them in the 1938 edition of *American Men of Science* are in Table 4.

The striking fact that emerges from the disciplinary self-definitions of our thirty-one "sex researchers" is that barely a third identified themselves as "physiologists." Even among these eleven men, four taught or worked in areas other than physiology. Huggins listed his first discipline as "surgery" and Hartman as "embryology." Two others, James Slonaker and Gregory Pincus, identified their field as "zoology, physiology."

The situation is further complicated by the fact that a number of CRPS grantees who called themselves physiologists and described their work as physiology in *American Men of Science* or in their CRPS grant proposals[10] did not publish in the *AJP*. (There were in all 104 CRPS grantees, only thirty-one of whom published in the *AJP*.) The most interesting of these are the seven people who labeled themselves physiologists in the 1938 edition of *American Men of Science* but who published none of their sex research in the *AJP*. This group of seven includes H. S. Liddell; Hooker, the editor of the *AJP*; and two of the three women who received CRPS grants, Evelyn Howard and Ruth Okey. Another three had mixed identities: W. R. Miles, who combined psychology and physiology, F. C. Koch, who combined physiology and physiological chemistry, and William U. Gardner, who combined physiology and anatomy. If one adds these seven to the eleven *AJP*-publishing physiologists, we see that only eighteen of the 104 CRPS grantees identified themselves as physiologists, eight of them jointly with a variety of other disciplines.

FUNDING LEVEL, STATUS, AND PRODUCTIVITY OF CRPS GRANTEES

The identity of the "sex researchers" with physiology becomes more problematic still if one looks at the related issues of funding, status, and publication rate of the thirty-one "sex physiologists" identified previously. The first two indices contradict one another. From the funding patterns of the CRPS,[11] one might conclude that the anatomists and zoologists enjoyed the greatest stature among the biological scientists who did sex research. At the CRPS the largest grants went to, and the most publications came from, those who identified themselves as anatomists and zoologists in *American Men of Science*. The pecking order in terms of grants and publications is shown in Table 5.

From Table 5, we see that the six anatomists and zoologists received $468,650 from the CRPS and had 808 publications. The total alloted for all 104 researchers by the Rockefeller philanthropies from 1922 to 1947 was $1.525 million and the total number of publications was about 1,800[12]; these

Table 3
Disciplines of the Thirty-One CRPS Grantees Published in *AJP*

Discipline	Name
Physiology	Philip Bard Henry Bazett S. W. Britton Walter B. Cannon Maurice Friedman Robert Oslund Samuel Reynolds James Slonaker (first affiliation, zoology) Gregory Pincus (first affiliation, zoology) Charles Huggins (first affiliation, surgery) Carl Hartman (first affiliation, embryology)
Zoology	Frederick Hisaw Frank R. Lillie (second affiliation, embryology) George Parker Gregory Pincus (second affiliation, physiology) James Slonaker (second affiliation, physiology) W. W. Swingle William Young
Anatomy	Edgar Allen George W. Corner R. S. Cunningham Herbert M. Evans Joseph E. Markee S. W. Ranson (also directed a neurology institute) Earl T. Engle, joint grant with P. E. Smith P. E. Smith (also had own grant) Franklin Snyder
Clinical Research (and Chemistry)	Sophie Aberle (ob/gyn, worked as anthropologist) Robert Frank (ob/gyn, joint grant with Reuben Gustavson) Reuben Gustavson (chemistry, also had own grant) Charles Huggins (surgery, physiology) Adolf Meyer (psychiatry, neurology, published with Hartman) John Rock (ob/gyn, not in 1938 or 1955 *American Men of Science*) Calvin Stone (psychology)

six anatomists and zoologists (a mere 6 percent of the grantees) thus gained about 30 percent of the funds and published almost half the papers listed in the bibliography by Aberle and Corner. There was, in short, a very uneven

Table 4
Self-Identified Physiologists

Name	Degree	Department
James Slonaker	Ph.D. 1896	Emeritus Professor, physiology (Stanford)
Walter B. Cannon	M.D. 1871	Higginson Professor, physiology (Harvard)
Carl Hartman	Ph.D. 1915	Resident Associate, embryology (Carnegie Institute)
Henry Bazett	M.D. 1919	Professor, physiology (University of Pennsylvania)
S. W. Britton	M.D. 1924	Professor, medicine (University of Virginia)
Robert Oslund	Ph.D. 1923 M.D. 1932	Practicing physician
Philip Bard	Ph.D. 1927	Professor, physiology (Johns Hopkins)
Charles Huggins	M.D. 1924	Instructor, surgery (University of Chicago)
S. R. M. Reynolds	Ph.D. 1931	Instructor, physiology (Long Island College of Medicine)
Maurice Friedman	Ph.D. 1928 M.D. 1933	Assistant Professor, physiology (University of Pennsylvania)
Gregory Pincus	Sc.D. 1927	Assistant Professor, general physiology (Harvard)

Table 5
Grants to Zoologists and Anatomists from the CRPS

	Grant	Number of Articles Published
Zoologists		
Frank R. Lillie	$146,040	309
Frederick Hisaw	32,000	70
Anatomists		
Herbert Evans	108,630	181
Edgar Allen	106,880	77
P. E. Smith/Earl T. Engle	75,100	171

division of labor and reward among our "sex researchers." The contribution of some zoology and anatomy departments was overwhelming.

It would be a mistake to conclude, however, that readers of the *AJP* saw the new sex research as part of anatomy or zoology. In numbers, the contributions of self-described physiologists, anatomists, and zoologists to the *AJP* were fairly evenly distributed. The nine anatomists contributed forty-four articles on sex to the *AJP*, the seven physiologists published forty-two articles, the four physiologists who also identified themselves as zoologists, embryologists, or surgeons contributed thirty-five articles, and the five other zoologists (one of whom bore the joint title of embryologist) published thirty-five. The six sex researchers in the medical group and one chemist published thirteen articles (Table 6).

Table 6
American Journal of Physiology:
175 Publications by 29 Grantees of the CRPS

Year	Number of Publications	Grantee	Author	(Number and Type of Publication)
1923	1	(1) Evans		abstract
1924	6	(2) Stone	Stone	2 articles
		(3) Lillie	Moore, Oslund	
			Moore, Quick	
			Oslund	
		(4) Slonaker	Slonaker	
1925	7	Slonaker	Slonaker	4 articles
		Stone	Stone	2 articles
			Stone, Miles	
1926	8	(5) Oslund	Oslund	
		Lillie	Moore	
			Quick	
		Evans	Evans, Burr	
			Teel	2 articles
		Slonaker	Slonaker	
		Stone	Stone	
1927	7	(6) Corner	King*	
		Evans	Evans, Hoagland	
		Slonaker	Slonaker	5 articles
1928	12	Evans	Evans	2 articles
		Lillie	Moore, McGee	
			McGee, Juhn, Domm	
			Gallagher	
		(7) Parker	Parker	
		(8) Smith/Engle	Engle, Mermod	
		Slonaker	Slonaker	3 articles
		Parker	Parker	
		(9) Allen	Allen	
1929	9	Evans	Evans, Simpson*	4 articles
		Lillie	Juhn,* Mitchell	
			Moore, Gallagher	
		Smith/Engle	Engle	
		Slonaker	Slonaker	
		(10) Frank/ Gustavson	Frank, Gustavson, McQueen, Goldberger	
1930	7	Lillie	Juhn,* Domm	
			Juhn,* D'Amour, Womack*	
		(11) Swingle	Nelson, Pfiffner, Haterius	
		(12) Gustavson	Kunde, D'Amour, Gustavson	
		(13) Hisaw	Hisaw, Leonard	
			Hisaw	abstract
		Slonaker	Slonaker	

Table 6—Continued

Year	Number of Publications		Grantee	Author	(Number and Type of Publication)
1931	19	(14)	Friedman	Reynolds	
			Bazett	Friedman	
				Reynolds	
				Weinstein, Reynolds, Friedman	
			Lillie	Moore, Samuels	
				Juhn*, Barnes*	
				Moore, Price*	
				Ihrke, D'Amour	
			Evans	Evans, Swezey*	
				Evans, Meyer, Simpson*	
				Evans, Simpson*	
		(15)	Smith	Shelesnyak	
			Slonaker	Slonaker	7 articles
1932	12	(16)	Markee	Markee	
			Smith/Engle	Smith	2 articles
				Severinghaus	
			Smith	Leonard	
			Friedman	Friedman	2 articles
			Hisaw	Leonard, Hisaw, Fevold	
				Martin	
			Evans	Evans	
				Reichert, Pencharz, Simpson, Meyer	
				Reese, McQueen-Williams*	
1933	15	(17)	Aberle	Aberle	
		(18)	Britton	Silvette, Britton	
			Britton		5 abstracts
			Hisaw	Fevold, Hisaw, Hellbaum, Hertz	
			Friedman	Friedman*, Friedman	
			Smith	White, Leonard	
			Lillie	Nelson, Smelser	
			Markee	Hinsey, Markee	
			Smith	Shelesnyak	
			Engle	Schockaert	
				Engle	
1934	8		Slonaker	Slonaker	
		(19)	Cunningham	Wolfe	
			Britton	Corey, Britton	
			Smith	Leonard	
			Engle	Leonard, Smith	
				Engle	
			Hisaw	Hertz, Hisaw	
				Fevold, Hisaw	

Table 6—Continued

Year	Number of Publications		Grantee	Author	(Number and Type of Publication)
1935	10	(20)	Reynolds	Sauer, Jett-Jackson, Reynolds	
				Reynolds, Kaminester	
			Allen	Allen, Diddle, Elder	
		(21)	Pincus	Burdick, Pincus	
			Hisaw	Greep	
				Lane	
		(22)	Bard		2 abstracts
			Slonaker	Slonaker	2 articles
1936	9		Allen	Allen, Diddle, Burford, Gardner	
			Bard		abstract
			Hisaw	Fevold, Hisaw, Greep	
			Gustavson	Halpern, D'Amour	
			Pincus	Pincus, Kirsch	
			Britton	Britton, Kline	
		(23)	Young	Dempsey, Hertz, Young	
			Reynolds	Reynolds, Kaminister	
				Fagin, Reynolds	
1937	7	(24)	Cannon	Reboul	abstract
				Reboul, Friedgood	
				Friedgood, Foster	
			Smith	Tyndale, Levin	
			Bard	Brooks	abstract
				Root, Bard	abstract
			Young	Dempsey	
1938	6		Lillie	Plagge	
			Bard	Brooks	2 articles
			Cannon	Friedgood, Bevin	
				Friedgood, Foster	
			Slonaker	Slonaker	
1939	6		Lillie	Browman	
			Reynolds	Reynolds, Foster	
			Cannon	Friedgood	
		(25)	Snyder		abstract
			Bard	Brooks, Lambert	
				Davis	
1940	12		Bard		2 abstracts
				Brooks, Gersh	
		(26)	Hartman	Ball*	
			Meyer		
			Pincus	Graubard, Pincus	
			Ranson	Dey, Fisher, Berry, Ranson	
			Reynolds	Reynolds, Foster	3 articles
		(27)	Rock		2 abstracts
			Smith	Jailer	

Table 6—Continued

Year	Number of Publications		Grantee	Author	(Number and Type of Publication)
1941	4	(28)	Huggins	Berg, Huggins, Hodges	
			Pincus	Graubard	
				Pincus	
		(29)	Ranson	Brookhart, Dey	
1942	4		Bard	Leonard, Brooks	
			Smith	Starkey, Leathem	
				Levin, Leathem, Crafts	
				Levin, Leathem	
1943	4		Huggins	Huggins, Vail	
			Lillie	Taber, Davis, Domm	
			Rock	Snodgrass, Rock, Menkin*	
			Smith	Levin	
1944	1		Smith	Levin	
1945	0				
1946	0				
1947	1		Bard	Root, Bard	

* Woman

The issue of status was not merely one of funding and number of publications. The available data on contemporary judgments of the scientists (from *American Men of Science*) and of their work (through publication in the *AJP*) leads to the conclusion that physiology had a higher status in American sex science during the interwar years than the more heavily funded areas of anatomy and zoology. The CRPS sought out as its early grantees those who had published in the *AJP*; the grantees themselves—even those who were anatomists and zoologists—became members of the APS to a far greater extent than their physiological counterparts sought external affiliations in other disciplinary societies. Anatomists in this group of "sex researchers" published in the *AJP* far more often than the physiologists among them published in anatomy journals.

WAS THIS RESEARCH SEX PHYSIOLOGY?

At the start of this chapter, I suggested that the research supported by the CRPS belonged to endocrinology and other parts of the domain of physiology because of its concern with control and function in vertebrate animals. What are we to make, then, of the fact that the official history of the CRPS, a brilliant and subtle discussion of that committee by Aberle and Corner in 1953, scarcely uses the word "physiology" in its description of the program of research sponsored by the CRPS? *Twenty-Five Years of Sex Research* discussed the need to create a science of sex that would explain the underlying biological

Table 7
Topics of Research Funded by the CRPS, 1927–1932

Topic	Author Published in *AJP*
Cycle and Hormones	Evans, Lillie, Hisaw, Frank, Gustavson
Sex Differentiation	Lillie, Swingle
Pituitary and Adrenal	P. E. Smith, Engle, Cunningham
Other Histological and Physiological Topics	Friedman, Markee, Parker, Slonaker
Insects	
Sex Behavior, Psychology	Stone
Anthropology	*
"Centers"	Lillie, Evans, Smith, Hisaw, Allen

* No *AJP* author published on this subject. [From Aberle and Corner, *Twenty-Five Years* (see n.1).]

mysteries of sex in animals before attempting to understand or alleviate the medical and psychological problems of man.[13] Its definition of sex was very broad: reproduction, sexuality, and sex dimorphism. The CRPS committee and its historians did not list the particular features of that biological program by discipline but rather by topic. Table 7 lists the topics funded by the CRPS from 1927 to 1932, the years of most active publication by the grantees.

Not all of these research programs produced publications in the *AJP*, but most of those that concerned nonhuman subjects or "lower forms" of life (as Corner put it) did. They included studies of the reproductive, sexual, or dimorphic systems of the body that could be analyzed in terms of control by the hormones or nervous system. By the 1930s when this research program had settled into sustained arguments about the sex hormones, the articles by anatomists, physiologists, and zoologists look very much alike. The most intense debate of the 1930s in the field of sex endocrinology, the debate about the number and function of the hypophysial hormones and their relation to the better known gonadal and adrenal hormones in laboratory animals, was led by anatomists Smith and Evans on the one hand, and zoologists Hisaw and Lillie on the other. Only two of the sex researchers who identified themselves as physiologists had projects that were part of this central research project. S. W. Britton was interested in the effect of adrenal hormones on sex, and Pincus studied the metabolism of the ovarian hormones in mammals. What was distinct about the research of the physiologists? They were more apt than the other sex researchers to correlate the physiology of sex with animal behavior (e.g., see the works of Bard, Cannon, Hartman and Meyer, and Slonaker), or with medically useful information about X-rays (Robert Oslund), prostate cancer (Huggins), hormones in ovulation (Maurice Friedman and Bazett), and uterine contraction (Samuel Reynolds).[14]

CONCLUSION

This survey of the meaning of the term "physiologist" as it applied to sex researchers in the United States in the 1920s and 1930s has led us to a number of seeming contradictions about the definition of that discipline in relation to sex research. These contradictions come, I think, from our expectations. The views of physiology are confusing if one expects to see a department, journal, or other institution populated only by people who chose to call themselves physiologists. This brief study, like other prosopographies of the period, suggests that American biomedical institutions were more complicated. Certainly Corner's history of the CRPS makes it clear that he, as an anatomist, felt comfortable with an absence of disciplinary labels.[15] One might even say that the field of sex research needed the blurred disciplinary boundaries that have been noted in this chapter. Those men and women who funded and supported the "sex researchers," like the sex researchers themselves, had a number of professional interests.

This field of interests in American sex research between the two World Wars deserves a larger study, but even now we can draw some conclusions about the "sex researchers" who published in the *AJP*. First the organizers of the CRPS called for basic medical science that was not only useful in education and practice, but also impeccable in terms of research standards. The CRPS established sex research as part of the concern of both the NRC and the Rockefeller Foundation with medical education. Their grantees worked largely in medical schools, whether they had M.D.s or Ph.D.s. This pattern applied to all twenty of the anatomists and physiologists who were grantees of the CRPS and who published in the *AJP*, whether their highest degree was an M.D. (8), Ph.D. (9), or both (3). The zoologists, though usually located in the faculty of Arts and Sciences, were also attracted to the medical schools, perhaps because—as one zoology student of Lillie put it—"there seems to be more demand for physiology and pathology."[16] The career patterns of those who published in the *AJP* suggest as much. Both Slonaker and Pincus transferred from departments of zoology to physiology. In 1937–1938 William Young was an honorary fellow at the Yale School of Medicine (along with two other CRPS grantees Edgar Allen and, a littler earlier, Aberle). At Yale, both Young and Swingle held appointments in biology rather than zoology while, at the University of Chicago, Lillie occupied the whole settee of the Division of the Biological Sciences after 1934.

This research had another very clear role, however: to solve problems of sex, real human problems. This cultural context for sex research is evident in Corner's history of the CRPS and in his own autobiography. In addition to discovering progesterone and chairing the CRPS, Corner contributed to sex education by publishing a highbrow book, *Hormones in Human Reproduction*,[17] that gracefully summarized the new scientific information on sex physiology in a way that allowed doctors and educators to enlighten their adult constituencies. Corner also provided a popular version of this enlightened expertise on human sexuality in his books for boys and girls. His *Attaining Manhood* and *Attaining Womanhood*[18] tried to make boys and girls understand the realities of their sexuality as revealed by the new sex research.

Corner's sex manual for girls also asked an informed citizenry to support the CRPS in its quest for enlightened sex mores.[19] More generally, the history of the CRPS by Aberle and Corner sought to promote the organization and support of biomedical research. Aberle's chapters on the functioning of the CRPS accurately describe that committee as an important model for the organization of large-scale government funding of biomedical science after

World War II. The committee was important and successful at the difficult task of developing an interdisciplinary field of research in an area that was not only culturally controversial but professionally competitive.

The word "interdisciplinary" was not explicitly deployed by Aberle and Corner in their history of the CRPS, but the phenomenon was important to the success of the committee. The new field of sex research, especially in the light of its definite if muted goals of social reform, needed to be comprehensive. Aberle and Corner recall how the CRPS wanted to get "young workers" committed to sex research, whatever their field.[20] As a contemporary observer of the new field, Corner emphasized the importance of the size and competence of the community rather than its disciplinary affiliations. This view was shared by Warren Weaver, one of the Rockefeller Foundation officers responsible for oversight of the CRPS and organizer of a 1934 conference on sex biology at Woods Hole, Massachusetts. In his report on that conference, Weaver enthusiastically proclaimed:

> One gets a vivid picture of the great advance which has been made under the committee. In 1900–1910, there were perhaps six investigators in all, making any progress on such problems. There are now well over six investigators at each of a considerable group of institutions. The undoubted leadership of the United States in endocrinology can be credited directly and indirectly, to the committee's activity.[21]

In Weaver's view of the new field, the goal was to build up a community of sufficient size and talent to solve the complex problems of the functioning and control of the sexual, dimorphic, and reproductive systems of higher organisms. As previously suggested, much of this new field hardly belonged to "physiology," strictly speaking, but it was physiological enough to benefit from the special advantage of that disciplinary affiliation for those who published in the *AJP*.

The advantages that accrued to physiologists in sex research were very real, as should be clear from our collective portrait of the "sex researchers" who published in the *AJP*. That portrait revealed, first of all, the prestige ascribed to physiology, to the APS, and to individual physiologists. Cannon was the apotheosis of that field for the early generation although Huggins, jointly a surgeon, was the only Nobel Laureate in this group. The APS was important to them and publication in its journal (the *AJP*) was important, especially during the early years of sex research. The anatomist Evans, for example, published sixteen articles in the *AJP* on his sex grant from 1926 to 1932, but none thereafter.

Anatomists like Corner and Evans felt no need to call themselves physiologists, but their work gained publication in the *AJP* because it was functional laboratory research on higher organisms. When this research became the basis for teaching in medical school education, its usual locus was the physiology segment of the curriculum.

By the 1950s, sex endocrinology enjoyed a prominent place in physiology textbooks for medical students.[22] In this way, physiology formed the stable base on which sex research was built. At the same time the physiological approach in sex research fit well with the reform of medical education that was then under way. Finally, the physiological approach fit well with the goal of reforming human sexual mores, a goal of great concern to the founders of the CRPS. In Corner's account of the history of the CRPS, and in the grant proposals to that committee, the concern with human sexuality was pictured as a natural extension of the physiology program. Corner thus lists the research of several of those who identified themselves as physiologists and who published in the *AJP* in his chapter on "Contributions to Knowledge of Neural and Psychological Aspects of Sex Behavior."[23] In the early 1930s, this research program included grants to Cannon and Bard at Harvard, and to the Hartman/Adolf Meyer team at Johns Hopkins.

Sexology per se, the study of human sex behavior and problems, developed as an autonomous field after World War II with the work of William Masters and others. Masters records that his mentor, Corner, gave him advice about the background needed for an individual doing research on human sexuality: "He should wait until he was at least forty before tackling sex research. He should first earn a scientific reputation in some other science. He should wait until he could secure the sponsorship of a major medical school or university."[24]

As a community interested in the advancement of sex research, the CRPS grantees surveyed here fit Corner's description very well. The "sex researchers" who published in the *AJP* were on the average thirty-nine years old. They held posts at major universities. Their vitae show that at some points in their careers eleven spent time at the University of Chicago, eight at Johns Hopkins University, seven at Harvard, six at Stanford, and three each at Yale, Columbia, and the University of California. In a sense, the whole field of sex research went through an apprenticeship in the basic medical and biological sciences before it went on to study *Human Sexual Response.*[25] This chapter, by focusing on the careers of key "sex researchers" who were funded by the CRPS and published in the *AJP*, suggests that the field of sex research was hardly coterminous with physiology. Yet its founders made dynamic use of both the prestige and fluid boundaries of that discipline to enlarge not only the scientific understanding of sex but also the interdisciplinary organization of the sciences in the American context.

NOTES

1. Sophie B. D. Aberle and George W. Corner, *Twenty-Five Years of Sex Research: History of the National Research Council Committee for Research in Problems of Sex, 1922–1947* (Philadelphia, PA: Saunders, 1953).
2. Edgar Allen, Charles H. Danforth, and Edward A. Doisy, *Sex and Internal Secretions: A Survey of Recent Research,* 2d ed. (Baltimore, MD: Williams and Wilkins, 1939).
3. Cf. Hans Lisser, "The Endocrine Society: The First Forty Years 1917–1957," *Endocrinology* 80 (1967): 5–28.
4. Diana Long Hall, "Academics, Bluestockings and Biologists: Women at the University of Chicago, 1892–1932," *Ann. NY Acad. Sci.* 232 (1979): 310.
5. I use the term "community of interest" to convey the shared professional and scientific concerns of these researchers without (here) addressing the question of their shared social characteristics or the sociology of their field of knowledge; Aberle and Corner, *Twenty-Five Years,* bibliography.

6. I want to thank Toby Appel at the American Physiological Society (APS) for access to and help in understanding the annual reports to the APS by the editor of the *American Journal of Physiology* (*AJP*) from 1923 to 1942. I also want to thank her for information about the thirty-six (out of 104) Committee for Research in Problems of Sex (CRPS) grantees who were members of the APS. There is a brief biography of Donald Russell Hooker by Charles W. Greene, "The American Physiological Society. History of the Second Quarter Century," in *History of the American Physiological Society Semicentennial 1887–1938* (Baltimore, MD: American Physiological Society, 1938), pp. 90–190. See also Wallace O. Fenn, ed., *History of the APS: The Third Quarter Century, 1937–1962* (Washington, DC: American Physiological Society, 1963), p. 12.
7. Aberle and Corner, *Twenty-Five Years*, bibliography.
8. *American Men of Science*, ed. J. McKeen and Jacques Cattell (New York: Science, 1938).
9. Ibid., 1955.
10. Aberle and Corner, *Twenty-Five Years*, appendix 7.
11. Ibid.
12. Ibid., appendix 6.
13. Aberle and Corner, *Twenty-Five Years*, chapt. 1 and 2.
14. Ibid., appendix 7.
15. Corner briefly alluded to the competitive interest that practitioners of anatomy and physiology had in functional studies at a time when anatomists had given up their old definition as morphologists; see *Clio Medica IV. Anatomy* (New York: Hoeber, 1930), final chapter.
16. Hall (Long) "Academics, Bluestockings and Biologists."
17. Corner, *The Hormones in Human Reproduction* (Princeton, NJ: Princeton Univ. Press, 1942).
18. Corner, *Attaining Manhood: A Doctor Talks to Boys About Sex* (New York: Harper and Row, 1938); idem, *Attaining Womanhood: A Doctor Talks to Girls About Sex* (New York: Harper and Row, 1939).
19. Ibid., 2d ed. (1952); new chapter on "The Scientific Study of Sex and Reproduction," pp. 92–104.
20. Aberle and Corner, *Twenty-Five Years*, p. 15: "to encourage younger workers to promise to shape their training for work in this field."
21. Warren Weaver, "Conference on the Biology of Sex." Files of the National Research Council: Sex Committee." Rockefeller Foundation Archives, Tarrytown, New York, f. 200.
22. Major physiology texts of the 1950s were in fact edited by endocrinologists with a heavy emphasis on the internal environment and on the endocrine basis of sex and reproduction. See, for example, H. M. Evans, "Forward," in Bernhard Houssay, *Human Physiology*, trans. Juan T. Lewis and Olive T. Lewis (New York: McGraw-Hill, 1951). Charles Best and N. B. Taylor, *The Physiological Basis of Medical Practice: A University of Toronto Text in Applied Physiology* (Baltimore, MD: Williams and Wilkins, 1937). Editions after 1950 dropped the reference to the University of Toronto. Arthur C. Guyton, *Textbook of Medical Physiology* (Philadelphia, PA: Saunders, 1956).
23. Aberle and Corner, *Twenty-Five Years*, chapt. 4.
24. Edward M. Brecher, *The Sex Researchers* (Boston, MA: Little, Brown, 1969), p. 285.
25. William H. Masters and Virginia E. Johnson, *Human Sexual Response* (Boston, MA: Little, Brown, 1966).

XII

Cardiac Physiology and Clinical Medicine? Two Case Studies

JOEL D. HOWELL

During the early twentieth century, orthodox American medicine incorporated the ideas and the ideology of the natural sciences, marking the beginning of a relationship that has continued to the present day. Medical schools started to devote ever-increasing hours to science, particularly physiology, and physicians began to perceive their task as the application of clinical science.[1] In this chapter I explore the historical justification for physicians' perception of physiology as a natural science valuable to medicine. In so doing I focus on the question: When, and how, did practicing American physicians come to see the physiology of the heart as clinically useful?

I consider regular physicians exclusively, a choice dictated by the desire to arrive at a manageable field for study, not by the belief that none but physicians delivered health care. Physiology was brought into the medical world of orthodox physicians earlier and more intensely than the world of other medical practitioners. Nurses eventually adopted the ideas and ideology of physiology; witness 1980s intensive care unit nurses for five minutes as they calculate base deficits, measure pulmonary artery wedge pressures, and diagnose and treat all manner of cardiac arrhythmias. In the early twentieth century, however, it was allopathic physicians who promoted, promulgated, and practiced physiological teaching in their newly scientific curriculum, and it was allopathic physicians who first saw physiology as useful in their daily work.[2]

Yet the two events—the growth of physiology teaching and the application of physiology to clinical medicine—were not nearly as closely linked as early reformers of medical education would have liked their audience to believe. Their polemics were advantageous to medical physiologists. These reformers convinced those in power that teaching physiology was useful, even essential, for young physicians, thus creating a place where physiologists could be employed, and where their research, as part of a medical school, could be seen as valuable. Eventually, of course, this physiological research was incorporated into patient care. Physiologists certainly promoted the connection: Ernest Starling once said, "The physiology of today is the medicine of tomorrow."[3] The question is how and when "tomorrow" came.

The dissemination of physiology from the scientist's laboratory to the practitioner's office holds broad relevance for the history of American medicine. One of the most significant transformations (social or otherwise) of medical practice has been its development into a field dominated by the rhetoric of the natural sciences and the ritual practices of physiological experimental medicine. Ideally one should follow this process at the level of the ordinary physician, the practitioner whose practices most directly affected and were most directly shaped by the American social system.[4] One needs to inquire not only when physiology was taken into account by prominent physicians at elite institutions, but also when physiological ideas or physiological instrumentation became a regular part of routine medical practice. This question forces the historian to peer inside the "black box" of medical practice and medical therapeutics, a box most social historians have not attempted to open.[5] Not to peer inside, however, dooms the historian to accepting at face value statements about the relevance of physiology made by its most ardent supporters, a problematic stance at best.

PHYSIOLOGY AND MEDICINE

How was, and how is, cardiac physiology seen as useful in medical practice? First, clinicians can perceive physiology as a guide to active intervention. In 1900 knowledge of the cardiac pressure tracing, although interesting to physiologists, was not perceived as relevant in caring for acutely ill patients. Patients' pressures could not be measured and no therapeutic intervention based on cardiac pressures had been described. Today physicians use a catheter placed in the central circulation to measure left ventricular filling pressure and thus obtain information used to guide therapy in cardiac or pulmonary failure. In this instance, a measurement derived from cardiac physiology is used to obtain a particular effect in a particular situation. If all observers agreed on the definitions of what constitutes cardiac physiology, and if all were convinced that such procedures are indeed common events in typical medical practice, this measure of perceived "usefulness" would surely satisfy the toughest critics. Thus one way in which physiology can be seen as useful is as a guide to the acquisition, justification, and implementation of medical therapy held to be effective within the context of contemporary medical thought.

A second function of cardiac physiology is diagnosis and prognosis. When people visit physicians they usually obtain a diagnosis, and with that comes a prognosis. "Tell me, doctor, will I get better? When?" Prognostication is far more common and central to medical practice than the actual treatment of illness, and cardiac physiology can be seen as useful because it aids in establishing a diagnosis and prognosis. For example, the diagnosis of myocardial infarction—a diagnosis that was first made frequently in the 1930s, using techniques derived from cardiac physiology—holds a far more ominous prognosis than another diagnosis sometimes made from similar symptoms, namely indigestion.

There is yet a third way in which cardiac physiology can be called useful within medical practice: when it serves as an explanatory or heuristic device, a way of conceptualizing the patient's cardiovascular system. Christopher

Lawrence, in exploring British perceptions of the heart, argues persuasively that conceptions of the heart changed during the late nineteenth and early twentieth centuries, from an anatomical to a physiological way of thinking.[6] Rather than asking where the lesion lay, early twentieth-century physicians began to ask what the functional capacity of the heart was—a physiological question related to asking what the pressures in the cardiac chambers were. Cardiac physiology was therefore seen as important because it changed the way that cardiac disease was conceptualized. This change was later used by American observers to endorse their accounts of medical progress. In 1940 Carl Wiggers, perhaps the most prominent American cardiac physiologist in the first half of the twentieth century, noted that there had been a change in attitudes toward disease and that medicine had entered a physiological period. He called this transformation the first pathway of progress taken by medicine, and thought it started around 1910.[7]

Wiggers also described a second pathway of progress, a change in how physicians approached the patient. They moved from taking an oral history to instrumentally measuring physiological parameters, a change that has been the subject of much debate in recent years.[8] For better or worse, physiology provided physicians with concepts with which to make that transition. Furthermore, this change in the conception of the heart was applied not only to people with diseased hearts, but also to healthy individuals.

However, for all three of these ways of justifying the usefulness of some element of cardiac physiology—that it changed therapy, that it allowed more accurate prognostication, or that it altered the way in which disease was conceptualized—one would like to establish how practicing physicians evaluated these criteria, and not to accept at face value the assertions of academicians who had much to gain by having physiology seen as clinically useful. Similarly, there is a great difference between one population asserting that a given device or way of thinking could be useful and demonstrating that another population thought that it actually was useful.

We have as yet no systematic study of how cardiac physiologists' ideas became a part of medical practice. The choice of sources for evidence about medical practice will influence the results one obtains. Traditional sources, medical texts and journals, reflect the theoretical approach of academic or elite physicians, not the actual behavior of the typical practitioners. When specific cases were described in medical journals, authors usually discussed exceptional cases rather than describing routine practice.

Case notes recorded by individual practitioners have a different set of limitations. Few such notes exist, and those that have been preserved often record the practice of a single notable physician. Hospital patient records offer another set of problems. Issues of bulk and of confidentiality have discouraged these records from being deposited in most archives. However, when available, hospital patient records can be extremely useful, illustrating how a number of physicians cared for patients with similar diagnoses.[9]

For this chapter I have used the case records of the Massachusetts General Hospital (MGH) as one indicator of medical practice. Starting in 1915, Richard Cabot led conferences based on a case presentation with subsequent discussion and pathological correlation.[10] Transcripts were first published as separate series, and later appeared (and continue to appear) in the *Boston Medical and Surgical Journal* and its successor, the *New England Journal of Medicine.* Obviously, these case records describe conferences at an elite institution, but at least they document practice and tell of specific actions taken in the care of particular, albeit highly selected, patients. Also, the transfer of ideas and practice from scientific laboratories into routine practice probably first occurred in elite institutions such as the MGH. Furthermore, the presentations were anything but spontaneous. House officers no doubt spent many long nights preparing their case history, knowing it would eventually be published in a leading medical journal. The perceived importance of these presentations may make the records even more valuable as an index of ideas about what physical signs or instrumental technologies were understood in physiological terms.

The questions asked on medical licensing examinations provide another source for ideas about the perceived usefulness of physiology. These are obviously politically charged documents and must be interpreted with care. Nonetheless, the material for which the student-physician was held accountable, and, no less important, the section of the exam in which it was placed, can tell us a great deal about how those who controlled the exam perceived the importance of various kinds of knowledge and technique.[11]

Rather than attempting a broad discussion of physiology within medicine, I offer a case study of two specific elements of cardiac physiology.[12] First, I discuss the electrocardiogram (ECG). (I shall use "ECG" to indicate both the machine and the tracing it produces; context should make the distinction clear.) The ECG is a graphic representation of the electrical action of the heart and is usually obtained from the body's surface. Originally a research laboratory instrument for cardiac physiologists, the ECG is now routine for any patient with heart disease, suspected heart disease, or serious illness of any kind. Figure 1 is a typical ECG taken in the early years of this century.

Second, I consider pressure tracings of the cardiac cycle, which measure hemodynamic pressure within a given cardiac chamber as the pressure changes over time. This information was often presented in graphic form. I include in this concept both the qualitative form of the venous pulse and the quantitative measurement of its magnitude. When physiologists integrated the graphic representations of the pressure of the four chambers of the heart into a single diagram, as in Figure 2, that diagram was seen to have enormous power to explain cardiac hemodynamics.

Both the ECG and cardiac pressure curves were considered central to early twentieth-century physiology. They were discussed at length in standard physiology textbooks, included on National Board Examinations, and dignified by the research and careers of many leading American physiologists. Although the ECG and pressure tracings presented what we now consider very different kinds of information—one electrical, one hemodynamic—both did so in graphic form, a form that to many symbolized the ideology of the natural sciences.

The differences between the historical development of technology associated with these two markers offer a chance to document two examples of how

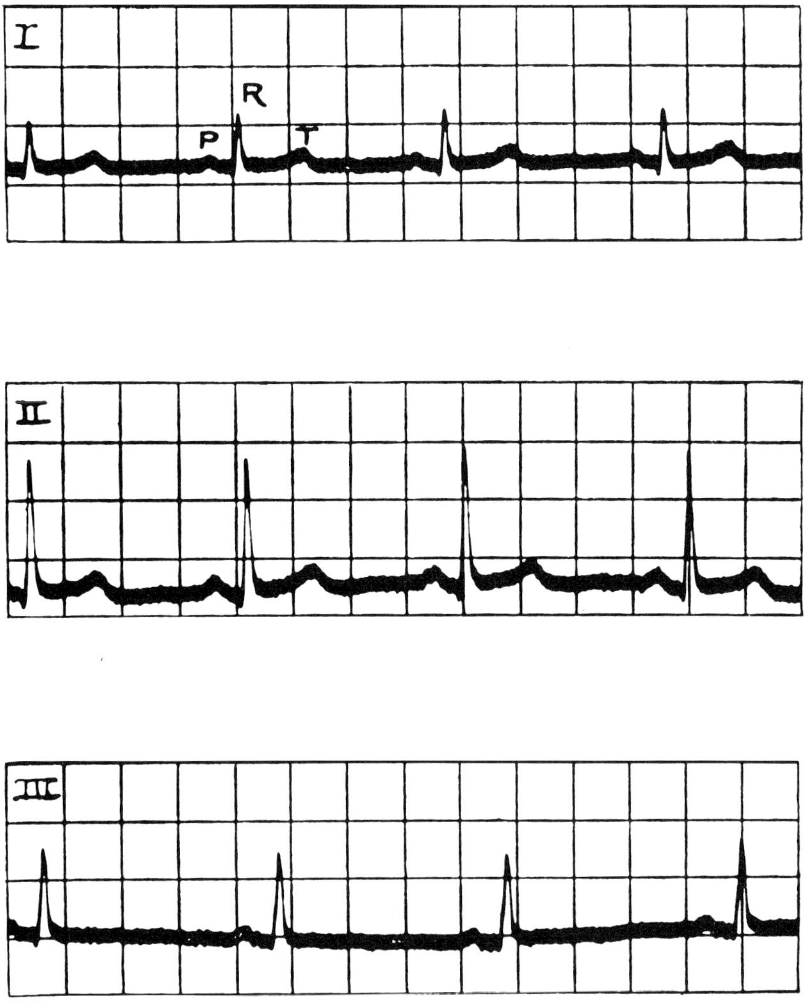

Figure 1: *Typical three-lead electrocardiogram tracing from healthy 40-year-old man. [Fredrick A. Willius, Clinical Electrocardiograms: Their Interpretation and Significance (Philadelphia, PA: Saunders, 1929), p. 20.]*

physiology came to be seen as an integral part of medical practice. One, the ECG, was introduced into clinical practice as a relatively mature instrumental technology. Although subsequent instrumental advances made ECG tracings easier to obtain, the technical quality of tracings taken by early machines is comparable with those taken today. From the outset, the ECG was a diagnostic technology seen as potentially applicable to human beings as well as to laboratory animals. The other, the tracing of pressure pulses in the cardiac cycle of the heart, was applied only in laboratory settings for almost a century after it was first proposed. Eventually, however, the measurement of cardiac pressure tracings became a routine event in clinical practice. Comparing these two elements of cardiac physiology can help us to better understand the relationship between physiology and medical practice.

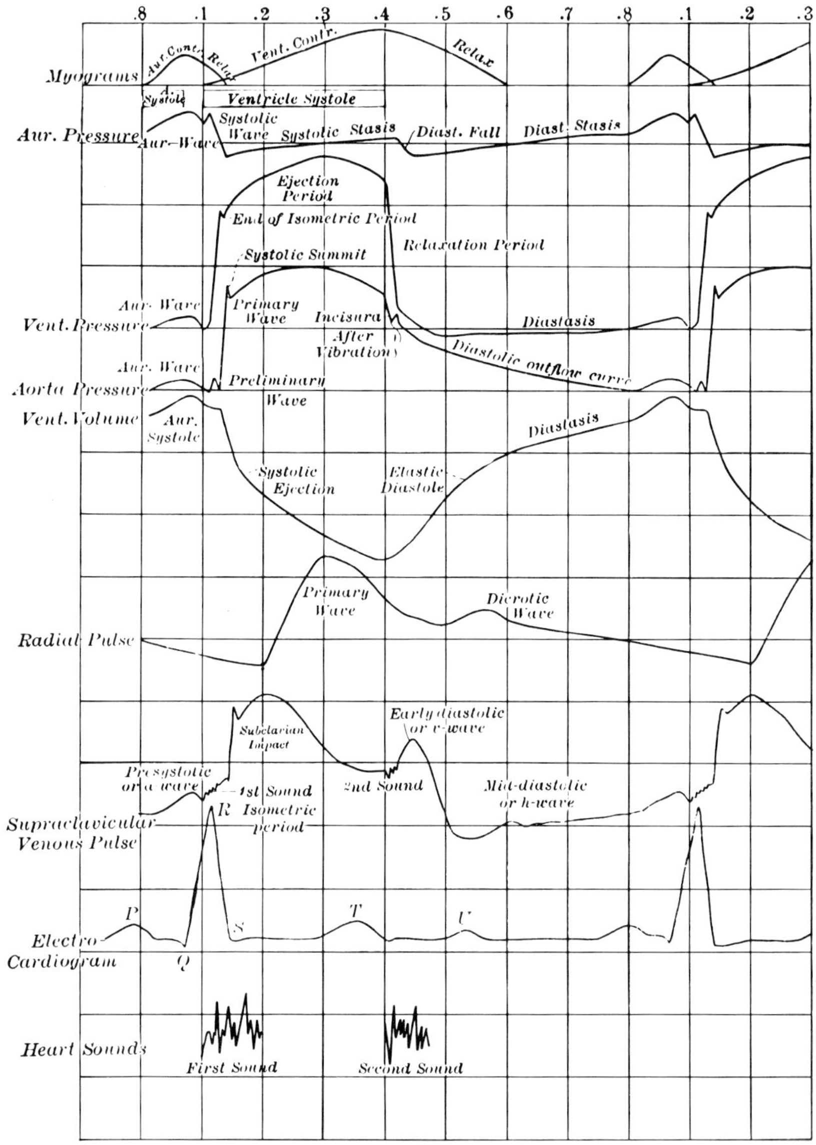

Figure 2: *Typical set of cardiac pressure curves showing pressure within chambers and other events thought important. [From Carl J. Wiggers, Modern Aspects of the Circulation in Health and Disease (Philadelphia, PA: Lea and Febiger, 1915), p. 98.]*

ELECTROCARDIOGRAM

The ECG was made in the early twentieth-century physiology laboratory, widely used in physiological experimentation, and soon discussed in almost every standard physiology text.[13] To some the ECG was the most important proof of cardiac physiology's importance for medicine, and those who advocated more and better physiology in the medical curriculum often claimed that every medical school needed to own an ECG machine.[14] Yet when the

ECG was first introduced, it was not perceived to have a significant impact on medical therapy.[15] Early twentieth-century physicians recognized no treatment for disease states that the ECG alone could diagnose. Now, however, the ECG is applied in a wide variety of therapeutic contexts. Physicians install pacemakers, defibrillate hearts, and decide whether or not to intervene in cases of coronary thrombosis—all largely on the basis of ECG data.

Our historical judgments about when the ECG was seen as useful will depend on how we choose to define utility. Let us begin with a definition of utility in terms of perceived effective therapy. By this standard the ECG was first seen as useful in the 1950s, with the advent of pacemakers to artificially stimulate a heart beating too slowly and defibrillators to restart a heart that had begun to fibrillate. As early as the 1930s a few physicians had suggested using electric shock to defibrillate the heart,[16] but it was not until 1947 that a patient was reported to have survived the use of countershock. Paul Zoll reported the successful use of external shock in the 1950s.[17] The first mention of external countershock in the MGH case records appears in 1960. This relatively quick application, within a decade, may be partly explained by the fact that Zoll did his pioneering work in Boston, at the Beth Israel Hospital and at the Harvard Medical School. By the 1970s, defibrillators were routinely carried on ambulances and used by paramedics.

Artificial external pacemakers were first described in 1952 and the implantable pacemaker was invented in 1959.[18] Pacemakers quickly became widespread (probably too widespread) and by 1975, approximately 260,000 patients had been treated with pacemakers. By 1985, approximately 500,000 were treated, and extrapolation of the current rate of growth would predict that by the year 2000 every American adult will have a pacemaker implanted! Before the early ECG could be perceived as a guide to definitive therapy by external defibrillation or by internal artificial pacemakers, further technological developments were needed—another story in itself.

What of diagnosis? Characteristic ECG findings for the diagnostic entities of atrial fibrillation or of heart block were described in the ECG's first decade. However, the ECG findings were not essential to the definition; both fibrillation and heart block had already been clinically described. In the early twentieth century, ECG users created their first unique diagnostic role in the diagnosis of myocardial infarction. Coronary artery disease was rarely diagnosed during life before 1918, although it was frequently described at autopsy. However, premortem diagnosis of the coronary disease increased dramatically following James Herrick's 1918 description of particular ECG changes in myocardial infarction.[19] The disease of myocardial infarction had become defined, in part, by ECG changes. Yet even at an elite hospital such as the MGH, the ECG was not a routine test for patients with suspected infarction as late as 1935. By 1940, however, long discussions about the diagnostic efficacy of the ECG tracing are found in the case records of patients presenting with chest pain. By around 1950 the ECG had become routinely used, as it is today, in the diagnostic evaluation of patients critically ill for whatever reason.

Medical licensing examinations started to test physicians' abilities to interpret an ECG tracing at about the same time. The National Board Examination was divided into two parts. The first part tested knowledge of what was seen as the basic sciences; this included a section on physiology. The second part tested knowledge of clinical skills; this included a section on medicine. The medicine section, not the physiology section, of the 1949 National Board Examination asked students to describe the typical ECG features of coronary occlusion. By placing the question in part two, the clinical portion of the exam, rather than in part one, which dealt with the basic sciences, those who wrote the exam implicitly emphasized that by 1949 the ECG was seen as a diagnostic tool with which all licensed physicians needed to be familiar.

So far this discussion has considered the two clinical contexts in which use of the ECG is easiest to document: treatment and diagnosis. It is far more difficult to identify when physicians used the cardiac physiology defined by the ECG to conceptualize the heartbeat. Physiologists used the ECG to redefine their image of the heart soon after the instrument's invention. By 1910, those who investigated either the mechanism of the heartbeat or cardiac arrhythmias applied the ECG intensively. During the first decades of this century, some clinicians, apparently, also wanted to take ECG tracings. The machine's tracings appear surprisingly early in American hospital records, long before the ECG was thought to have clinical relevance for diagnosis or therapeutics. The Pennsylvania Hospital started to request ECG tracings from a nearby hospital (the hospital of the University of Pennsylvania) around 1920. The Pennsylvania Hospital purchased its own ECG machine in 1921, and tracings were soon taken from patients with a wide variety of diagnoses.[20] Similarly, case records from the MGH show frequent ECG tracings by the 1920s, usually with interpretations by Paul Dudley White.[21] In neither of these sets of records does one usually find claims for the diagnostic or therapeutic usefulness of the ECG. The tracings were recorded out of curiosity, taken because there was a new machine around to use. Although we need to know more about when the ECG was used at small, rural hospitals and in office practice, we may conclude that by the 1920s at least a few physicians used the ECG to conceptualize the heart in a new way.

CARDIAC PRESSURES

It has been known for centuries that the heart generates not only electrical impulses but also pressure waves that propel blood around the body. One of the first attempts to measure the force exerted on blood by the heart came in 1710 when Stephen Hales introduced a gun barrel down the carotid artery of a horse, thus measuring the arterial pressure by a relatively direct method.[22] More than a century later, in the 1840s, Claude Bernard again attempted to measure the pressure of the blood from within the heart.[23] By the end of the nineteenth century Auguste Chauveau and Jules Marey had succeeded in tracing pressure curves from inside a horse's heart, and Adolf Fick had described a theoretical method for measuring cardiac output that would eventually require blood samples taken from within the central circulation.[24] Early in the twentieth century interest in pressure changes in the heart escalated. Starling described his now famous "Law of the Heart" between 1912 and 1920, although, as W. Bruce Fye has pointed out, it was a generation later before Starling's insights were seen to have clinical relevance.[25] In 1915,

Wiggers described the form taken by the pressure curves of the four chambers.[26]

X rays, first described in 1895, soon enabled physicians to place catheters into a specific cardiac chamber. The original concept has now become the Swan-Ganz catheter, a triple (or more) lumen tube found in almost every American hospital and used to measure left ventricular filling pressures from within the pulmonary circulation.[27] Physicians use measurements to decide how much fluid should be administered. Swan-Ganz catheters are also seen as valuable when used to measure the cardiac output, using the method described by Fick in the nineteenth century, or when measurements of oxygen content permit the diagnosis of abnormal communications between the various cardiac chambers. (Recent models come equipped with pacing electrodes, thus linking perceived use of the Swan-Ganz catheter with the perceived usefulness of the ECG previously discussed.) By the early 1980s the measurement of cardiac pressures with a Swan-Ganz catheter was widely considered valuable, although some observers called it a "cult phenomenon."[28]

To trace how cardiac pressure curves came to be seen as useful by physicians is difficult because so many technical factors were required before pressure curves led directly to specific therapy.[29] For example, surgical repair of cardiac lesions requires precise preoperative anatomical investigation, usually by contrast aided X-ray study.[30] Other specific therapies were not considered before the advent of new pharmacological agents, or new machines such as the heart-lung machine.

The absence of therapies, however, mattered little to early twentieth-century physiologists, who were quite fond of cardiac pressure curves. Wiggers's graphs of the pressures in the four chambers, like that shown in Figure 2, were quickly incorporated into almost every major physiology text. Clinicians, however, could directly evaluate only one chamber's pressure, namely, the right atrial pressure, manifest as the central venous pulse and visible in the jugular vein of the neck. The pulse's absolute magnitude was understood to indicate the filling pressure for the right side of the heart, a pressure predicted by Starling's Law to control the output of the right ventricle. The qualitative form of the pulse wave reflected the heart's rhythm. That form could either be observed with the naked eye or recorded by the direct mechanical linkage of the polygraph, an instrument more commonly used in Great Britain than in the United States.[31] Although effective therapy for either elevated pressures or abnormal pulse waves was not described in the 1910s, the venous pulse could still have been seen as clinically valuable, either for diagnosing abnormal cardiac rhythms or heart failure, or for understanding the heart's action in a physiological way.

Obviously, it was in the medical physiologists' interest to emphasize the clinical importance of the venous pulse. They succeeded in incorporating the cardiac pressure curves into medical school teaching, as well as placing relevant questions on the basic sciences portion of the national examinations. Starting with the first set of National Board Examinations in 1922, questions pertaining to the cardiac cycle were included in part one of almost every examination.[32] These included such queries as: "Describe the venous pulse"; "Explain the causes of the closure and opening of the valves of the heart: Give the phases in the cardiac cycle"; and "Describe the effect of changes in the heart beat on the volume output of the heart."

Were these physiological ideas accepted as important concepts by medical practitioners? Interestingly, cardiac pressure curves were incorporated into medical practice far more slowly than the ECG. There are no records of the venous pulse in either the Pennsylvania Hospital records through 1927 or in the MGH case records from the 1930s. A few physicians specifically interested in studying heart disease preached the value of the venous pulse for diagnosis of heart disease, although well into the 1940s they decried the fact that observations were far too infrequently made in practice.[33]

By the mid-1940s, several investigators had succeeded in measuring intracardiac pressures in human beings. Werner Forssmann passed a tube into the right side of his heart in 1929; this feat was prompted by the search for a better way to infuse medications, not by the desire to measure physiological parameters.[34] His experiment went largely unnoticed. In a series of studies in 1941 motivated by the desire to find improved means of measuring physiological parameters, Andre Cournand and Dickinson Richards demonstrated that right-heart catheterization could be safely performed in humans.[35] By 1945 James Warren and Eugene Stead had used right-heart catheterization to establish a clinical diagnosis of potential prognostic importance: atrial septal defect.[36]

Cardiac catheterization of the 1940s and 1950s provided two different kinds of information. First, it gave anatomical information about the heart, information that Fye argues was uppermost in the minds of investigators such as Agustin Castellanos, George Robb, and Israel Steinberg.[37] Such anatomical information was realizable in two ways: (1) injection of contrast material outlined vascular structures, and allowed direct visualization of the structures of interest; and (2) measurement of intravascular pressures and oxygen saturation allowed physicians to deduce the location of abnormal communications between cardiac chambers. By measuring intracardiac pressures and blood oxygen saturation, physicians also derived functional physiological information and could derive the cardiac output using Fick's nineteenth-century method. By the 1950s physicians started to apply this data to determine whether or not the heart was failing, and if so, how best to aid that damaged heart.

An increased belief in the clinical relevance of pressure measurements was reflected on licensing examinations. The medical section of the 1950 National Board of Medical Examiners test asked the student to describe circulatory adjustments following surgical closure of a patent ductus (a common congenital abnormality that results when a normal vascular connection between two major blood vessels, the aorta and pulmonary artery, fails to close after birth). This question, one that required discussion of changing pressures and blood flows after birth, reflects both the perceived availability of surgical repair and the increased emphasis on physiological cardiology.

Routine use of cardiac catheterization to measure intracardiac pressures and diagnose anatomical abnormalities is noted in MGH case records starting around 1960. Measurement of intracardiac pressures was probably not very common at smaller, less prestigious hospitals until at least a decade later.[38] Massachusetts General Hospital case records reveal that, along with this increased use of technology-based cardiac measurements, by the later 1960s house officers were increasingly likely to document the venous pulse in cardiac examination. Until then the intellectual construction of the heart and cardiac lesions does not seem to have been informed by ideas about cardiac pressures.

Since the 1960s both the ECG and the measurement of cardiac pressures have been a central part of a more general medical trend toward continuous measurement of physiological parameters, measurement usually carried out in some form of intensive care unit. First established in 1962, coronary care units were based on the idea that "some hearts were too young to die." The perceived ability to save those dying hearts came largely through technology for immediate countershock of ventricular fibrillation, a technology whose application was guided by ECG monitoring.[39] In addition to constant ECG monitoring, coronary care units quickly incorporated measurement of pulmonary artery pressures. By 1976 almost all hospitals with 300 or more beds had coronary care units, as did 57 percent of all community hospitals with 200 or more beds.[40] Certainly by the start of the 1980s, physiological measurements—and cardiac physiology as I have elected to define the term—were widely accepted as useful in clinical medicine.

CONCLUSION

I have briefly sketched how two innovations in cardiac physiology came to be seen as clinically useful for twentieth-century medical practice. The ECG produced graphic tracings that looked scientific and therefore fit well with ideology of early twentieth-century American medicine. Moreover, taking an ECG produced information at no risk to the patient. From the 1920s physicians perceived the ECG as useful for organizing information about the heartbeat because it was easy to take and it fit well with contemporary ideas about what medicine should be. It may also have been thought a useful tool for diagnosing myocardial infarction at about that same time. By the 1940s and 1950s the ECG had become widely accepted as a necessary tool for treating patients with heart disease.

On the other hand, while physiologists in the 1910s were intensely interested in cardiac pressure tracings from the laboratory, such tracings could not be obtained from human beings until well into the 1940s. Pressure measurements were not commonly used to care for patients with heart disease until the devices for obtaining them became safe, or at least easy for physicians to use. Lacking easy techniques and measuring devices, the concept of cardiac pressure curves was not used as a heuristic device before the 1940s. Thereafter, accurate measuring devices, effective surgical intervention, and the need for contrast-mediated anatomical localization all helped cardiac pressure tracings appear useful to clinicians.

Ultimately, to answer the question I asked at the start of this chapter, about the practice of medicine by typical physicians, will require analysis of case records from a wide range of practitioners. Other markers, perhaps drawn from other communities, will no doubt give different perceptions of usefulness. I have pointed out the need to ask the question at this level, lest one be left at the mercy of those observers who were most eager to promote basic laboratory science as useful to clinicians—those whose careers were based on basic science within the context of the medical school.

The tension between the cognitive nature of laboratory science and the practical imperatives of clinical practice is common to most of twentieth-century medicine.[41] A particular physiological innovation may be seen as useful by physicians on several levels. Whether or not it will be so perceived depends on the nature of medical practice at that time. The physician-patient interaction is mediated by a wide range of social, economic, and political factors related to, but not the same as, those factors that inform the production of scientific knowledge.

This chapter has discussed how ideas and knowledge about cardiac physiology acquired power in the practice of medicine. Initially, power came through changing physicians' understanding of the heart. Later, physicians saw physiological knowledge as useful in giving the power to predict the future. Finally, the knowledge and techniques of laboratory physiology were seen as useful guides to therapy.

The ultimate success of physiology may lie in its ceasing to be "physiology" at all and being called clinical medicine instead. Depending on how one defines the term, cardiac physiology, as exemplified by the ECG and cardiac pressure curves, generally came to be seen as useful by American physicians sometime between 1920 and 1960. The mechanism of that process deserves closer examination.

I wish to thank Janet Golden and Christopher Lawrence for useful suggestions. The preparation of this paper was supported in part by a grant from the Henry J. Kaiser Family Foundation, Menlo Park, California.

NOTES

1. For a general review, see John Harley Warner, "Physiology," in *The Education of American Physicians*, ed. Ronald L. Numbers (Berkeley: Univ. of California Press, 1980). The best work on the history of American medical education is Kenneth Ludmerer, *Learning to Heal* (New York: Basic, 1985). For a critical evaluation of recent historiography see Daniel M. Fox, "The New Historiography of American Medical Education," *Hist. Educ. Q.* 26 (1986): 117–124.
2. For a representative essay, see Carl J. Wiggers, "The Interrelations of Physiology and Internal Medicine," *J. Am. Med. Assoc.* 91 (1928): 270–274. Gerald L. Geison discusses the push for physiology in "Divided We Stand: Physiologists and Clinicians in the American Context," in *The Therapeutic Revolution*, ed. Morris J. Vogel and Charles E. Rosenberg (Philadelphia: Univ. of Pennsylvania Press, 1979).
3. Quoted by Eugene F. DuBois in "Fifty Years of Physiology in America: A Letter to the Editor," in *The Excitement and Fascination of Science* (Palo Alto, CA: Annual Reviews, 1965), p. 86. Another noted physiologist, Paul F. Cranefield, was somewhat less certain of physiology's impact on medicine in "Microscopic Physiology Since 1908," *Bull. Hist. Med.* 33 (1959): 275.
4. This ideal has been largely defined by the Annales school. See Fernand Braudel, *Afterthoughts on Material Civilization and Capitalism* (Baltimore, MD: Johns Hopkins Univ. Press, 1977).
5. For a review of recent historical analysis of medical science see Warner, "Science in Medicine," *Osiris* 1 (1985): 37–58.
6. Christopher Lawrence, "Moderns and Ancients: The 'New Cardiology' in Britain 1880–1930," *Med. Hist. Suppl.* 5 (1985): 1–33.
7. Wiggers, "Pathways of Medical Progress," in *Blood, Heart and Circulation*, ed. Forest Ray Moulton (Washington, DC: Science, 1940), pp. 1–9.
8. Stanley Joel Reiser, *Medicine and the Reign of Technology* (Cambridge, UK: Cambridge Univ. Press, 1978).

9. Joel D. Howell, "Early Use of X-ray Machines and Electrocardiographs at the Pennsylvania Hospital: 1897 Through 1927," *J. Am. Med. Assoc.* 255 (1986): 2320–2323.
10. W. B. Cannon, "The Case Method of Teaching Systematic Medicine," *Boston Med. Surg. J.* 142 (1900): 31–36; Thomas Franklin Williams, "Cabot, Peabody, and the Care of the Patient," *Bull. Hist. Med.* 24 (1950): 462–465; Benjamin Castleman and H. Robert Dudley, Jr., Preface to *Clinicopathological Conferences of the Massachusetts General Hospital: Selected Medical Cases* (Boston, MA. Little, Brown, 1936–1960), pp. vii–ix.
11. Robert Joy has suggested that the use of machines by life insurance examiners and by physicians who certified aviators can be considered as other indicators of perceived usefulness.
12. I am not attempting to say anything about the development of cardiology as a specialty (or a subspecialty of internal medicine). There can be little doubt that some relationship exists between clinical cardiology and cardiac physiology. The nature of that relationship, however, awaits further clarification. Also, before the mid-twentieth century, most practicing physicians were general practitioners, not specialists. To define the use of cardiac physiology in terms of specialists' practice would take us away from the level of ordinary practice and quickly put us into the level of elite, academic researchers.
13. George E. Burch and Nicholas P. DePasquale, *History of Electrocardiography* (Chicago, IL: Yearbook, 1968); John Burnett, "The Origins of the Electrocardiograph as a Clinical Instrument," *Med. Hist. Suppl.* 5 (1985): 53–76.
14. For example, C. C. Guthrie, "Twenty Years Progress in Physiology and Pharmacology: Including Present Status and Outlook to the Future," in *Proceedings of the Annual Congress on Medical Education, Medical Licensure, Public Health and Hospitals* (Chicago, IL: American Medical Association, 1925), pp. 13–21.
15. Howell, "Early Perceptions of the Electrocardiogram: From Arrhythmia to Infarction," *Bull. Hist. Med.* 58 (1984): 83–98.
16. W. Bruce Fye, "Ventricular Fibrillation and Defibrillation: Historical Perspectives with Emphasis on the Contributions of John MacWilliam, Carl Wiggers, and William Kouwenhoven," *Circulation* 71 (1985): 858–865; Kouwenhoven and Orthello R. Langworthy, "Cardiopulmonary Resuscitation: An Account of Forty-Five Years of Research," *Johns Hopkins Med. J.* 132 (1973): 186–193.
17. Paul M. Zoll, et al., "Termination of Ventricular Fibrillation in Man by Externally Applied Electric Countershock," *N. Engl. J. Med.* 254 (1956): 727–732.
18. Victor Parsonnet and Marjorie Manhardt, "Permanent Pacing of the Heart: 1952–1976," *Am. J. Cardiol.* 39 (1972): 250–256; Peter C. Hanley, et al., "Two Decades of Cardiac Pacing at the Mayo Clinic (1961 Through 1981)," *Mayo Clin. Proc.* 59 (1984): 268–274.
19. R. L. Levy, H. G. Bruenn, and D. Kurtz, "Facts on Diseases of the Coronary Arteries Based on a Survey of the Clinical and Pathological Records of 762 Cases," *Am. J. Med. Sci.* 187 (1934): 376–390.
20. Howell, "Early Use of X-Ray Machines and Electrocardiographs"; and Howell, "Machines and Medicine: Technology Transforms the American Hospital." In press.
21. For biographical details, see Paul Dudley White, *My Life and Medicine* (Boston, MA: Gambit, 1971).
22. For a general discussion of the history of measuring cardiac pressures and output, see William F. Hamilton and Dickinson W. Richards, "The Output of the Heart," in *Circulation of the Blood: Men and Ideas*, ed. Alfred P. Fishman and Richards (New York: Oxford Univ. Press, 1964), pp. 71–126. Also, Stephen Hales, *Statistical Essays: Containing Haemastatistics* (1733; Reprint ed., New York: Hafner, 1964).
23. Andre Cournand, "Historical Details of Claude Bernard's Invention of a Technique for Measuring the Temperature and the Pressure of the Blood Within the Cavities of the Heart," *Trans. NY Acad. Sci.*, 2d ser., 39 (1980): 1–14.
24. Edward Shapiro, "Adolf Fick—Forgotten Genius of Cardiology," *Am. J. Cardiol.* 30 (1972): 662–665; T. Doby, "History of Cardiovascular Catheterization," *Clio Med.* 12 (1977): 185–188.
25. Fye, "Ernest Henry Starling, His Law and Its Growing Significance in the Practice of Medicine," *Circulation* 68 (1983): 1145–1148; see also Mark D. Altschule, "Invalidity of Using So-Called Starling Curves in Clinical Medicine," *Perspect. Biol. Med.* 26 (1983): 171–187.
26. Wiggers, *Reminiscences and Adventures in Circulation Research* (New York: Grune & Stratton, 1958).

27. H. J. C. Swan and William Ganz, "Hemodynamic Monitoring: A Personal and Historical Perspective," *Can. Med. Assoc.* 121 (1979): 868–871.
28. Eugene D. Robin, "The Cult of the Swan-Ganz Catheter: Overuse and Abuse of Pulmonary Flow Catheters," *Ann. Intern. Med.* 103 (1985): 445–449.
29. For a discussion of the many advances needed, as well as an excellent chronology that goes far beyond the dates listed in the title, see Julius H. Comroe, Jr. and Robert D. Dripps, *The Top Ten Advances in Cardiovascular-Pulmonary Medicine and Surgery 1945–1975* (Washington, DC: Government Printing Office, 1977).
30. Fye, "Coronary Arteriography—It Took a Long Time," *Circulation* 79 (1984): 781–787.
31. James Mackenzie, *The Study of the Pulse* (Edinburgh: Young J. Pentland, 1902).
32. For example, the National Board of Medical Examiners questions in the Physiology Section of part one (the basic sciences) for February, 1922; September, 1922; and almost every exam thereafter. *National Board of Medical Examiners Bulletin No. 8: Examination Questions Parts I and II*, 1922–1925.
33. For example, George R. Hermann, *Synopsis of Diseases of the Heart and Arteries* (St. Louis, MO: Mosby, 1944), p. 68.
34. A. Werner Theodor Otto Forssmann, "Die Sondierung des Rechten Herzens," *Klin. Wochenschr.* 8 (1929): 2085–2087, 2287; and Werner Forssmann, *Experiments on Myself: Memoirs of a Surgeon in Germany* (New York: St. Martin's, 1974).
35. Cournand and Hilmert Albert Ranges, "Catheterization of the Right Auricle in Man," *Proc. Soc. Exp. Biol. Med.* 46 (1941): 462–466; Richards, "The Contributions of Right Heart Catheterization to Physiology and Medicine, with Some Observations on the Pathophysiology of Pulmonary Heart Disease," *Am. Heart J.* 54 (1957): 161–171.
36. Warren describes this event in J. Willis Hurst, "History of Cardiac Catheterization," in *Coronary Arteriography and Angioplasty*, ed. Spencer B. King III and John S. Douglas (New York: McGraw Hill, 1985), pp. 1–9.
37. Fye, "Coronary Arteriography," 783–784.
38. Louise B. Russell, *Technology in Hospitals: Medical Advances and Their Diffusion* (Washington, DC, Brookings Institution, 1979), documents that larger hospitals adopted intensive care units, where such measurements were performed, more rapidly than smaller hospitals.
39. Hughes W. Day, "History of Coronary Care Units," *Am. J. Cardiol.* 30 (172): 405–407.
40. Russell, *Technology in Hospitals*, p. 43.
41. Russell C. Maulitz, "Physician versus Bacteriologist: The Ideology of Science in Clinical Medicine," in Vogel and Rosenberg, *Therapeutic Revolution*, pp. 91–108.

XIII

Instruments and an Independent Physiology: The Harvard Physiological Laboratory, 1871–1906

MERRILEY BORELL

This essay concerns the new vision of experimental physiology that was successfully fostered in American universities at the end of the nineteenth century. Integral to this new vision was a deep faith in the experimental method, a belief in the ultimate accessibility of life phenomena to analytical methods, and a recognition of the need for investigators to develop special tools and techniques by means of which the complicated phenomena of life might be analyzed into simpler components. This intellectual transformation involved the creation of a new social context for the production of physiological knowledge: research and teaching laboratories in which students investigated individual problems and learned how to question nature.[1]

In the laboratory, students learned observational skills, surgical techniques, and the experimental method. Physiological laboratories were equipped with instruments designed to extend the senses and enhance the investigator's powers of discrimination. The complex phenomena associated with living organisms were to be analyzed into sequences of cause and effect. Ludwig's revolving drum recorder or kymograph, adapted from instruments used in the physical sciences, was the prototype for other devices successfully introduced into physiological research in the period 1850–1870. The teaching laboratory became the site for the use of these instruments and the focus for the transmission of knowledge derived from them.

The introduction of recording techniques allowed investigators to separate and measure transient physiological events.[2] Graphic registration of processes such as muscle contraction and nerve conduction gave researchers insights into phenomena inaccessible merely by the dissecting knife. Newly confident of their method, these investigators promised clinical medicine understanding and, ultimately, control of disordered physiological processes. These optimistic, if premature, claims gave physiology a unique power and status within the medical sciences. At the end of the nineteenth century, clinical observations acquired a deeper meaning to the extent that they could be correlated with phenomena exposed by laboratory investigation. In 1898, Harvard's associate professor of physiology, William Townsend Porter, triumphantly proclaimed:

> Physiology is the most highly developed rational discipline in medicine,—not a merely descriptive science like anatomy,—and is well adapted to train the mind in scientific procedure, in the setting of problems for research, in the criticism of methods and results, and in the tests which lay bare shallowness,—matters of great moment to men who shall practise an applied experimental science in the midst of quackery, illusion and pretense.[3]

This paper examines the rise of physiology in the Harvard Medical School and the creation of the "first American physiological laboratory for the use of students."[4] By analyzing the development of the laboratory created by Harvard physiologists Henry Pickering Bowditch (1840–1911) and William Townsend Porter (1862–1949), it follows the importation of the European physiological tradition to America and examines the flowering of that tradition in the American context.

The roles of Bowditch and Porter in the professionalization of American physiology are well known.[5] Bowditch was one of the founding members and second president of the American Physiological Society (Fig. 1). Porter financed and edited the *American Journal of Physiology* from its first volume in 1898 through its thirty-third in 1914 (Fig. 2). Less well understood are both Bowditch's and Porter's contributions to the process by which the new experimental techniques associated with the rise of European physiology were effectively incorporated into the American scientific and medical traditions. Both men studied abroad: Bowditch in France and Germany, notably with

Figure 1: *Henry Pickering Bowditch, second from right, at Cambridge University (1898) with other recipients of honorary degrees. Left to right: Marey, Dohrn, Golgi, Haeckel, Hubrecht, Kühne, Bowditch, and Kronecker. [From photo taken by F. J. Allen, professor of physiology at Birmingham, U.K. Courtesy of Harvard Medical Archives, Francis A. Countway Library of Medicine, Harvard Medical School (hereafter, CL-HMS).]*

Figure 2: *Portrait of William Townsend Porter. [From The Medical School of Harvard University Class of 1906. Courtesy of CL-HMS.]*

Carl Ludwig (1816–1895); Porter, at several laboratories in Germany. Their apprenticeships served well to transmit the European research ideal to America. Yet, while both men borrowed heavily from the program and goals of German physiology especially, in their lifetimes they helped create a laboratory-based curriculum for students that was distinctively American, emphasizing hands-on experience with research instruments at the introductory level. Their individual careers highlight the changing role of physiology within nineteenth-century American medicine and underscore the mechanism by which new laboratory techniques introduced in Europe were institutionalized in America.

Bowditch created the physiological laboratory at Harvard in the 1870s. Porter reorganized it in the 1890s, creating "routine student laboratory experiments."[6] Porter also initiated the manufacture of "Harvard" physiological apparatus and founded the Harvard Apparatus Company. He did this to provide inexpensive and reliable apparatus for large classes and to eliminate American dependence on European equipment. In fact, the Harvard Apparatus Company shaped both the tools and character of American physiological research. Standard research instruments were greatly simplified and mass-produced for the specialized market that arose from the increased importance of physiology in the medical and scientific curricula of American universities. More than that, the Harvard Apparatus Company eventually supplied European as well as North American laboratories, influencing the conduct of research and teaching throughout the world.

CONTINENTAL MODELS FOR BOWDITCH'S PHYSIOLOGICAL LABORATORY

The enormous value of instruments to the progress of physiological research was already evident in the 1860s following the phenomenal rise of French and German physiology. When English physiologists William Sharpey (1802–1880) and Michael Foster (1836–1907) upgraded physiological teaching in Britain, they imported Continental techniques. These methods spread rapidly as physiological laboratories were established at each of the major British universities.[7]

During this period of the creation of laboratories and research schools throughout Europe, Bowditch worked in Ludwig's new institute in Leipzig. In fact, Bowditch's account of the horseshoe-shaped "Neue Physiologische Anstalt" appeared in 1870 in both the *Boston Medical and Surgical Journal* and the British journal *Nature.* Ludwig's institute was composed of three wings, one each for microscopy, chemical studies, and experimental physiology. Each section was well equipped with specialized apparatus. Bowditch observed with no small measure of admiration:

> It will thus be seen that abundant facilities are here offered, not only for learning the existing state of physiological science, but also for becoming familiar with the manner in which physiology is at present studied in Germany. The patient, methodical, and faithful way in which the phenomena of life are investigated by the German physiologists not only inspires great confidence in their results, but encourages one in the hope that the day is not far distant when Physiology will take its proper place as the only true foundation of Medical Science.[8]

Bowditch subsequently brought German laboratory apparatus back to America in order to equip the new physiological laboratory at Harvard. In 1871, as assistant professor, he set up an attic laboratory in the old buildings of the medical school on North Grove Street, where, consistent with the medical school reform then in progress[9] and with the policies then being developed by Harvard's President Charles Eliot, he helped to "innoculate" the university with "German science."[10] The annual report for 1870–1871 recorded the new emphasis on laboratory work: "in the important subjects of anatomy, physiology, chemistry, and pathological anatomy, laboratory work will be substituted for, or added to, the usual didactic lectures. Every student will have his place and time in the anatomical and chemical laboratories and in the microscope room..."[11]

To support this new emphasis, in the following year "the chemical laboratory was greatly enlarged, and a new laboratory for microscope work was built."[12] Although physiological experiments were not yet formally required, the 1870–1871 report noted, "The appointment of an assistant Professor of Physiology and the equipment of a physiological laboratory has put that department of instruction upon a much better footing than ever before."[13]

Bowditch's physiological laboratory soon became established within the new three-year medical curriculum as a "research laboratory open to qualified medical scientists."[14] His former student Frederick W. Ellis recalled in 1938, "The main room of the old laboratory, which was not a large one, had for its principal furniture a table for experiments, a sink, a workbench and foot lathe and a tub for frogs. A smaller room which was shared with Quincy, who had just been appointed an assistant in histology, contained cases for instruments and a few books."[15]

In this setting, very limited practical instruction was given to beginning students; experimental technique was only a minor part of Bowditch's elementary course. To quote from Ellis again:

> The course in physiology in the Harvard Medical School was a complete refutation of the view that real instruction in science cannot be given to students without individual work by them in a laboratory.... The elementary laboratory is desirable and useful, but its results depend largely upon the kind of student who enjoys its advantages. . .our class did have an infinitesimal amount of practical instruction in the attic laboratory. We were divided into squads, which were summoned to the laboratory twice during the first year. I remember that on the first occasion that I enjoyed this privilege Garland [Bowditch's assistant] showed us how to make the classic nerve-muscle preparation from the frog's leg, and that on my second visit Bowditch was on hand to initiate us into the mysteries of graphic registration on the smoked drum.[16]

Although the laboratory experience served a minor role in Bowditch's elementary classes, he did encourage independent research by advanced students.[17] He was very interested in the advancement of experimental technique. In 1879, in the second volume of the new British *Journal of Physiology*, he published details of special apparatus designed for his laboratory, describing a mechanism for artificial respiration adapted from kymograph clockwork, a dog holder, a cannula for observing the vocal cords, and unpolarizable electrodes. In an accompanying article, James J. Putnam (1846–1918), neurologist and clinical instructor at Harvard, described a simplified and inexpensive pendulum myograph, which he had modified from the design of the German psychologist Wilhelm Wundt (1832–1920).[18]

In 1883, a larger physiological laboratory was opened on the second floor of the new Harvard Medical School building at Boylston and Exeter Streets.[19] Physiology occupied about one-quarter of the floor, and chemistry the other three-quarters. The physiological laboratory was well designed and carefully equipped with specialized instruments from Germany (Fig. 3). The adjoining amphitheater or lecture hall was patterned after that attached to the private physiological laboratory of Johannes Czermak (1828–1873) in Leipzig.[20]

Bowditch shared his lecture hall with the chemistry department. Here demonstrations were presented to large classes from the lecture table at the front of the room. The table was completely equipped with pipes for water, air, gas, and waste; it also possessed binding posts for electrical experiments. A hood, pneumatic trough, and sand baths were available nearby. The central part of the table was movable: experimental demonstrations in both physiology and chemistry could be prepared and wheeled in separately. Related microscopic demonstrations were set up in the mezzanine gallery of the lecture room.

In physiology's general laboratory next to the lecture hall, six to eight students could be accommodated at tables for "practical courses or for special work."[21] Two researchers could work at the Ludwig-style kymograph, which was equipped with a dust-proof protector and had nearby storage for operating instruments. A varnishing apparatus based on the design of Hugo Kronecker (1839–1914) of Berlin was employed to preserve the tracings.[22]

A large central "tower" or interrupter case held apparatus for regulating "the galvanic current" (Fig. 4). This was connected by wires to a battery closet in the gallery that supplied current to other apparatus, including a large Ruhmkorff coil.[23] A seconds pendulum could be put in to regularly interrupt the circuit, thus providing stimuli of a known rate. A new interrupter with platinum contacts, based on a model described by Christian Bohr (1855–

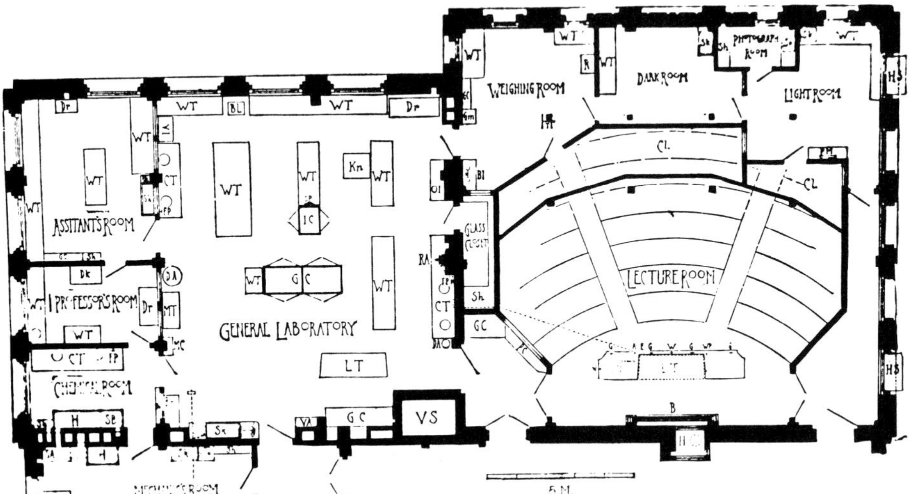

FIG. 1.—GENERAL PLAN OF LABORATORY AND LECTURE-ROOM.

Lettering to plans of laboratory.—*BC*, battery-closet (gallery); *BL*, blast-lamp; *Bl*, balance; *Bn*, basin; *CC*, clothes-closets; *Cl*, closet (under seats); *CT*, chemical table; *DA*, digestion apparatus; *Dk*, desk; *Dr*, chest of drawers; *FP*, filter-pump; *GC*, glass case; *Gm*, galvanometer; *H*, hood; *HS*, heliostat shelf; *IA*, injection apparatus; *IC*, interrupter case; *Kn*, kymographion; *L*, lathe; *LT*, lecture-table; *MC*, meat-cutter; *MT*, mercury-table; *OI*, operating instruments; *PC*, portfolio-case; *PM*, pendulum myograph; *R*, refrigerator; *RA*, respiration apparatus; *SA*, soldering apparatus; *SB*, steam-bath; *SE*, steam-engine; *Sh*, shelves; *Sk*, sink; *SP*, seconds pendulum; *ST*, saw-table; *T*, telescope; *VA*, varnishing apparatus; *VS*, ventilating shaft; *WB*, work-bench; *WT*, working-table. *Lecture-room.*—*A*, air & gas; *B*, blackboards; *E*, electricity; *G*, gas; *PT*, pneumatic trough; *W*, water; *WP*, waste-pipe.

Figure 3: *Floor Plan of the Harvard Physiological Laboratory at Boylston and Exeter Streets, 1883–1906. [From Science, 1884 (see n. 19).]*

1911) in 1882, had been constructed by the German instrument maker Baltzar and imported from Leipzig. It was described in a feature on the laboratory published in *Science* in 1884:

> In this a metallic cylinder turned by clockwork carries two sets of pegs (like the pins in a musical box), which strike against levers, and thus break contacts in the trough below. The pins of each series are so set that the contact is broken in one lever a little sooner than in the other, and consequently is still broken in the latter when the former closes. In this way a simple change of the wires from the induction apparatus permits the short-circuiting of the opening or the closing induction shock at pleasure. By an ingenious arrangement a cog-wheel can be shifted so as to give the cylinder a very slow or a rapid motion, as desired: the series of pegs thus do double work, and permit the great range of interruptions already mentioned.[24]

A clock that Bowditch had described in 1871 had been constructed by Zachariae, also of Leipzig. The article in *Science* explained:

> The clock...has a revolving pendulum, and a set of toothed wheels, which interrupt the electric current at intervals of one, two, three, four, five, ten, fifteen, twenty, thirty, or sixty seconds. The duration of the interruptions may be anything less than four seconds. By using a relay these may be changed into closures of corresponding length and interval.[25]

The laboratory also had an acoustic interrupter modified from the design of Julius Bernstein (1839–1917), "in which a steel bar of variable length (deter-

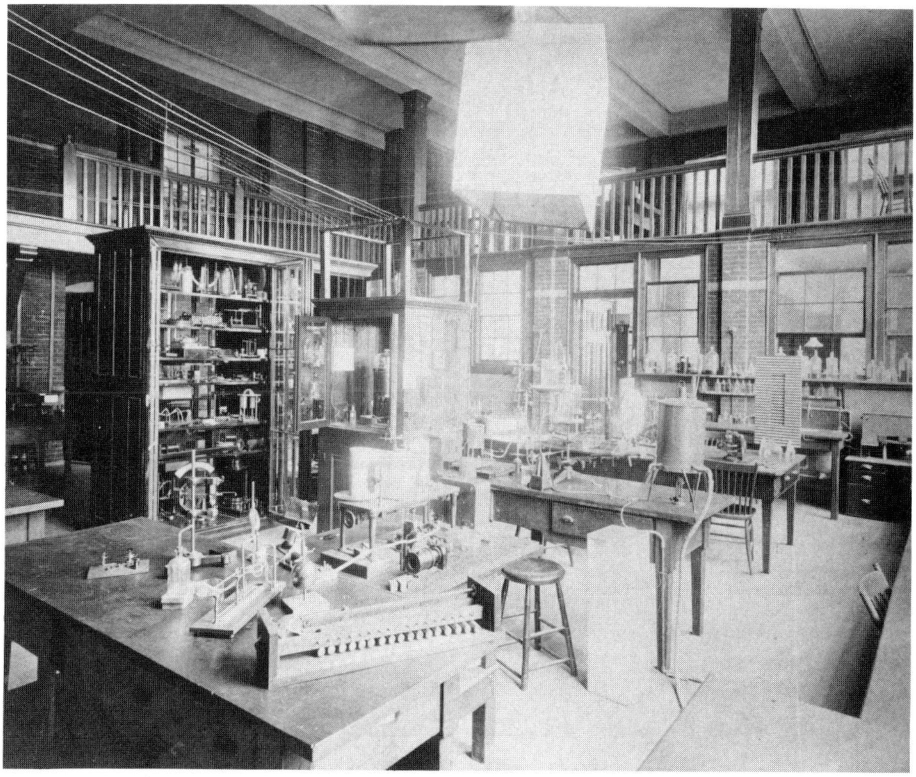

Figure 4: *Interior of the general laboratory. Note glassware cupboard and interrupter case or "tower" at rear; inductorium and kymograph among apparatus in foreground. [Photo courtesy of CL-HMS.]*

mined by a sliding clamp) is kept swinging by a temporary magnet above, while a platinum point makes and breaks in a cup of mercury below."[26] Several other homemade interrupters were available for use in stimulation studies.

The respiration apparatus, also developed by Bowditch, used kymograph clockwork to provide interrupted bursts of air for respiration experiments with dogs, cats, and rabbits. It could supply suction as well. Other prominent features of the laboratory were the mercury table, a meat-juice press, injection apparatus, a water bath, and a constant-temperature chamber (38°–60° C) made of two boxes with an intervening water layer heated by a Bunsen burner. These were standard laboratory items that Bowditch had noted in his description of apparatus in Ludwig's institute.

Separate rooms adjoining physiology's general laboratory accommodated Bowditch and his assistant, as well as chemical storage, instrument maintenance, and other supporting activities. The nearby mechanic's room had a hood, lathe, and saw table activated by a steam engine. The weighing room contained a balance, a Wiedemann's reflecting galvanometer, and a refrigerator. The dark room was used for optical experiments and would eventually contain a Thomson's high-resistance astatic reflecting galvanometer to be used for studying electromotive phenomena in muscle. In this instrument, a mirror on the galvanometer needle reflected a beam of light from a paraffin lamp onto a scale, thereby magnifying tiny electrical currents.[27] The photography room adjoined a lightroom that held the pendulum myograph made by Putnam. The intention was eventually to provide space in the gallery above the general laboratory for courses in "practical" physiology, which, the account in *Science* noted, "are given for the students of the first year in as large sections as can be conveniently managed."[28] The purpose of the laboratory was clear:

> It is intended to serve primarily as a laboratory of research, and secondarily as an adjunct to the lectures on physiology in the preparation of suitable apparatus and experiments. Courses in "practical physiology" are also given in the laboratory to the class in sections of a convenient size, but no instruction in "biology" is contemplated.[29]

PORTER'S LABORATORY METHOD FOR TEACHING PHYSIOLOGY

When Porter arrived ten years later, despite the vigorous advocacy of experimental physiology at Harvard, elementary classes were "still [being] taught in lectures illustrated by occasional experiments."[30] In 1893, Bowditch, by then dean of the Harvard Medical Faculty, invited Porter to fill the vacancy in the physiology department that had been created by the departure of William Henry Howell (1860–1945) to the physiology department at the Johns Hopkins Medical School. Porter's "task was to assume responsibility and reorganize the teaching of physiology."[31] Hands-on experimental work, hitherto only occasionally used in teaching, was to be dramatically increased.[32]

Porter announced the reform of the Harvard physiology curriculum in 1898 and 1900 in the *Boston Medical and Surgical Journal* and the *Philadelphia Medical Journal*.[33] In these public presentations, he criticized didactic teaching, rote learning, and "memory drill." He emphasized the need for laboratory work in physiology to train medical students in the scientific method:

> Practical work in physiology has ... been kept back by the erroneous ideas that the cost of apparatus and other materials is prohibitory, that medical students cannot master the details of exact experimentation, that delicate apparatus cannot be trusted in their hands, and that instruction to the extent required cannot be given to large classes because the course will become too complicated to be carried out.[34]

Moreover, medical teachers "fail[ed] to see that medical training should be 'for power' and only secondarily for information."[35] This training could be accomplished by:

> Deal[ing] so far as possible with the phenomena themselves and not with the descriptions of them. Where the fundamental experiments cannot all be performed, fill the gap with the original protocols from the classical sources. Associate facts which the student can observe for himself with those which he cannot observe. Use as the basis of professional instruction, where practicable, the facts and methods to be used by the student in earning his living. Teach the elements by practical work.[36]

In Porter's course, lectures were secondary; they came after the experiment. They were primarily useful late in the course to "show the student the historical development of scientific problems, the nature of scientific evidence, and the canons of criticism that help to sift the wheat from the chaff of controversy."[37] This approach contrasted with the other "practical" courses elsewhere, for example, those developed in England in biology and physiology.[38]

Porter's new physiology course was taught for four months in the first year of the medical curriculum, following anatomy. It was held from 9 A.M. to 1 P.M. every day during the second term. The afternoons were devoted to physiological chemistry. Instruction in physiology was divided into three main parts.

Weeks 1–5:

> Part I, of five weeks' duration, provides thorough experimental work in some limited field. In this, the student should acquire the point of view, the general physiological method, training in technic, and a complete knowledge of one or more tissues to serve as an introduction to the physiology of the remaining tissues. There can be little doubt that the physiology of muscle and nerve should be chosen for this purpose.

Weeks 6–12:

> Part II, of about seven weeks' duration, comprises carefully arranged fundamental experiments giving in turn the elements of each field in physiology except that of nerve and muscle which has just been studied.

Weeks 13–16:

> In Part III, covering the remainder of the term of sixteen weeks, the instruction is divided into special courses on the physiology of the eye, ear, larynx, digestion, the spinal cord, the innervation of the heart, etc.[39]

Neuromuscular physiology became the central focus for the new method. Porter argued:

> It is the most fully developed chapter in physiology and is well adapted to train the mind in habits of exact experimentation and close reasoning. Moreover, the physiology of muscle and nerve is in large measure the physiology of all living tissues, so that a man learned in this one field is in effect already acquainted with the general principles of physiology.[40]

AMERICAN SYSTEM OF MANUFACTURE: QUANTITY PRODUCTION OF QUALITY INSTRUMENTS

The "development of the mind rather than the imparting of information"[41] was to be supported by intensive laboratory training requiring specialized equipment, but the first class to be taught by "the laboratory method" consisted of 208 men.[42] This required 104 setups for experiments. Imported Baltzar kymographs would have cost $200 each, as these were made to order one at a time in Leipzig.[43] Porter thus invented a new kymograph: "The old kymograph, costing $200, was too heavy to be moved easily. The new one did essentially the same work, cost $16, and could be lifted in one hand."[44] Porter also invented an inductorium. Both the kymograph and the inductorium were highly praised. Indeed, Max Von Frey (1852–1932) of the Physiological

Institute at Würzburg and disciple of Ludwig later wrote Porter: "I especially prize the inductorium as all the connections are exposed to view, which is a great saving of time for the students. The two kymographs have approved themselves highly and I am astonished that such a useful instrument can be sent out at so small a price."[45] Memoranda written in the early 1930s emphasized Porter's strategy: "New apparatus was invented; old apparatus was redesigned; the strictest regard was paid to economy of weight and to the smallest possible 'number of operations.'... Reliable instruments were turned out at one-tenth the previous cost. Wholesale laboratory teaching became possible."[46]

Such accounts preserved in Porter's papers stress that his method of manufacture was American: "American quantity production requires economy of material, the fewest operations, and the making of large numbers of identical pieces at one time. The saving in price [per item] is magical...."[47] The surplus that resulted from quantity production was sold at cost to other teaching institutions, which had previously relied upon imported apparatus or in-house "mechanics." The 1904 Harvard Physiological Apparatus Company catalog explained:

> Physiological apparatus has theretofore been made upon the "model" plan, each piece for itself, without regard to the number of "operations" required, and with little or no thought as to the subsequent maintenance of the apparatus in good condition. A large laboratory class requires a great supply of apparatus....The sound training of large classes in physiology depends absolutely upon the invention of apparatus designed with reference (1) to its "manufacture," (2) to its storage and issue, and (3) to its maintenance in good condition.
>
> By "manufacture" is meant technically the making of a large number of "parts" for the same apparatus consecutively, by preference upon special lathes and with special tools. The number of steps or "operations" should be the fewest possible, as each operation must be repeated many times. The labor of "setting" a turret lathe which will make eight operations with eight different tools on the same piece of metal is as great for one piece as for a thousand. Astonishing economies may also be secured by the use of special automatic devices.
>
> Effective storage and issue demand that the apparatus be compact, and that important parts be protected so that they may not be injured when the apparatus is handled rapidly by persons of small experience.
>
> Economic maintenance requires that the apparatus should be constructed with special reference to durability and cleanliness.
>
> With all this, the apparatus must be admirably exact, serving the student for the repetition of classical experiments and satisfying the investigator for all the ordinary needs of research.[48]

Ease of manufacture, storage, and maintenance all affected instrument design. The catalog emphatically noted, "the supply of apparatus of this type bears the same relation to the advance of physiology that the commissariat bears to the advance of an army."[49]

The first sales catalog of Harvard Physiological Apparatus instruments appeared in 1900.[50] The instruments were initially manufactured by the Waltham Clock Company and sold by the Chicago Laboratory Supply and Scale Company. The Harvard kymograph with "accurate clock-work motor" and aluminum sheath then listed for $17.60. It was featured alongside electrical apparatus, including a "Harvard" induction coil with a swivel track to allow greatly reduced current, an interrupter, electrodes, keys, and signal magnet, as well as equipment for myographic studies, that is, a femur clamp, moist chamber, muscle warmer, muscle lever, ergograph, and several other items.

Porter's arrangement with the Waltham Clock Company did not succeed. He felt the instruments were badly made and too expensive because of the introduction of a middleman.[51] Instead, from 1901 "the Mechanics of the Harvard Physiological Laboratory" produced the instruments. The preface to the 1901 catalog explained Porter's intent:

> The student of physiology should perform the classical experiments on which science rests. The laboratory teaching of physiology should be brought within the reach of every school. These ends require that apparatus of precision be designed upon lines permitting its manufacture in larger quantities at small cost. It is believed that the apparatus here offered shows progress in this direction. It is well made, inexpensive, trustworthy, always in stock.[52]

Porter's vision of the role physiology could play in the reform of American medical education required mass manufacture of quality instruments and hands-on experience for all students. Apprenticeship in a research laboratory by advanced students would be preceded by experience in the teaching laboratory. Here all beginning students would learn the techniques, procedures, and modes of reasoning that underlie physiological research. This vision was obviously shared by Porter's colleagues elsewhere. He later recalled: "The Harvard Physiological Apparatus was welcomed everywhere. The success was embarrassing. The demand outgrew the very limited shop in the Physiological Laboratory. The work became in a small way a business, though an educational and not a commercial business."[53]

In March 1904, the propriety of the university selling apparatus was questioned. The Corporation of Harvard University requested removal of the manufacture of apparatus from university property, and the Harvard Apparatus Company was formally established as an independent company. To Porter, President Eliot wrote:

> I hope you will not feel that the Corporation does not recognize the value of your labors in improving the method of teaching physiology. Personally I believe very firmly in the changes you have introduced not only into our Medical School but into many others; but I agree with the Corporation in thinking it important to disconnect this indispensable manufacture of apparatus from the Medical School establishment both now and in the future. This manufacture has always been the only case of the sort in the entire University. It has been an exception of an undesirable and even dangerous sort as regards the financial management of the University and the relation of the University to public taxation. It is this action which has determined the present action of the Corporation.[54]

To Bowditch, he emphasized: "It does not mean, however, any distrust of the value of Dr. Porter's labors in revolutionizing the teaching of physiology in medical schools."[55] Three days later Eliot wrote to Francis Blake, the telephone inventor, requesting his assistance, noting "... you will, in my opinion, render a real service to medical teaching, and also help a man who has done an excellent piece of work during the past four years at much personal sacrifice and against formidable opposition."[56] Funds to start the company were secured through Blake, philanthropist Augustus Hemenway, and three Harvard professors.

Porter's innovations in the production of laboratory apparatus, like his curricular reforms, were announced immediately to American physiologists. At the annual meeting of the APS in 1899, Porter reported on "Apparatus for Laboratory Work by Large Classes."[57] Two years later, he displayed "Harvard Physiological Apparatus."[58] Porter was not alone in his interest in improving teaching apparatus. Warren Plimpton Lombard (1885–1939), a former Bowditch student and then professor of physiology at the University of Michigan, also demonstrated apparatus "intended for the use of students"[59] during these years, as did E. T. Reichert from the University of Pennsylvania, with whom Porter had once studied.[60]

From 1902 to 1904, Porter continued to exhibit physiological apparatus that he had developed for teaching.[61] So did several other physiologists, some of whom like Porter began to write laboratory manuals for beginning students. At this time, Porter published both a laboratory text, *An Introduction to Physiology* (1901 and 1906) and a guide to the Harvard course entitled *Physiology at Harvard* (1902, 1903, 1904, 1905, and 1906).[62] Lombard's *Directions for Laboratory Work in Physiology for the Use of Medical Students* appeared in 1906.[63]

Porter's effort to reform the teaching of a science by introducing regular laboratory instruction was not unique. The vital role of laboratory training in chemistry and physics education had already been established at Harvard by this time; nor was Porter's campaign new to physiology. Lombard was accomplishing a similar task at Michigan, and for similar reasons. In the preface to his laboratory manual, Lombard emphasized:

> [Students] are made to feel that the capacity to observe and correctly interpret the results of an experiment, is of even more importance than the ability to make an experiment successfully, and that one who has trained his powers of observation and has learned to accurately report the phenomena he has witnessed, can make a reliable diagnosis and keep a trustworthy casebook.[64]

In 1904 Winfield S. Hall, professor of physiology at Northwestern University Medical School, observed in his text, *A Manual of Experimental Physiology*:

> The general method of presenting the subject of physiology is the same as that followed in all of the other experimental sciences, viz., the laboratory method, according to which the student is led to discover for himself certain facts and to formulate from his collected data conclusions which represent fundamental principles of the science...the modern laboratory method of instruction, cultivates in the student the capacity of close and accurate observation; it affords an opportunity for valuable practice in the systematic recording of the observations; it develops the power of logical thought in drawing tenable conclusions from the observed data; and, in the formulation of conclusions, it stimulates the ability to express the thoughts in concise and unambiguous terms. In the second place, practice of this kind makes the student independent and furnishes him with just the mental equipment needed for later life in whatever line his activities may be directed. If such an education is more important for one profession than another, the medical profession is certainly the one in which its importance is greatest. The physics, chemistry, and biology studied in preparation for medicine, and the

physiology, pharmacology, and pathology of the medical course should give the student a most admirable equipment for dealing with the complex problems of clinical practices.[65]

When Porter began the production of inexpensive precision instrumentation for use by large numbers of students and placed primary emphasis on experiment in physiological teaching, the supporting pedagogical arguments were already in place. Reform could be accomplished rapidly and cheaply, even at small institutions, now that reliable, well-made, inexpensive apparatus could be secured from an American manufacturer. Porter's initiative rapidly secured physiology's place as an independent experimental science in American universities.

APPARATUS FOR TEACHING ELEMENTARY PHYSIOLOGY

For the new Harvard curriculum in physiology, Porter had devised several specialized teaching instruments himself, all of them adapted from research instruments then in use. The Harvard Physiological Apparatus catalog of 1901 included a capillary electrometer, a circulation scheme, an ergograph, an eye box (or artificial eye) and optical lantern, a moist chamber, and a signal magnet that Porter had designed.[66]

The capillary electrometer ($2) relied on the changing surface tension of mercury, that is, a meniscus displaced by changing potential, to indicate the current of injury in muscle and to measure electrical potential in nerve, muscle, and heart. Several variants of the original Lippmann instrument were in use at the time. The circulation scheme ($3.50), illustrating the mechanics of circulation, simulated cardiac action by means of a pump and valves and simulated peripheral resistance by means of channels in a piece of bamboo. The receiving tambour of a sphygmograph ($1) recorded "time relations of the changes in the pressure within the pump," while mercury manometers measured arterial and venous pressure. With this apparatus, the physical principles of circulation could be demonstrated. Many such schemes had been described. The Harvard Apparatus model, monitored by means of a sphygmograph tambour and kymograph drum (in this catalog approximately $12), was based on physiological techniques then approximately four decades old (see Figs. 5, 6, and 7).

The ergograph ($1.50) recorded isotonic and isometric contractions and monitored fatigue of the index finger abductor. Porter's artificial eye ($3.00) with interchangeable diaphragms and lenses simulated the physics of vision, while an optical lantern ($3.75) supplied light. Light rays were made visible in the apparatus through the use of Japanese incense. Willy Kühne (1837–1900) had designed an earlier device, the water-filled artificial eye manufactured by Jung of Heidelberg, then in use in Britain.

The moist chamber ($1.75) used for muscle preparations consisted of a porcelain plate covered by a glass plate in which a nerve-muscle preparation could be placed and monitored or stimulated by means of unpolarizable potter's clay boot-shaped electrodes. The signal magnet ($1) used in circuit

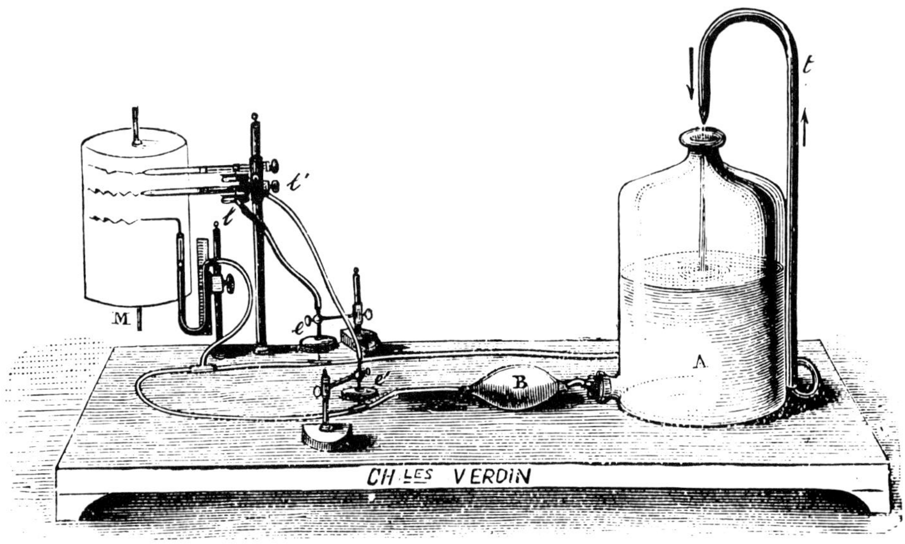

Figure 5: *Circulation scheme of Leon Frédéricq. A, venous system; B, ventricle; tubing connected to mercury manometer represents arteries and peripheral resistance. Marey tambours (e and e′) record pressure-pulse and rate of propagation of pulse wave. Heart motions are simulated by contraction of bulb (B) by hand. [From Stirling, Outlines of Practical Physiology, 4th ed., 1902 (see n. 27).]*

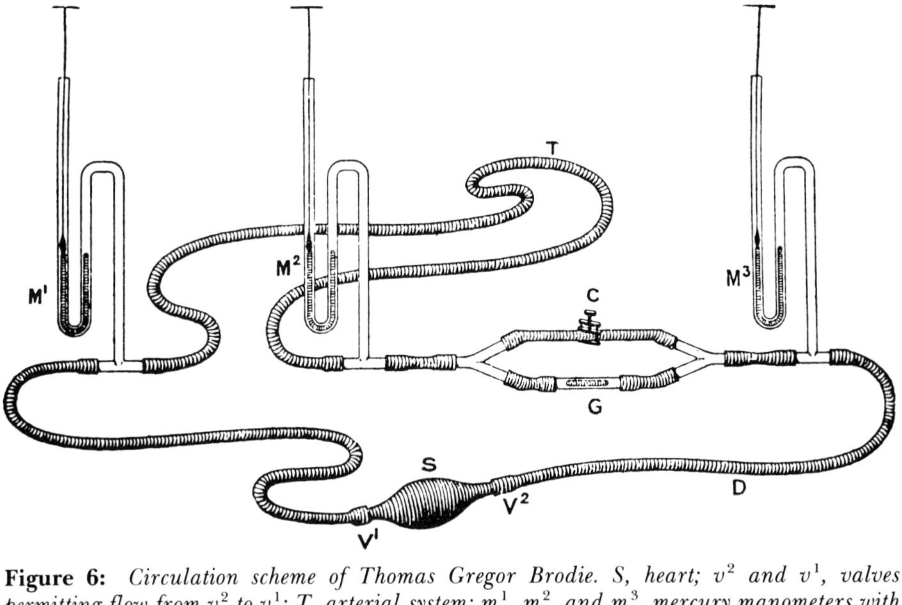

Figure 6: *Circulation scheme of Thomas Gregor Brodie. S, heart;* v^2 *and* v^1*, valves permitting flow from* v^2 *to* v^1*; T, arterial system;* m^1*,* m^2*, and* m^3*, mercury manometers with recording floats; C, screw clamp; G, tubing packed with glass wool; D, venous system. [From Brodie, Essentials of Experimental Physiology, 1898 (see n. 68).]*

with a tuning fork ($.50) could record 100 double vibrations per second. In the primary circuit of the inductorium ($5), it recorded the make and break of the current.

Other apparatus produced by the mechanics of the Harvard physiological

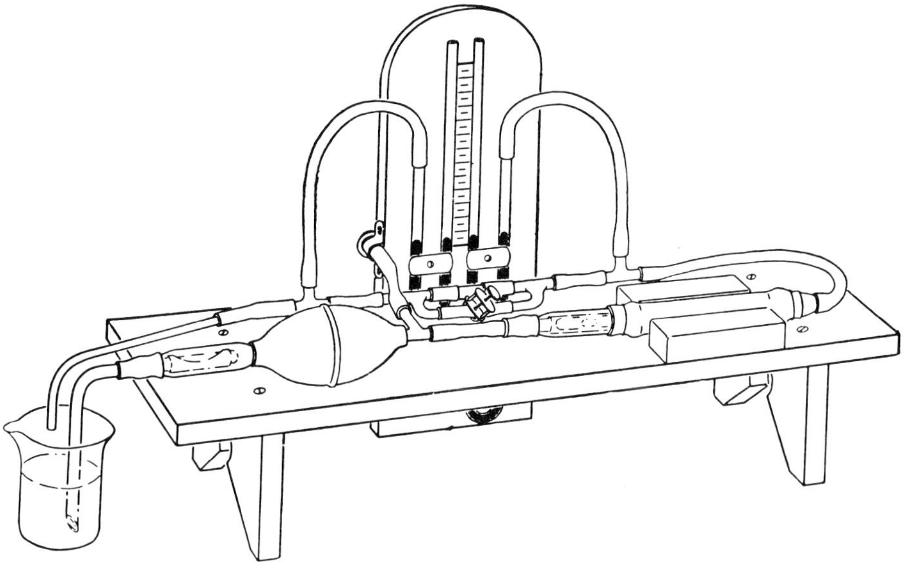

Figure 7: *Circulation scheme manufactured by Harvard mechanics showing mounting of pump, valves, and elastic tubing. Bamboo simulates peripheral resistance. Side branch with clamp simulates dilatation. Mercury manometers measure arterial and venous pressures. Sphygmograph tambour (not shown) monitors pulse in the aortic tube. [From Lord Electric Company, "Harvard Physiological Apparatus," 1901 (see n. 52).]*

laboratory included attachments for electrical, kymographic, myographic, and sphygmographic work. Electrical devices included platinum electrodes, keys (short-circuiting, simple contact, and rocking), the induction coil, interrupter wheel, and two types of rheochord. Myographic equipment included the frog board, the femur clamp, adjustable plate, muscle warmer, lead weights, and work adder, while the mercury manometer, sphygmograph tambour, tuning fork, and several miscellaneous items were adaptable to multiple uses in the elementary laboratory. All were "modified forms of familiar apparatus."[67]

Porter's course text, *An Introduction to Physiology*, described how to use these instruments to perform basic operations in physiology. Students stimulated muscle and nerve and recorded the response. They studied irritability, conductivity, and change of form in muscle; electromotive phenomena in muscle and nerve; and the mechanics of the circulation and the innervation of the heart and blood vessels. From Porter's text, we are able to assess the application of these teaching instruments and to compare them with apparatus in use during the preceding decades.

Comparable British texts by physiologists William Stirling (4th ed., 1902), A. D. Waller and W. Legge Symes (1905), and T. G. Brodie (1898) indicate the increasing importance of routine laboratory instruction at this time.[68] Such manuals were by no means new to physiology. These texts evolved from other handbooks of the last quarter of the nineteenth century, such as Cyon's *Methodik der physiologischen Experimente und Vivisectionen* (1876) and Sanderson, Foster, Klein, and Brunton's *Handbook for the Physiological Laboratory* (1873).[69] However, the widespread use of clear plates showing apparatus in most manuals (Figs. 8 and 9) suggests both the increasingly entrenched position

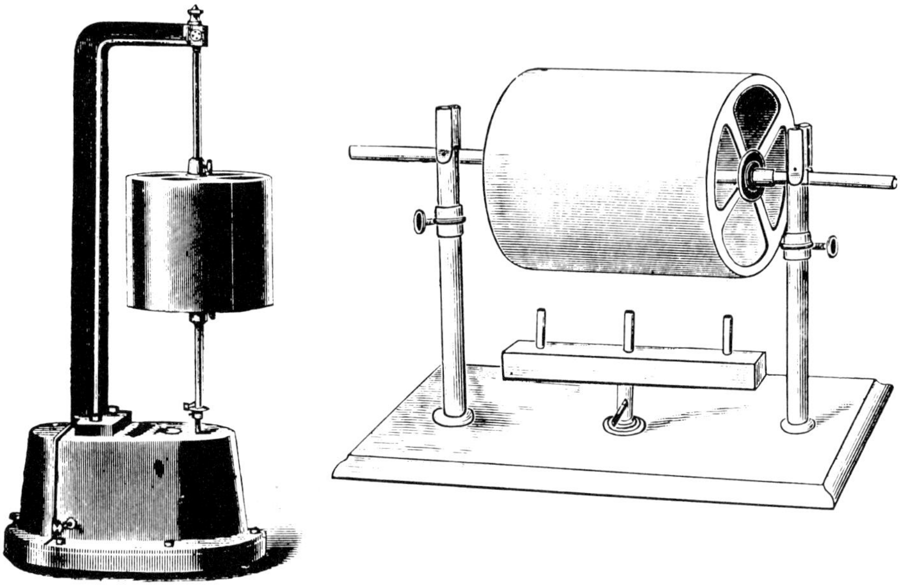

The kymograph. Drum support for use in smoking the kymograph drums.

Figure 8: *Kymograph and apparatus for smoking paper. [From Hall, Manual of Experimental Physiology, 1904 (see no. 65).]*

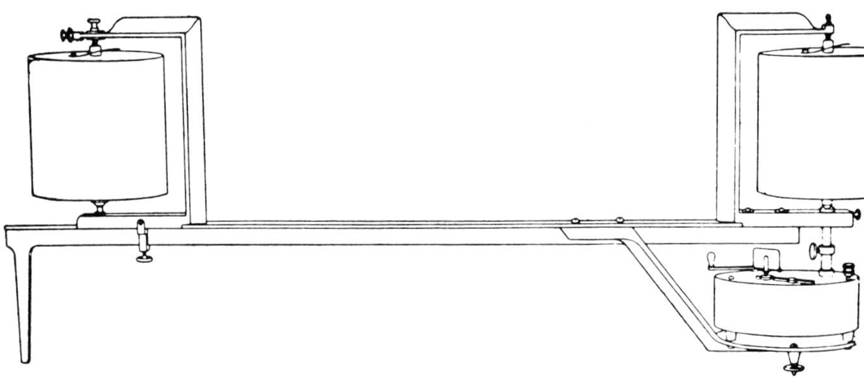

Figure 9: *Adjustable long-paper kymograph manufactured by Harvard Apparatus Company. [From Porter, Introduction to Physiology, 2d ed., 1906 (see n. 62).]*

of physiological instrumentation in the laboratory, and the need for clear, concise, practical instructions at a time when laboratory techniques were being taught to large numbers of students, rather than individually or in small groups by apprenticeship.

The most complete and historically useful of the laboratory manuals written in English at the end of the nineteenth century is that of William Stirling (1851–1932), Brackenbury Professor of Physiology and Histology at Owens College and professor at Victoria University, Manchester, England. The fourth edition of his *Outlines of Practical Physiology* appeared in 1902. It was subtitled a "Manual for the Physiological Laboratory including chemical and experimental physiology with reference to practical medicine." Stirling's work incorporated useful data from other recent texts including that of Porter and was profusely illustrated. From it we can survey the array of techniques and instruments in use at the turn of the twentieth century. Four hundred and sixty-five illustrations present the techniques and data of experimental physiology. Of these, less than twenty percent are devoted to chemical physiology. The text portrays physiological instruments manufactured in both Europe and North America and provides a list of thirty-three instrument makers.[70] Such an extensive presentation of laboratory technique and instruments in competing texts raises the question of the extent to which Porter's contributions to teaching laboratory physiology were really unique.

The increasing range of devices that had become useful to physiological research was already clear in the 1860s. In 1868 the French physiologist E.-J. Marey (1830–1904) had provided clear descriptions of the kymograph, myograph, sphygmograph, cardiograph, sphygmoscope, hemodromograph, and polygraph, as well as other apparatus, for his students at the Collège de France.[71] These laboratory devices had been complemented in the latter part of the century by the invention of various plethysmographs, oncometers, ergographs, and a wide range of electrical apparatus; all were well represented in Stirling's text. From 1876, physiology as a fundamental problem area of biological science to be explored by experiment had been developed in the United States by Henry Newell Martin (1848–1896) at Johns Hopkins on the Foster-Huxley model.

In 1903, in the second edition of his *Physiology at Harvard*, Porter confronted criticisms regarding his originality:

> The traditional method of teaching physiology consists of a systematic course of lectures illustrated by occasional demonstrations. For thirty years or more, especially in England this didactic teaching has been further illustrated by certain experiments performed by students themselves. Laboratory experiments, therefore, have long been a valued part of the instruction in physiology in many universities. When the new method of teaching was introduced in the Harvard Medical School...it was said that this was only doing in a large way that which had been done in a small way for many years.[72]

Porter emphasized the goals behind his new method:

> The new method is not an extension of the old. It is fundamentally a different process. The old method is chiefly didactic. The new is a systematic course of experiment and observation by the student himself. In the old the student rests upon the dictum of the professor and the text-book. In the new he relies upon the fundamental experiments done with his own hands. In the old his experiments follow the lecture and attempt to verify its statements. In the new the lecture follows his experiments and discusses them in relation to the work of other observers. In the old the stress is upon didactic teaching. In the new the stress is upon the observation. Under the old method students ... used to ask, "Who is the authority for that statement?" Under the new they ask, "What is the experimental evidence?"[73]

Here was a pedagogical revolution of great significance that relied upon intensive laboratory training for beginning students. Porter repeatedly emphasized that in his course, experiments were intended "to illustrate physiological experimentation rather than to disclose step by step the groundwork of the science of physiology."[74] The Harvard Physiological Apparatus Company had been created to make this opportunity available to each and every student. Not only Harvard students, but students across the country benefited from the availability of the tools and techniques of modern physiology. Porter had earlier noted: "The way of the physiologist is not peculiar. The method of getting a real education is the same from the kindergarten to the specialist. The principle is to train 'for power,' to use President Eliot's phrase, and not primarily for information."[75]

TEACHING OF ELEMENTARY LABORATORY PHYSIOLOGY AT HARVARD, 1900–1906

The second edition of Porter's booklet *Physiology at Harvard* appeared in 1903. The preface explained: "This book is written to explain a new method of teaching, sound in theory and feasible in practice; to provide the Harvard Medical School with a precise account of the work done by each student in Physiology; and to create for students and instructors alike a working plan by which they may find their way unvexed through much detail."[76] This little book of 101 pages explained Porter's method and goals, detailed the first-year course, the advanced and summer courses, as well as opportunities available for independent research. The standard laboratory apparatus being used was also listed in full.[77] Included were the manometer, kymograph, sphygmograph, and plethysmograph for recording mechanical phenomena; and the batteries, inductorium, electrodes, keys, rheochord, galvanometer, and capillary electrometer for providing stimuli and recording electrical events. Along with the thermometer and microscope, and some chemical apparatus, these instruments formed the central apparatus for study and analysis of physiological phenomena.

Many of these instruments had been invented in Europe in the 1850s and 1860s, but their designs had been improved by later workers, and their applications extended to study a wide range of physiological events. Phenomena that had not originally been accessible to earlier workers had since become objects of intense study. The spectacular progress in neuromuscular and cardiovascular physiology that underlay much of Porter's course resulted from the application of these research tools.

Many of these physiological devices had also been tested as possible clinical tools immediately after their invention. However, the invasive nature of most physiological instruments meant that each device had to be modified to be used for clinical purposes. By the end of the century, as physiologists and physicians had developed distinct goals, their instruments reflected different requirements. This conflict over purpose had been evident to those who attempted to apply the new instrumentation to medicine even in the 1870s.[78] In the 1880s and 1890s, physiologists had become increasingly concerned with their own questions, and the disjunction between pure and applied experimental medicine had become more apparent. By the time of Porter's push for experimental physiology, physiology had become a discipline to train the mind in exact reasoning rather than the royal road to either diagnostic precision or effective therapy. By 1900, physiology as a discipline had become removed from clinical medicine, even though physiology as an intellectual pursuit was becoming intimately tied to clinical concerns. After the turn of the century, physiologists increasingly studied problems arising from clinical observations and whole organism studies. They were busy establishing base lines and standard curves that would impinge on medical practice only much later.

Porter's course was required in the first year of the medical curriculum, accompanied by physiological and pathological chemistry. It followed four months of study in anatomy, histology, and embryology. Additional work could be undertaken formally in the fourth year. The research laboratory was open at any time to qualified students, and research from such investigations was often published in the *American Journal of Physiology.* It is interesting that Porter excluded from his course most diagnostic instruments except the ophthalmoscope and sphygmograph. Such artificial schema as the eye box and circulatory scheme provided the student with simulated objects for study. The separation of physiology from practical medicine was evident in the content of his course and would prove to be a hindrance to its continuation.

The first-year laboratory course accommodated approximately 200 students, and Porter began to devote increasing attention both to it and to the production of teaching apparatus. This decision was also to have an adverse impact upon his later career. His efforts to develop both the teaching of experimental physiology at Harvard and instrumentation for that purpose were to have unexpected results. Porter was not promoted to professor of physiology on Bowditch's retirement.

AMERICAN PHYSIOLOGY IN TRANSITION

In a letter of 31 March 1906 to President Eliot, Porter announced his impending resignation and recounted his accomplishments, noting that he "[had] made Harvard methods of teaching Physiology known among physiologists in all parts of the world."[79] In an attachment to this letter, Porter claimed that the *American Journal of Physiology*, which he managed and edited, "[had] doubled the volume of physiological research in this Country and [had] given American physiology a new standing." Moreover, Harvard physiological apparatus was "used in 162 laboratories in 35 States and Territories, and in 43 laboratories in 15 foreign countries including Syria, Russia, Japan and Australia." By this means "73 different pieces of physiological apparatus" had been made "accessible to educational institutions." Porter himself had invented twenty-two of them.[80]

Notwithstanding these achievements, Porter had been criticized for neglecting his research.[81] Perhaps more importantly, the new course to which he had devoted so much energy was criticized for its high failure rate,[82] and Porter's innovations in physiological apparatus had proved to be a burden to the university.[83] Porter's method for revolutionizing the teaching of physiology at Harvard had not developed as he had intended. In 1906, Walter B. Cannon (1871–1945) was appointed professor of physiology. Porter instead became professor of comparative physiology. In the same year the medical school formally moved to the new buildings at Longwood Avenue. Porter had been a major force in designing the laboratory there,[84] but he was no longer to run the elementary teaching laboratory.

Under Cannon, the teaching of physiology was refocused. As the annual report for 1906–1907 noted:

> During the past year several changes were made in the teaching of Physiology. The laboratory experiments were rearranged in such order that the student learns first of what activity an organ or tissue is capable, next how certain factors condition or modify that activity, and finally what may be the effect of the activity. Where possible, experiments combining physiological training and the use of instruments valuable for clinical investigation were introduced.[85]

In addition, the curriculum was altered to give students more time for reading and for repetition of experiments. Courses in physiology and biological chemistry were also coordinated.[86] In the words of the annual report for 1907–1908: "A new laboratory manual has been written by Dr. Cannon, based on seven years' experience in laboratory instruction, and the members of the next class will have the new directions and experiments in their work."[87] From 1908 to 1909, it was noted that "the use of new laboratory directions has rendered the instruction more orderly and developmental than it hitherto has been."[88]

Porter's vision of physiology had been that of an experimentalist. Promotion of that ideal, however, had been fraught with difficulties. Porter sought to train men in rational science and quantitative work, but his students were accustomed to memorizing instead of reasoning, as he pointed out in a letter of 1904 to President Eliot:

> The majority of the men now entering the School cannot observe phenomena nor record them accurately; still less can they reason upon their observations. The boy's natural power of observation and his lively curiosity regarding the world about him are not infrequently dulled by years of memorizing in the secondary schools. Four additional years in college without a rational science often complete the blunting of the special powers required for the successful study of physiology. Moreover, the student's first half-year in the Medical School is devoted chiefly to memorizing, for which he is well prepared."[89]

In 1921, Porter's special contributions to the development of physiology became apparent in debates over whether the APS might accept ownership of the Harvard Apparatus Company. As then APS President Lombard wrote to Porter:

> I think that the chief objections centered about the doubt whether as time went on suitable men could be found to look after the business. The factory would have to be located at some one point, and there would be no surety that a man like yourself, with knowledge of the requirments [sic] of physiologists, inventive, yet conservative, and of good business capacity, could be found in that region; or if there were such a man, it would be doubtful whether he would be willing to take from his teaching and research the time which would be necessary to handle such an enterprise properly.[90]

Lombard further noted: "All of the members of the Council expressed their recognition of the fact that the Harvard Apparatus Company has been of the very greatest value to physiological teaching and research, and their desire that it should continue to be of use."[91]

As evidenced by his founding of both the Harvard Apparatus Company and the *American Journal of Physiology*, Porter was well aware of the need for building support institutions in North America that nurtured individual inquiry and independent research. By creating such support institutions himself, Porter sought to bring to American physiology the independent status enjoyed by European physiologists. Yet even as professional independence for physiology was being achieved in this country, physiologists were individually confronting the perennial issues surrounding their discipline's intellectual and economic relationship to medicine.

In other words, the eventually successful campaign for the separation of physiology from anatomy and medicine in the nineteenth century was being followed by a reconsideration at the turn of the twentieth century of the mutual benefits enjoyed by each intersecting discipline. Physiology itself was about to be threatened by the separation from its domain of fields like biochemistry and endocrinology.[92] Moreover, physiologists were beginning to discover that the more interesting problems yet to be solved lay in the common ground between clinical medicine, experimental physiology, and physiological chemistry. The regulation of physiological processes, for example, could be studied by the techniques and apparatus already in hand in these neighboring fields.[93]

CONCLUSIONS

Science is simultaneously a craft, an intellectual discipline, and a social enterprise. Physiology, as a foundational subject of medicine, is also an applied science. Porter's appointment to the chair of comparative physiology is one indication of the extent to which he had neglected clinical concerns in his efforts to promote American physiology.[94] In the United States, as even Porter must now have realized, the advancement of physiology was intimately linked to medicine. Physiology was not only expected to train "for power" but also to provide the material fruits of scientific medicine, to make rational therapy possible. As professor of physiology at Harvard, Cannon reclaimed that goal, seeking to promote the link between laboratory and clinic and to enhance physiology's role in providing the rational underpinning for diagnosis and therapy.[95]

Even with the rise of the elementary teaching laboratory, individual research beside experienced workers would continue to provide the primary mechanism for the transmission of new research techniques in physiology. Physiological research was a craft to be learned at the bench by apprenticeship. This had been the goal of the laboratory or "workshop" method, as Porter later called it.[96] The elementary teaching laboratory taught the essentials of the craft by introducing large numbers of students to the basic techniques, instruments, methods, and modes of reasoning that underlie experimental analysis of physiological events.

By the first decade of the twentieth century, physiology had become the model discipline for exploitation of the experimental method in the life sciences. The methods and material supports for that work were firmly established. Students were being introduced to the art of experiment and the science of physiology in institutions across the continent, in laboratories equipped for research with instruments designed by research physiologists and manufactured for teaching purposes (Fig. 10). Horace Davenport has called the laboratory "the way to the social and institutional life of Physiology."[97] Porter in 1900 was acutely aware of its intellectual role. The techniques and tools, theoretical foundations, and institutions of physiology are mutually supportive aspects of the same enterprise. They arose jointly and must be considered as a unit in any historical analysis of the rise of modern physiology.

The laboratory method has been a central feature in the process of the self-definition of American physiology. In the opening years of this century, its social value was asserted in the often repeated injunction to train "for power," echoed in different institutional contexts by Porter and his colleagues.[98] Cannon, too, observed in his laboratory manual published in 1910:

> To the student of biology and the student of that part of applied biology included in medicine, the laboratory method of learning the functions of the tissues and organs of the animal body has demonstrated its superiority over other methods, not only in impressing vividly the facts of observation, but also in demanding a discipline which furnishes power to go forth with assurance in independent research.[99]

Figure 10: *Harvard Physiological Laboratory, Boylston Street, 1905. Students from Dental Section using Harvard Physiological Apparatus kymographs. Note addition of electric lights. [Courtesy of CL-HMS.]*

Nineteenth-century laboratory instruments, the tools of both research and education, provide an important record of the rise of that method. The elementary physiology laboratory, equipped with mass-produced but specialized precision apparatus, taught basic research skills to those generations of American investigators who would shape the emerging ideology of twentieth-century scientific medicine.

This research was supported by a grant from the National Science Foundation (RII-8503650) through its Visiting Professorships for Women program. The author wishes especially to thank the Department of the History of Science at Harvard University and curators Richard Wolfe and Clark Elliott and their staffs at the Harvard Medical Archives, Francis A. Countway Library, and the Harvard University Archives, Pusey Library, respectively, for assistance in locating historical materials relating to the development of the Harvard Physiological Laboratory. The excellent research assistance of Hughes Evans is gratefully acknowledged, as are the insightful observations and questions of Atsushi Akera, Hughes Evans, Jason Harry, Charles Millstein, Sam Murrell, and Rob Spaulding, all of whom participated in the author's conference course at Harvard, "Instrumentation and the Rise of Modern Physiology." Speical thanks are due to A. Clifford Barger, Adele Clarke, Deborah Coon, Gail Hornstein, Naum Kipnis, Clark Sawin, Howard Segal, Arleen Tuchman, and Carroll Williams for their very helpful comments on an earlier draft of this paper. The skillful and timely word processing of Carol Troyer-Shank is gratefully acknowledged. The early stages of this work were supported by the Bakken Library in Minneapolis.

NOTES

1. J. Schiller, "Physiology's Struggle for Independence in the First Half of the Nineteenth Century," *Hist. Sci.* 7 (1968): 64–89; and Karl E. Rothschuh, *History of Physiology*, trans. and ed. Guenter B. Risse (Huntington, NY: Krieger, 1973). For a lucid summary of the main features of the emergence of physiology as an independent experimental science, see Gerald L. Geison, *Michael Foster and the Cambridge School of Physiology: The Scientific Enterprise in Late Victorian Society* (Princeton, NJ: Princeton Univ. Press, 1978), pp. 13–47.
2. Merriley Borell, "The Kymograph and the Origins of the Graphic Method," *Elec. Q.* 7 (Summer, 1985): 2–3; idem, "Extending the Senses: The Graphic Method," *Med. Heritage* (March/April, 1986): 114–121; L. A. Geddes, "Perspectives in Physiological Monitoring, *Med. Instrum.* 10 (1976): 91–97; and idem, "Clinical Engineering and the Background of Interdisciplinary Engineering, *Med. Instrum.* 9 (1975): 239–249.
3. W. T. Porter, "The Teaching of Physiology in Medical Schools," *Boston Med. Surg. J.* 139 (1898): 647–652, on 649.
4. William H. Howell and Charles W. Greene, *History of the American Physiological Society Semicentennial 1887–1937* (Baltimore, MD: American Physiological Society, 1938), p. 14.
5. Ibid., see esp. pp. 10–15, 171–173, and 201–206, as well as chapter I by Robert Frank, chapter II by W. Bruce Fye, and chapter III by Alejandro Laszlo in this book. For recent scholarship on Bowditch, see Everett Mendelsohn, "Henry Pickering Bowditch," in *Dictionary of Scientific Biography*, ed. C. C. Gillispie, vol. 2 (New York: Scribner, 1970), pp. 365–368, and Fye, "Why a Physiologist?—The Case of Henry P. Bowditch," *Bull. Hist. Med.* 56 (1982): 19–29. On Porter, see Fye, "Acute Coronary Occlusion Always Results in Death—Or Does it? The Observations of William T. Porter," *Circulation* 71 (1985): 4–10, and A. Clifford Barger, "The Meteoric Rise and Fall of William Townsend Porter, One of Carl J. Wiggers' 'Old Guard,'" *Physiologist* 24(5) (1982): 407–413.
6. Barger, "Meteoric Rise," p. 409.
7. On this process see Geison, *Michael Foster*, esp. p. 78, and Horace W. Davenport, "Physiology, 1850–1923: The View from Michigan," *Physiologist* 24, suppl. 1 (1982): 26 and 83. Davenport notes: "Sharpey began to correct the deficiencies in laboratory teaching in 1867 by inviting Michael Foster to become Professor of Practical Physiology. Together Sharpey and Foster

toured continental laboratories, and they successfully transplanted foreign physiological practice to England. Foster's textbooks, beginning with the second edition, are filled with pictures of physiological apparatus and descriptions of techniques. Burdon Sanderson, who succeeded Foster at University College, edited an even more copiously detailed handbook...Foster took the methods to Cambridge, Burdon Sanderson took them to Oxford, and Schäfer took them to Edinburgh," (p. 26).

8. The description of Ludwig's laboratory, which opened in 1869, appeared in both the *Boston Med. Surg. J.* 82 (1870): 305–307, on 307, and in *Nature Lond.* 3 (1870–1871): 142–143, on 143. For full details of Bowditch's travels and career, see chapter I by Frank and chapter III by Laszlo in this book. Timothy Lenoir has prepared a new study of the origins of Ludwig's institute at Leipzig.
9. *Annual Reports of the President and Treasurer of Harvard College 1870–71* (Cambridge, MA: University Press, 1872), pp. 18–22 (hereafter, HAR).
10. Frederick W. Ellis, "Henry Pickering Bowditch and the Development of the Harvard Laboratory of Physiology," *N. Engl. J. Med.* 219 (1938): 819–828, on 820. Also see Fye, "Why a Physiologist?", pp. 26–27, esp. n. 44, for details of Bowditch's expenditures for equipment. On the role of Eliot, see Kenneth M. Ludmerer, "Reform at Harvard Medical School, 1869–1909," *Bull. Hist. Med.* 55 (1981): 343–370.
11. HAR, (1870–1871), p. 21.
12. HAR, (1871–1872), p. 25.
13. HAR, (1870–1871), p. 22.
14. Ellis, "Henry Pickering Bowditch," p. 820. For details see Fye, "Why a Physiologist?", p. 27, n. 45.
15. Ellis, "Henry Pickering Bowditch," p. 821.
16. Ibid., p. 822. George Minot Garland served as assistant in physiology from 1877–1881.
17. Ibid., p. 827. Ellis noted "[Bowditch's] work and interest was in that division of physiology that was in closest contact with that of the master in the Leipzig laboratory." For a list of Bowditch's publications, as well as publications from the physiological laboratory between 1873 and 1886, see Fye, "Henry Pickering Bowditch: A Case Study of the Harvard Physiologist and His Impact on the Professionalization of Physiology in America" (MA thesis, Johns Hopkins University, 1978).
18. H. P. Bowditch, "Physiological Apparatus at the Harvard Medical School," *J. Physiol. Lond.* 2 (1879–1880): 202–205; and James J. Putnam, "Description of a Modified Pendulum-Myograph," *J. Physiol. Lond.* 2 (1879–1880): 206–208.
19. Thomas Francis Harrington, *The Harvard Medical School*, vol. 3 (New York: Lewis, 1905), p. 1073. A useful account of the change in physical setting is found in *The Harvard Medical School, 1782–1906* (Boston, 1906); on the Department of Physiology, see pp. 87–95; on the present medical buildings, see pp. 177–183 and 185–193. A full description of the Boylston Street laboratory is provided in "The Harvard Physiological Laboratory," *Science* 4 (1884): 128–135. The reprint copy (Boston, MA: Franklin, 1884) held in Pusey Library, Harvard, is signed "with the author's compliments, J. W. W." [Joseph Weatherhead Warren].
20. [J. W. Warren], "Harvard Physiological Laboratory" (reprint), p. 5. Czermak's amphitheater opened in 1872.
21. Ibid., p. 7. The General Laboratory measured 10.8 m by 9.6 m. The students worked at "chemical tables" furnished with shelves, chemicals, bowls, and filter pumps. Hot water was also available in the laboratory.
22. Ibid., p. 19. Kronecker later became president of the Marey Institute, an organization dedicated to the standardization and control of physiological methods. Cf. Rothschuh, *History of Physiology*, p. 252; and n. 29, this chapter.
23. Named after Heinrich Daniel Ruhmkorff (1823–1887), the Ruhmkorff coil is an induction coil in which the secondary coil is not movable. Cf. W. A. Newman Dorland, *The American Illustrated Medical Dictionary* (Philadelphia, PA: Saunders, 1947), pp. 340 and 1272. For details, see J. M'Gregor-Robertson, *The Elements of Physiological Physics: An Outline of the Elementary Facts, Principles and Methods of Physics; and Their Applications in Physiology* (Philadelphia, PA: Lea, 1884), pp. 38–42.
24. "Harvard Physiological Laboratory" (reprint), p. 12.
25. Ibid., p. 13.
26. Ibid. The Bernstein- and the Bohr-style interrupters were used by Bowditch and Warren for

their article, "Plethysmographic Experiments on the Vasomotor Nerves of the Limbs," *J. Physiol. Lond.* 7 (1886): 416–450.

27. Sir William Thomson's (Kelvin's) galvanometer is described in Silvanus P. Thompson, *Elementary Lessons in Electricity and Magnetism*, rev. ed. (Chicago, IL: Thompson & Thomas, 1906), pp. 168–170. It was devised for signaling through submarine cables. On its use in physiology for the study of the demarcation and action currents, see William Stirling, "Outlines of Practical Physiology (London: Griffen; Philadelphia, PA: Blakiston, 1888), pp. 186–192 and 4th ed. (London: Griffen, 1902), pp. 311–319. On Wiedemann's galvanometer, see E. Cyon, *Methodik der physiologischen Experimente und Vivisectionen* (Giessen, Germany: Ricker'sche Buchhandlung and St. Petersburg, Russia: Ricker, 1876), pp. 350–362 and Gustav Wiedemann, *Die Lehre von den Wirkungen des Galvanischen Stromes*, Zweite Auflage, Erste Abteilung, Zweiter Band (Braunschweig, Germany: Vieweg, 1873), pp. 227–231. For a comparison of the two galvanometers, see M'Gregor-Robertson, *Physiological Physics*, pp. 99–110. I am grateful to Naum Kipnis for directing me to several of the references cited in this section.
28. "Harvard Physiological Laboratory" (reprint), p. 22.
29. Ibid., p. 9. The teaching of "practical" biology had been instituted in the United States in 1876 by the appointment of Henry Newell Martin at the Johns Hopkins University. Martin, a student of Michael Foster, was to equip his laboratory with English apparatus. In his *History of the Semicentennial*, p. 9, Howell recalled that: "[Martin] had galvanometers, rheotomes and recording apparatus of all kinds and I remember a particularly massive pendulum myograph made after the design of Fick, which was securely bolted to the wall in a room all its own and made a very impressive appearance particularly when we were showing visitors through the laboratory....It played the role of the telescope of an astronomical laboratory." Howell recalled that most of this apparatus had been purchased from the Cambridge Scientific Instrument Company, a company founded by one of Foster's students, Albert George Dew-Smith (1848–1903). For details of the origins of this company, see "Fifty Years of Scientific Instrument Manufacture," *Engineering* (11 May 1945): 361–363; (25 May 1945): 401–403; (15 June 1945): 461–463; and (29 June 1945): 501–502. Also see "List of Apparatus Exhibited by the Cambridge Scientific Instrument Company, Ltd," *Proc. 4th Int. Physiol. Congr. Cambridge, UK, 23–26 August 1898, J. Physiol. Lond.* 23 Suppl. (1899): 73–74. At this meeting E.-J. Marey proposed formation of a body to standardize physiological recording instruments. See E.-J. Marey, "Nécessité de créer une commission international pour l'unification et le contrôle des instruments inscripteurs physiologiques," ibid., p. 6–7. The commission was composed of Bowditch, Foster, von Frey, Hürthle, Kronecker, Marey, Mislawsky, Mosso, and Weiss. I am grateful to Karen Johnson and Hughes Evans for several of these references on British instrument manufacture.
30. "Concerning the Harvard Apparatus Company," unsigned typescript and n.d. [ca. 1933–1934] (hereafter, "Concerning the Company"), William Townsend Porter Papers, Harvard Medical Archives, Francis A. Countway Library of Medicine, Harvard Medical School, Boston, MA (hereafter, CL-HMS), GA71, box 3, and a longer document with a similar title called "Concerning the Harvard Apparatus Company, Memorandum Accompanying Letter of Dr. William T. Porter to the Members of Harvard Apparatus Company, Incorporated," 9 August 1934, Porter Papers, CL-HMS, GA71, box 3, (hereafter, "Memorandum"). In this memorandum, presumably written by Porter, the limitations of earlier attempts at practical instruction are described, chiefly the need to teach the class in sections and do experiments out of sequence with the didactic lectures.
31. Barger, "Meteoric Rise," p. 409. Porter had been professor of physiology at the St. Louis Medical College and chief surgeon at the St. Louis Hospital.
32. Alfred C. Redfield, "The Laboratories of the Harvard Medical School," *Methods Probl. Med. Educ.* 3 (1925): 19–34. Some student experiments were done under Warren (assistant 1881–1987 and instructor 1887–1991) and under Howell (1892–1893). Cf. ibid. p. 24, n. 3. Porter reorganized instruction in 1899–1900.
33. W. T. Porter, "Teaching of Physiology in Medical Schools," and idem, "The Teaching of Physiology," reprinted from the *Phila. Med. J.*, special issue on Medical Education (1900), 16 p. [hereafter, "Teaching of Physiology (1900)"].
34. Ibid., p. 5.
35. Ibid. On Eliot's and Bowditch's promotion of this general attitude in teaching, see Ludmerer,

"Reform at Harvard," pp. 348–349. In "Teaching of Physiology in Medical Schools," p. 649, Porter attributed the phrase training "for power" to Eliot.

36. Porter, "Teaching of Physiology (1900)," p. 6.
37. Ibid., p. 7.
38. Geison, *Michael Foster*, p. 133. In Thomas Henry Huxley's course at South Kensington, London, an hour's lecture was followed by four hours of laboratory work. Porter's method of instruction proved to be difficult. Redfield later explained: "Experience accumulated to show, however, that the new method was not entirely satisfactory. From the first the impossibility of covering the entire subject by the objective experiment had been recognized, but physiology itself was growing so rapidly that the futility of attempting to organize the course on this basis became more and more apparent," ("Laboratories," p. 25).
39. Porter, "Teaching of Physiology (1900)," p. 10.
40. Ibid.
41. Ibid.
42. "Concerning the Company," p. 1.
43. "Memorandum." This document is more detailed than that just cited in n. 42. The two disagree slightly as to the precise number of students enrolled.
44. Ibid., p. 2.
45. Von Frey to Porter, 3 July 1904, "Extracts from Letters Regarding the Harvard Apparatus," Porter Papers, CL-HMS, GA71, box 3.
46. "Concerning the Company," pp. 1–2.
47. "Memorandum," p. 3. Porter was referring to the "American System" of manufacturing, based on machine-produced uniform interchangeable parts. On its origins, see Merritt Roe Smith, "Eli Whitney and the American System of Manufacturing" in *Technology in America: A History of Individuals and Ideas*, ed. Carroll W. Pursell, Jr. (Cambridge, MA: MIT Press, 1981), pp. 45–61; Robert S. Woodbury, "The 'American System' of Manufacturing," in *Technology and Social Change in America*, ed. Edwin T. Layton, Jr. (New York: Harper and Row, 1973), pp. 47–63; and Otto Mayr and Robert C. Post, eds., *Yankee Enterprise and the American System of Manufactures* (Washington, DC: Smithsonian Institution Press, 1981). I thank Howard Segal for guiding me to this literature.
48. "The Harvard Apparatus Company" (Boston, MA: Harvard Apparatus Company, 1904), pp. iii–v of this catalogue. A copy is preserved in the Charles W. Eliot Papers, box 238, Harvard University Archives, Pusey Library, Harvard University (hereafter, PL-HU).
49. Ibid., p. v.
50. Chicago Laboratory Supply and Scale Co., "Catalogue of the 'Harvard' Physiological Apparatus" (Chicago, IL: Windsor & Kenfield, 1900).
51. See Porter to Eliot, 31 March 1906 letter, "Statement of Work Accomplished," p. 2, Eliot Papers, box 238, PL-HU.
52. Lord Electric Company, "The Harvard Physiological Apparatus; Made or Assembled by the Mechanics of the Harvard Laboratory of Physiology under the direction of Professor W. T. Porter," (Boston, MA: Lord Electric Company, 1901), p. iii (hereafter, "Apparatus").
53. "Memorandum," p. 3.
54. Eliot to Porter, 15 March 1904, Porter Papers, GA71, box 2, CL-HMS.
55. Eliot to Bowditch, 15 March 1904, Porter Papers, GA71, box 2, CL-HMS.
56. Eliot to Blake, 18 March 1904, Porter Papers, GA71, box 2, CL-HMS; reproduced in "Memorandum," pp. 4–5. The Porter Fellowship of the American Physiological Society (APS), derived from the Harvard Apparatus Company profits, was established in 1921. The transfer of the Harvard Apparatus Company to the APS was discussed in 1920–1921 and in 1929. When the Society declined, a foundation was set up (on 15 June 1934), as a "non-profit educational body, selling only to Physiologists and to persons themselves engaged in physiological work." See "Incorporation," p. 5 of undated typescript [ca. 1944], Porter Papers, GA71, box 3, CL-HMS.
57. Twelfth Annual Meeting, Yale University, New Haven, CT, 27–29 December 1899, *Am. J. Physiol.* 3 (1899–1900): xxxii.
58. Porter, "Demonstration of Apparatus," Fourteenth Annual Meeting, University of Chicago, Chicago, IL, 30–31 December 1901 *Am. J. Physiol.* 6 (1901–1902): xxv. Porter showed the following: "Adjustable plate, aortic cannula, circulation scheme, femur clamp, platinum electrodes, capillary electrometer, artificial eye, frog board and clips, inductorium, rocking

key, short-circuiting key, simple key, optical lantern, light muscle lever, heavy muscle lever, moist chamber, square rheochord, signal magnet, sphygmograph tambour."

59. Warren P. Lombard, "Demonstration of Apparatus," *Am. J. Physiol.* 6 (1901–1902): xxiv–xxv. Lombard showed an ergograph, designed by Thomas A. Storey of Leland Stanford Jr. University, California, as well as a platinum mercury key, an upright rheochord, a graphite rheostat, a short contact key, a modified form of Fitz's belt pneumograph, a ball and socket joint to connect the disk on a tambour with a lever, a celluloid lever, a spring support for a glass slide, plaster of Paris electrodes, and a method of moistening the skin for excitation of human nerve and muscle.

60. At the Thirteenth Annual Meeting, held at Johns Hopkins University, Baltimore, MD, 27–28 December 1900, Reichert presented a student's laboratory table, a new rheotome, and a rheochord. See *Am. J. Physiol.* 5 (1901): xvii.

61. Porter, "New Inductorium, Kymograph, Heart Lever, Heavy Muscle Lever, and Square Rheochord," Fifteenth Annual Meeting, Washington, DC, 30–31 December 1902, *Am. J. Physiol.* 8 (1902–1903), xxxv–xli; Porter, "An Improved Kymograph," "Respiration Scheme," and "Muscle Warmer," Sixteenth Annual Meeting, Philadelphia, PA, 29–30 December 1903, *Am. J. Physiol.* 10 (1903–1904): xxxix–xliii; Porter, "An Exhibition of Physiological Apparatus," Seventeeth Annual Meeting, Philadelphia, PA, 27–28 December, 1904, *Am. J. Physiol.* 8 (1905): xxxvii.

62. Porter, *An Introduction to Physiology* (Cambridge, MA: University Press, 1901); 2d ed., 1906; and Porter, *Physiology at Harvard* (Cambridge, MA: University Press, 1902); 2d ed., 1903; 3d ed., 1904; 4th ed., 1905; 5th ed., 1906. (I have been unable to locate copies of the 4th and 5th editions.) In addition Porter published several small pamphlets called "Experiments for Students in the Harvard Medical School."

63. On the evolution of Lombard's course see Davenport, "View from Michigan," esp. pp. 61–67.

64. Lombard, *Directions for Laboratory Work in Physiology for the Use of Medical Students* (Ann Arbor, MI: George Wahr, 1906) p. iv.

65. Winfield S. Hall, *A Manual of Experimental Physiology for Students of Medicine* (Philadelphia, PA: Lea Brothers, 1904), p. 2. See also his earlier work, *A Laboratory Guide in Physiology. Chapters I and II. On Circulation and Respiration* (Chicago, IL: Keener, 1896). This emphasis on laboratory training was first initiated at Harvard in chemistry. Charles Loring Jackson and Gregory Paul Baxter recalled in their retrospective on chemistry at Harvard [see Samuel Eliot Morison, *The Development of Harvard University Since the Inauguration of President Eliot 1869–1929* (Cambridge, MA: Harvard University Press, 1930), p. 269] "To fit men for places in the chemical industry we taught them how to solve chemical problems by practical work in research." Reform in physics followed: beginning in 1886–1887, candidates for admission were examined in elementary experimental physics. This requirement, continuing at Harvard until 1927, stimulated the teaching of experimental physics in schools. [Ibid., pp. 261 and 285–286; and Edwin H. Hall and Joseph Y. Bergen, *Text-book of Physics, Largely Experimental on the Basis of the Harvard College "Descriptive List of Elementary Physical Experiments"* (New York: Henry Holt, 1895), pp. iii–iv.] For a historical analysis of these events see Albert E. Moyer, "Edwin Hall and the Emergence of the Laboratory in Teaching Physics," *Phys. Teach.* 14 (February 1976): 96–103. I am grateful to Clark Elliott for directing me to this literature.

66. Lord Electric Company, "Apparatus," p. iii. The instruments are described in the catalogue. Brief accounts of variant models are found in Stirling, *Outlines*, 4th ed., 1902 and M'Gregor-Robertson, *Physiological Physics*, as well as Porter's texts.

67. See Lord Electric Company, "Apparatus," pp. iii–iv.

68. A list of reference works in practical physiology appears in Stirling, *Outlines*, 3d ed. (Philadelphia, PA: Blakiston, 1905), p. 373, and Stirling, *Outlines*, 4th ed., 1902, pp. 588–589. For comparable texts, see Augustus D. Waller and W. Legge Symes, *Exercises in Practical Physiology Part II. Exercises and Demonstrations in Chemical and Physical Physiology.* (London: Longmans, Green, 1905); and T. G. Brodie, *The Essentials of Experimental Physiology for the Use of Students* (London: Longmans, Green, 1898). Stella V. F. Butler has been studying the institutionalization of physiology in British universities.

69. E. Cyon, *Methodik* and *Atlas zur Methodik der physiologischen Experimente und Vivisectionen* (Giessen, Germany: Ricker'sche Buchhandlung; St. Petersburg, Russia: Ricker, 1876). The atlas included data on instrument makers and prices. On the impact of the British handbook

by Sanderson, Foster, Klein, and Brunton, *Handbook for the Physiological Laboratory* (London: Churchill, 1873), see Richard D. French *Antivivisection and Medical Science in Victorian Society* (Princeton, NJ: Princeton Univ. Press, 1975), pp. 47–50. Note the list of instruments and suppliers cited in the American edition: J. Burdon Sanderson, ed., *A Handbook for the Physiological Laboratory* (Philadelphia, PA: Blakiston, 1884), pp. 574–575.

70. Stirling, *Outlines*, 4th ed., 1902, pp. xii–xxiv and 607–608.
71. E.-J. Marey, *Du mouvement dans les fonctions de la vie: leçons faites au Collège de France* (Paris: Germer Baillière, 1868).
72. Porter, *Physiology at Harvard*, 2d ed., pp. 1–2.
73. Ibid., p. 2.
74. Ibid., pp. 2–3.
75. Porter, "Teaching of Physiology in Medical Schools," p. 649.
76. Porter, *Physiology at Harvard*, 2d ed., preface.
77. Ibid., pp. 81–90. Students used an average of forty-five frogs each, as well as a rabbit and a tortoise.
78. For example, Marey was not accepted into the Académie des Sciences at his nomination in 1876 because "... Mr. Marey is a scientist who is too far removed from the investigation of man's diseases for us to propose him as the most worthy representative of French medicine in your section ... to replace Mr. Andral." From the report of Léon Gosselin (15 May 1876) translated in Eighth European Congress of Cardiology, *The Graphic Method and Cardiovascular Research: Étienne Jules Marey (1830–1904)* (Cachan, France: Medtronic, JLR Arts Graphiques, 1980), p. 13.
79. Porter to Eliot, 31 March 1906, Eliot Papers, box 238, PL-HU.
80. Ibid., "Statement of Work Accomplished," p. 2, n.d.
81. Ibid., Porter to Eliot, 31 March 1906 letter.
82. Porter to Eliot, 19 December 1904 letter, Eliot Papers, box 238, PL-HU. For details, see Barger, "Meteoric Rise," pp. 412–413.
83. See J. D. Greene to C. F. Adams, 10 October 1906, in Porter Correspondence, Eliot Papers, box 238, PL-HU, regarding proprietary rights over equipment used by the Harvard Apparatus Company.
84. Porter to Eliot, 31 March 1906 letter, "Statement of Work Accomplished."
85. HAR, (1906–1907), p. 163.
86. Ibid.
87. HAR, (1907–1908), p. 173.
88. HAR, (1908–1909), p. 176.
89. Porter to Eliot, 19 December 1904 letter, p. 8. This letter concerns the high failure rate in physiology. Cf. Barger, "Meteoric Rise," p. 412.
90. Lombard to Porter, 5 January 1921 letter, Porter Papers, GA71, box 3, CL-HMS.
91. Ibid.
92. See Robert E. Kohler, *From Medical Chemistry to Biochemistry: The Making of a Biomedical Discipline* (Cambridge, UK: Cambridge Univ. Press, 1982) and chapter VI by Toby Appel in this book, especially the section on the Federation of American Societies of Experimental Biology. I have found the work of Adele Clarke and that of the Tremont Research Institute in San Francisco especially helpful in understanding the intersection and segmentation of scientific disciplines.
93. The growing interest in such problems is evident in the early volumes of the *Am. J. Physiol.* and in the career of Cannon.
94. Porter's appointment was primarily a research appointment. In 1906–1907 (HAR, p. 164), he gave informal lectures and demonstrations, and his weekly presentations were doubled because of student interest.
95. Saul Benison and A. Clifford Barger, "Walter Bradford Cannon," in *Dictionary of Scientific Biography*, ed. C. C. Gillispie vol. 15 (Suppl. 1) (New York: Scribner, 1978), p. 71–77; Saul Benison, A. Clifford Barger, and Elin R. Wolfe, "Changing of the Guard," *Harv. Med. Alum. Bull.* 56 (1982): 34–41. Benison, Barger, and Wolfe are now completing their forthcoming book, *Life and Times of Walter Bradford Cannon: A Documentary History.*
96. "Memorandum," p. 2, "The workshop method of teaching large classes was at that time [before the development of Harvard Apparatus] impossibly expensive." This phrasing emphasizes the role of craft and apprenticeship in physiological training.

97. Davenport, "View from Michigan," p. 74.
98. A similar intellectual revolution had been accomplished nearly fifty years earlier in the German universities. Its goals were embodied in a similar manner by Helmholtz's phrase, "Wissen ist Macht." For a superb analysis of this process in Germany in the 1850s and 1860s, see Arleen Tuchman, "Science, Medicine and the State: The Institutionalization of Scientific Medicine at the University of Heidelberg." (Ph.D. diss., University of Wisconsin—Madison, 1985), esp. pp. 221–249. As Dr. Tuchman pointed out during a recent visit (February 1986), the correspondence between the sequence of events in the institutionalization of physiology in Germany and America is striking. I am grateful to her for sharing with me her observations, interpretations, and forthcoming manuscript, "From the Lecture to the Laboratory: The Institutionalization of Scientific Medicine at the University of Heidelberg."
99. Walter B. Cannon, *A Laboratory Course in Physiology* (Cambridge, MA: Harvard Univ. Press, 1910), p. 1.

XIV

Research Materials and Reproductive Science in the United States, 1910–1940

ADELE E. CLARKE

In order to observe or produce the phenomena they study, all working scientists must obtain and manage research materials.[1] In the United States at the turn of this century, the shift of emphasis from descriptive morphological approaches to experimental physiological approaches in the life sciences radically altered scientists' needs for research materials and had significant consequences for the organization of such materials.[2] This chapter focuses on the consequences of that shift for the acquisition and organization of research materials in the case of reproductive physiology during the first half of this century.

By 1900, physiology had emerged as an independent professional discipline in the United States, as it had a generation or two earlier in Europe. Its central problem structure had been defined and delimited, and American physiologists could claim their own society, journal, and university departments. The terms "physiological" and "physiologist" were now used as identity claims for a variety of reasons under a variety of conditions. Physiological approaches, combining a set of processual questions about nature with experimental methods, were spread throughout the life sciences, far beyond the discipline of physiology itself and its official university departments.[3] As the research agenda in American biology, medicine, and (soon after) agriculture began to extend beyond structure and form—the traditional preserve of anatomists and morphologists—entirely new lines of scientific work emerged, several of which were developed into new disciplines or specialties.[4]

The new approaches were found in a variety of institutions, old and new. Several of the new departments, research institutes, sponsoring organizations, and marine biological stations, and museums have already been subjected to scholarly examination.[5] However, historians and sociologists of science have almost entirely ignored the emergence of new infrastructural organizations to support the altered agenda for research in the life sciences—among them, biological supply firms, technical equipment industries, animal houses, and research colonies.[6] In this chapter, special attention is given to new infrastructural arrangements for supplying a variety of live and fresh materials for the experimental investigation of mammalian reproductive physiology. Interestingly, many of the reproductive scientists who now adopted experimental approaches in their research had been trained not as physiologists, but as zoologists or anatomists (see chapter XI by Long in this book). They, and "reproductive biochemists" soon after, quickly integrated and absorbed physiological experimental approaches into their routine working procedures.[7]

In 1979, Charles Rosenberg called for a historical "ecology of knowledge" that would include disciplines and institutional contexts.[8] The larger aim of this chapter is to suggest that a still richer ecology of knowledge would result from giving closer attention to such issues as the organization of research materials and the development of instrumentation and techniques. This approach would enlarge our historical appreciation for the day-to-day practices of scientists and bring them within the purview of recent literature on the sociology of work, organizations, and professions.[9] We would thereby expand our understanding of the social and material conditions of scientific research and their consequences for the production of scientific knowledge.[10] Just as local institutional, disciplinary, and departmental conditions may shape scientific work, infrastructural organization may also be consequential. Research is, after all, constrained or enhanced by the accessibility, cost, organization, and pacing associated with specific materials, instruments, and techniques.[11]

MATERIAL CONSEQUENCES OF PHYSIOLOGICAL APPROACHES

In terms of research materials, the shift of emphasis from descriptive morphological work to experimental physiological approaches had three major consequences: (1) increased demand for live and fresh materials; (2) increased demand for large quantities of the same species rather than limited numbers of specimens of several different species; and (3) development of colonies for on-site access to desired research materials.[12]

Life scientists who adopted the new experimental approaches in their research had to explain and justify the changing nature of their work to a variety of audiences in order to secure support for their research and related materials.[13] In 1914, for example, Frank R. Lillie, chairman of the Department of Zoology at the University of Chicago, wrote to the president of the university in a budget request for a vivarium—a facility for housing live animals for research:

> The study of living animals has replaced to a considerable extent the study of their dead bodies, and a vivarium is as much a need of the modern zoological laboratory as a museum. . .zoology is becoming an experimental science. A department that is well equipped for the older descriptive zoology and embryology, as ours is, may yet be deficient in facilities for pursuing the newer lines of work which must necessarily claim most of our interest. The development of experimental methods of studying zoological problems has been gradual; it is no sudden fashion but a normal development. And it is as certain that it is a permanent direction for research for zoology as for chemistry. A science cannot retreat from the vantage ground of experimental methods.[14]

Lillie's request illustrates in striking fashion some of the new and very different requirements for materials, space, and laboratory organization that scientists and their institutions faced because of the new experimental approaches.[15] These new requirements and their implications for the daily work of the scientist deserve some specification and elaboration here.

DEMAND FOR LARGE QUANTITIES OF SAME-SPECIES MATERIALS

Nineteenth-century morphologists and comparative anatomists generally sought to acquire representative samples of a wide variety of species, mainly for descriptive interspecies comparative work. They might then look for homologues (parallel structures) across species, seek interspecies characteristics for finer classificatory distinctions, and so on. The goal in acquiring materials for morphological work was, therefore, to secure one or at most a few specimens of each species. These specimens were generally (although not invariably) dead and complete organisms. A single specimen could theoretically be utilized for hundreds of studies, given adequate organization of access to it.[16]

The new experimental approaches entailed very different requirements for research materials. Both live and fresh cadaver materials became essential. Moreover, large quantities of such materials were typically sought, and often from the same species. Two general considerations lay behind these new requirements for research materials. First, live and fresh materials were needed because biological processes and their histological and cytological manifestations could rarely be adequately observed in dead or preserved specimens; preservatives and stains often altered materials in damaging ways.[17] Second, large numbers were needed because the new standards of research included an attempt to confirm results through adequate sample size. In particular, it was hoped that one might be able to control "accidental" variations between individual specimens by using large quantities of homogeneous materials, preferably from the same species.[18]

The crucial if unexamined assumption here was that basic biological processes were parallel in different species.[19] Scientists who made this assumption naturally focused on the physiological processes rather than the species or material per se. The goal was to work out a particular process in one species, developing a model or analogue that would subsequently be refined across species.[20]

GAINING ACCESS TO MATERIALS

The means of gaining access to mammalian materials changed dramatically with the shift in goals from traditional descriptive work to the new experimental approaches. In the older tradition, the primary means of access consisted of collecting single or limited numbers of each specimen, preserving them, and establishing collections of a wide variety of specimens.[21] Collecting expeditions were the major means of acquiring morphological research materials.[22] The fundamental social processes were (1) the development of and reliance upon both amateur and professional collectors in desired locations to obtain specimens[23] and (2) the creation of professional scientific networks to share and/or pass along desired specimens. Such networks enhanced access to research materials both geographically and in terms of the range of collected species.

The new experimental approaches entailed drastically different means of gaining access to research materials. Life scientists, and especially reproductive biologists, now sought reliable supplies of fresh and live materials, a task that was in fact a large and time-consuming part of their daily work.[24] They relied on five major means of access to appropriate mammalian materials for their reproductive research: (1) collecting exotic animals and specimens (materials native to foreign habitats, and/or rare to the environment in which the research is conducted, and/or difficult to maintain alive); (2) using mundane (local, easily accessible, and easily maintained) materials such as domestic animals (e.g., swine and cattle) or other routinely available materials; (3) using medically supplied materials (e.g., aborted and stillborn embryos, surgically removed organs of the reproductive system, etc.); (4) using animals and specimens obtained from biological supply companies; (5) establishing and maintaining on-site research colonies.

At first, the new breed of reproductive biologists and medical researchers primarily collected and used exotic and mundane materials from the field. Acquiring such materials was generally a do-it-yourself project for the research scientists themselves. Technical assistance was usually very limited (consisting mainly of students) and at times nonexistent. The selection of materials was therefore often drastically limited to what came easily and cheaply to hand. As Robert M. Yerkes commented in 1916, "Most investigators are either impelled or compelled by circumstances to work on easily available and readily manageable organisms."[25]

There was thus a catch-as-catch-can ethos about the selection of and access to experimental materials at the turn of the century. Indeed, the new band of experimental biologists imitated the practice of their more traditional predecessors by developing quite elaborate amateur and professional networks to enhance their access to materials, even though they sought live and fresh materials as opposed to dead specimens. At times, such "networking" led to joint acquisitions and research.

Such networks remained important even as life scientists came increasingly to rely on biological materials suppliers, medically-generated materials, and specifically designed biological colonies. Patterns of access to and use of particular materials were often passed down from teacher to student or from one scientist to another at a given research site. In such cases, and indeed more generally, there was a tendency to continue to use a single animal material once a stable means of access to it had been achieved and as the intricacies of its maintenance had been mastered. As their organizational commitments became entrenched,[26] scientists often explored a variety of research problems by using an already available animal rather than seeking new materials for new problems. This is no small wonder, since the quest for access to appropriate materials represented such a major investment of resources in terms of time and energy as well as money. The availability of and familiarity with particular biological materials surely did much to shape the pace, direction, and even the content of research in reproductive physiology and no doubt in other fields as well.

COLONIES AS MEANS OF ACCESS TO RESEARCH MATERIALS

The task of acquiring large quantities of live materials was challenging. Here agricultural scientists involved in reproductive and other physiological research had distinct advantages. Most agricultural institutions maintained herds of various animals as a matter of course in their efforts to find and teach improved methods of animal production. Scientists had fairly easy access to these herds, which were supported by state and federal funds. Moreover, well-established agricultural scientists could also often make use of the private herds of farmers and ranchers through coordination with extension agents of the United States Department of Agriculture.[27] Inspired in part by such agricultural models, biomedical scientists gradually articulated their need for large numbers of a given species immediately at hand for ongoing research.

In 1892, Charles Otis Whitman, first chairman of the Department of Zoology at the University of Chicago, put forward one of the earliest proposals in the United States for a major biological colony for research purposes (although his proposal was not published in *Science* until 1902).[28] Whitman was a transitional figure; raised on descriptive morphology and experimental zoology, he nonetheless encouraged the adoption of physiological approaches at Chicago and at the Marine Biological Laboratory at Woods Hole.[29] Toward this end, he proposed the establishment of animal and botanical colonies of both experimental and field types for multiple purposes at one site.

Whitman's proposal contains this striking expression of the shifting priorities and approaches of American biologists at the turn of the century:

> The biological laboratories of to-day, in design, equipment and staff, are almost exclusively limited to the study of *dead* material. Living organisms may find a place in small aquaria or vivaria, but they are reserved, as a rule, not for study, but for fresh supplies of dead material. It is no disparagement of the laboratory to point out a broad limitation in its ordinary functions and the pressing need of new facilities for observation and experiment on *living* organisms.[30]

Whitman further insisted that:

> The functions to be fulfilled by a farm are . . . [now] prescribed by the . . . deeper and broader needs of pure research on *living* organisms. . . . The biology of today . . . has not too much laboratory, but too little of living nature. The farm will certainly do much to mend this great deficiency.[31]

Like later advocates of biological colonies, Whitman argued for the creation of such farms partly on practical grounds, predicting the development of useful animal and plant hybrids. He also assumed that such colonies would encourage the wider adoption of one of the most significant features of the new scientific agriculture—advanced record keeping and monitoring.[32] Such modern farming methods, Whitman insisted, could provide scientists with the "history of the material to be investigated." These "exact records . . . will thus render a most important service to laboratory records."[33]

The leading models for Whitman's project were agricultural experimental stations, the Experimental Stations of the Carnegie Institution of Washington, D.C., and Luther Burbank's establishment in Santa Rosa, California.[34] Saying that "there is no satisfactory name for the new plant [i.e., institution] required for such work," Whitman settled on the term "Biological Farm," capturing the notion of plants and animals under cultivation.[35] He proposed that several different university departments should contribute to and use the large colony simultaneously. His proposal thus stands out in the history of materials acquisition as an early call for a multicontributor, multipurpose, multiuser biological colony—a new institutional form.

During the first quarter of this century, American reproductive biologists established ambitious colonies of a wide variety of research materials. They developed two major types of colonies: research-only colonies (in which additional materials were purchased or obtained as needed) and research-and-production colonies (in which some animals were reserved and maintained for

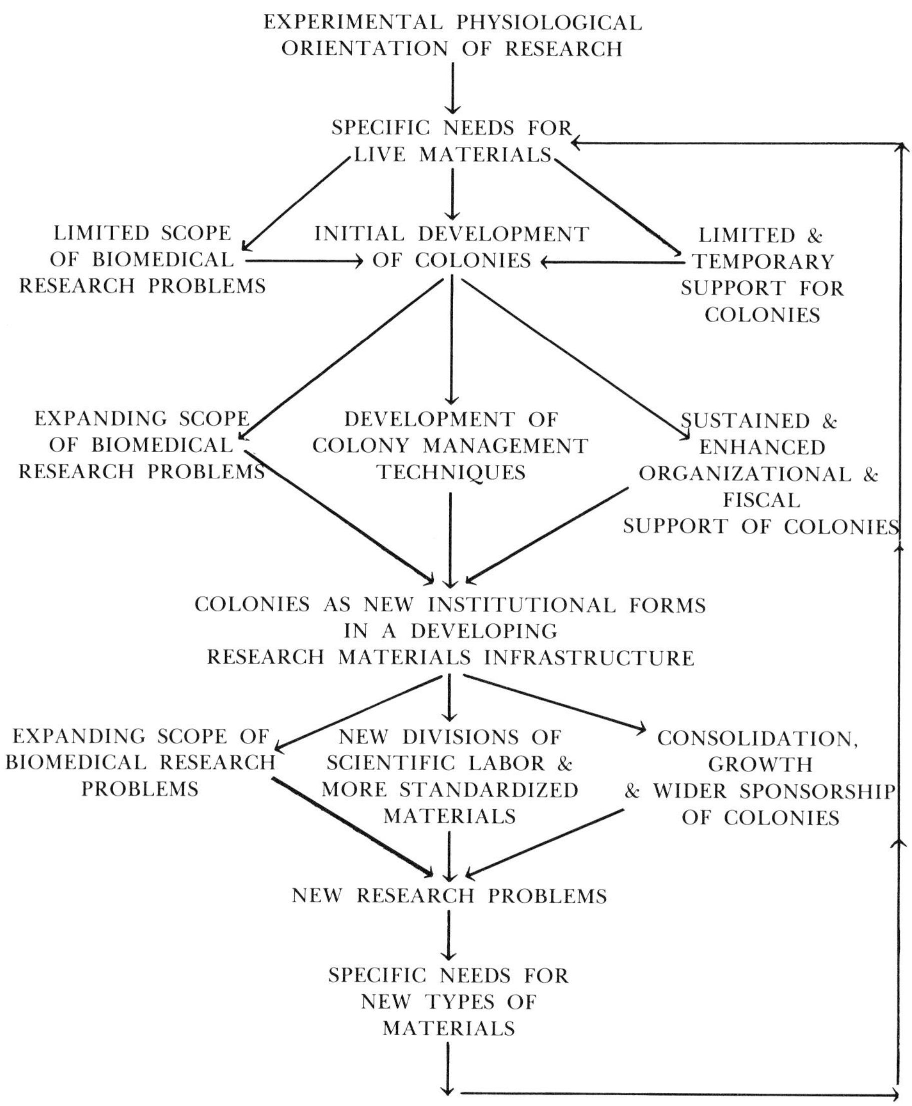

Figure 1. *Preliminary chronology of colony development.*

breeding, while experimental supplies were also made available). Research-and-production colonies were usually more expensive to operate, since breeding could be difficult to achieve and larger numbers of animals were needed to meet both goals. The duration of gestation also affected production-colony costs; rapidly reproducing animals such as guinea pigs and rats were less expensive to colonize than primates, for example.

Figure 1 provides a preliminary conceptual flow chart of the development of biological colonies. Further research is needed to clarify the nature of such relations in specific contexts and circumstances.

HISTORICAL ILLUSTRATIONS

The rest of this chapter illustrates the analytic themes previously identified by discussing patterns of access and use of specific mammalian materials in reproductive science ca. 1910–1940. Table 1 presents a classification of the major types of materials and lists specific examples from reproductive science that will be discussed in the next section.

MUNDANE FRESH SPECIMENS

A variety of both domesticated and wild animals are mundane materials in the sense that they are native and/or local and can be easily obtained. In this section, I discuss access to three domesticated animals commonly used in reproductive science: swine, cattle, and horses.

SWINE

Starting in the 1890s, pigs or swine were used in mammalian reproductive biological research in Franklin Mall's laboratory at Johns Hopkins. As Florence Sabin, Mall's biographer, noted in 1934:

> The anatomical laboratory in Baltimore was situated only a block away from an abattoir, where a large number of pigs were killed each day. It has often been said that the embryo pigs which were so abundantly obtained from this source made Mall's department.[36]

Table 1
Access to Major Types of Physiological Materials For Reproductive Research

	Major Types of Materials	
	Mundane materials	Exotic materials
Fresh specimens	Pig and cow ovaries and uteri Cow urine Mares' blood	Human embryos Human ova
Live materials	Cats and dogs "Laboratory animals" (rats, mice, guinea pigs)	Opossums Nonhuman primates

Mall's students later filled about twenty-five chairs in anatomy in the United States, and many of those students utilized pig embryos in their research, notably including Herbert Evans, George L. Streeter, Florence Sabin, and George W. Corner.[37]

Corner thus inherited access to and a pattern of use of a particular mundane material from Mall. By 1910, Corner was investigating the embryological development of the pancreas in pig embryos. He then became Mall's assistant in anatomy at Hopkins, where he first studied the development of corpora lutea in the ovaries of sows, partly as a means of determining the age of early human embryos. Specifically, Corner hoped that the stage of development of the corpora lutea could be determined in relation to the stage of development of the embryos in pregnant pigs and that this data could then be applied to humans who could not be studied directly, since appropriate human ovarian and embryonic material was "very scarce."[38] The study failed to solve the problem of dating early human embryos, but Corner was nonetheless delighted in the long run:

> I had chanced to begin my study of the mammalian reproductive cycle with a species whose cycle turned out to be so comprehensible as to constitute a model against which to compare the more complex cycles of other species on which I was later to work.[39]

Corner again used mundane materials from sows in his ovarian cycle research from 1915–1921. According to his sometime colleague Carl Hartman, "many's the day which he and his assistants spent in the local packing house laboriously searching for appropriate stages through the entrails of freshly slaughtered sows."[40] Corner's autobiography is full of reminiscences of making and re-making arrangements with local pork slaughter houses or abattoirs to obtain their discarded reproductive organs in a fresh and timely fashion. As Corner moved among several universities during this early part of his career, he became familiar with slaughter houses in Baltimore, Berkeley, and Rochester.

In 1919, Corner was doing research on sows to examine the developmental cycle of corpora lutea in the ovary over their twenty-one day estrus cycle. This investigation required three sequential weeks of frequent observations, tagging the animals at various stages of the cycle and then slaughtering them. In his autobiography, Corner noted, "Paradoxically, I could do the work far more easily and less expensively on the outskirts of urban Baltimore than on the [University of California, Davis] farm."[41] This was true at that time because, due to the high costs of proteins and fats during World War I, pig farms had been established near urban garbage dumps and citizens were encouraged to save kitchen scraps to feed the pigs. Corner easily gained access to pigs for his research at one such establishment. Then:

> The Johns Hopkins University. . .went into the pork business, with me as its agent, by purchasing the [twenty-two] sows for $800. I had them hauled to Hohman's slaughterhouse. . .near the medical school . . . Hohman's butchered the lot at the end of a morning's run, slowing the pace of the killing floor routine so that I could collect the ovaries and uteri systematically. The university lost a hundred dollars or so in the deal.[42]

Corner spent the summer of 1920 with Charles Davenport at the Carnegie Institution's Station for Experimental Evolution in Cold Spring Harbor, New

York. A student there went daily by train to and from a New York City abattoir (a trip of several hours) to obtain an adequate number of fresh sow uteri (1000 over the summer) at varying stages of pregnancy for another of Corner's research projects.[43]

Corner was hardly alone in the quest for such research materials. His sometime student and colleague Karl M. Wilson, for example, also secured sow uteri from a "near-by abattoir, the genitalia being procured entire for examination."[44] Wilson described as follows some of the constraints and opportunities that were provided by such research materials:

> There are certain obvious disadvantages connected with the use of such material. The sexual history of the animal is, of course, unknown, and, furthermore, many of the animals are very young, and these doubtless have rather irregular cycles during the early period of their sexual activity. . . . On the other hand, these obvious disadvantages are at least partially compensated for in the abundance of material which one is able to obtain for study.[45]

Thus swine became a commonly used research material in the development of modern reproductive physiology, due to both their ready availability and their appropriateness for the pursuit of embryological and reproductive problems. Research on sows was apparently curtailed by the removal of abattoirs from locations convenient to scientists as refrigerated trucks and trains became widely used.[46]

Cattle Fetuses and Reproductive Organs

Scientists also commonly obtained fresh cattle materials for reproductive research from the killing-room floors of local abattoirs. For scientists at the University of Chicago, for example, the stockyards of the meat-packing industry were a major materials resource. During World War I, Lillie, chairman of the Department of Zoology at Chicago, went so far as to seek draft exemption for a technical assistant, who "has done all of our collecting in the stockyards and in various localities for many years."[47] The technician made "at least two regular trips a week to the stockyards, and frequently additional trips" for which Lillie requested a department car in 1919.[48] The famous Chicago stockyards were relatively close, and the Department of Zoology had connections with nearby farms as well. Lillie identified both the stockyards and the farms as sources of freemartin cattle embryos for his classic paper on this topic.[49] The stockyard resources thus serve as a striking example of industrial contributions to reproductive science that are often difficult to specify.[50]

Cow Urine and Mare Blood

After ca. 1925, the core of the problem structure of reproductive science shifted from physiological to endocrinological research.[51] The biochemical nature of endocrinological research then generated a new set of needs for fresh materials. Once certain reproductive hormones were detected in urine, stable sources of urine were sought for endocrinological research.[52] In agricultural institutions, cattle herds became a key source.[53] Graduate and undergraduate student/technicians collected the urine by placing pails under the cows and then tugging on the cow's labia to induce urination.[54] Distilling the urine to obtain the desired hormones remained a highly-scented project until the development of radioimmunoassay techniques.

Pregnant mares came into demand once gonadotropins were detected in their blood.[55] Adequate supplies of blood could be obtained routinely without harm to the mares. Agricultural scientists obviously had the initial advantage in terms of access to pregnant mare serum gonadotropins. Reseachers at the University of California, Davis, for example, later sold their patents on the process to pharmaceutical companies through the university, and royalties were fed back into local research budgets.[56]

These examples illustrate that even in the case of mundane and relatively accessible materials, reproductive scientists spent considerable time and energy in the quest for appropriate research materials.

EXOTIC FRESH SPECIMENS

More exotic materials were also sought for reproductive research. In particular, highly prized human embryos and ova were obtained largely from cadavers and surgically removed organs and tissue. Access to such materials was uneven during the period 1910–1940. This result was due not only to the contingent nature of death and surgery but also to inadequate preservation methods and a lack of any systematic means of distributing specimens and cadavers. The following examples illustrate how access to such materials was achieved, how scientists organized their acquisition, and how surgical events were restructured to provide desired materials for research.[57]

HUMAN EMBRYOS

As late as 1938, E. Scipiades stated that "The science of human embryology is still incomplete, for the reason that research in this field has been hampered by two obstacles."[58] The first obstacle was poor microscopy and inadequate histologic techniques. The second was the scarcity of young and adequately preserved human embryos.

The most ambitious early effort to collect and study human embryos was that initiated by Mall at Johns Hopkins in the 1890s. Although he began by collecting research materials independently, Mall's collection was subsequently sponsored by the medical school at Hopkins, where he was chairman of the Department of Anatomy. When he became director of the Department of Embryology of the Carnegie Institution in 1913, Mall transferred his collection there.[59]

Shortly after he arrived at Hopkins, Mall made an appeal to physicians through the *Bulletin of the Johns Hopkins Hospital* to send him embryos derived from miscarriages. He noted that "Poor specimens . . . are better than none at all, therefore in all cases the ova [sic] should be preserved, even if there be but little hope to obtain a good specimen."[60] He then provided instructions for the preservation of the embryos in alcohol. Ten years later Mall reported that "The collection consists now of 208 specimens . . . over one-half of which were sent to me by physicians residing in Baltimore."[61] In return for such donations of research materials, Mall and his assistants often sent reports on the embryos to physicians who sought information on the cause of miscarriage. Fairly elaborate forms were filled out on each embryo collected.[62]

In 1903, Mall again called for contributions to his collection of human embryos:

> Although embryologists have emphasized again and again the importance of preserving carefully early human embryos for study, it is necessary from time to time to remind physicians not to throw away the valuable material which is constantly coming into their possession. There are numerous questions which may be answered were there an abundance of specimens at hand, and they can be procured only through the cooperation of physicians in active practice.[63]

In the same article, Mall also published a list of the embryo specimens with the names of the physicians who had contributed each one!

A decade later, in 1913, Mall made a formal "Request to the Graduates of the Johns Hopkins Medical School for Embryological Specimens."[64] Instructions for preservation were included and messenger service for embryo pick up was offered for the Baltimore area. Two similar letters were sent out from the Hopkins Department of Anatomy by Evans at about this time. In one such letter, Evans wrote: "Our success depends absolutely on you. . . . And so we are asking you to remember us; to give us anything you can collect from an abortion case."[65] Evans' letter reflects some refinements in preservation techniques:

> It must be preserved fresh. If the case happens in town and a telephone is at hand [number to call provided] let us send for it immediately at any hour of the day or night for we want to use special fixatives. A few hours' delay is serious and shows in the appearance and staining of all the tissues afterwards.

After Mall's death in 1917, efforts to obtain embryos were continued by Streeter, who succeeded him as director of Embryology at the Carnegie Institution. Today this embryo collection of the Carnegie Institution is housed at the University of California, Davis, where it is used in conjunction with the Primate Research Center.

Comparative embryology did not, however, become a focus of the Baltimore group until about 1920.[66] A. H. Schultz noted of his early years as research associate of the Carnegie Institution's Department of Embryology at the Johns Hopkins Medical School:

> I obtained an abundance of fetuses of different human races but I found very little interest *in* and no material *of* nonhuman primates to satisfy my main desire for the sort of comparative studies which I had learned to appreciate in Zurich.[67]

Hospital surgeries and local physicians were routine sources of human embryos and fetuses.

About 1924, while at Rochester Medical School, Corner became interested in obtaining human embryological materials from the operating room. He spoke of this goal to the chief surgeon and others there. Such informal requests were the regular means of acquisition. In his autobiography, Corner gives this firsthand account of the receipt of one such embryo:

> One evening about nine o'clock Dr. Jameson had called me to say that he was operating on a woman for a ruptured tubal pregnancy. . .hastening to the hospital I found the patient still on the table. The specimen he had saved for me, a small blood clot, was in a basin on the steam radiator, where he had ordered it put to keep warm. I had arrived just in time to save it from cooking. I took it out to my room in the animal house, to which the night watchman admitted me.[68]

This turned out to be a perfect embryo. G. W. D. Hamlett carefully reported a similar case of receiving a tubal embryo: "I received the unopened ovum, still in the injured tube, shortly after the operation and less than three hours after the tube's rupture."[69]

Fetal materials of multiple conceptions were particularly prized because of their rarity, multiple placentation, and the potential interrelation of the fetuses.[70] In 1935 A. W. Diddle and T. H. Burford reported extensively on premature quadruplets (which they said occur only once in every 654,455 labors). They wrote as follows of two such pairs of twins and two placentas: "Unfortunately, the placentae were not weighed nor preserved with the fetuses."[71] Surgeons providing materials were, in this instance as in the Corner case, not necessarily aware of the particular needs of scientists or of appropriate means of preserving such precious specimens.

Human Ova

Scientists who sought unfertilized human ova could not easily rely on physicians to supply medical wastes. Fallopian tubes removed for pathology were not a promising source because of the pathology itself, and it was considered clinically inappropriate to remove healthy tubes.[72] Stimulated by primate research,[73] and developing their technique initially on dogs, Edgar Allen, J. P. Pratt, Q. U. Newell, and L. J. Bland reported in 1930 on a joint program through which they had recovered unfertilized ova from women. Since it was "important to tap abundant sources of material," they initiated a cooperative investigation at two hospitals (one in Detroit and the other in St. Louis), "utilizing all available cases from both these services."[74]

The method that Allen and his associates had developed allowed ova to be flushed in situ from normal fallopian tubes during necessary abdominal surgery "without danger of injury, which might result later in obstruction."[75] The procedure involved clamping off the cervix, directly injecting a saline solution into the uterus, and permitting the fluid to escape from one fallopian tube at a time while collecting it in a series of watch glasses. Ultimately materials were obtained from over ninety operations, with sufficiently normal tubes on at least one side in thirty-five cases. These cases yielded five identifiable and successfully sectioned tubal ova, along with several dubious specimens. The corpora lutea that were believed to be associated with the ova were also excised from the ovaries for examination.[76]

In another restructuring of surgical events, Arthur T. Hertig and John Rock reported in 1939 on the "feasibility of recovering human embryos by the simple device of scheduling unavoidable hysterectomies according to the fertile period, as worked out on the monkey."[77] This line of work, which Hertig and Rock pursued for fifteen years, was dubbed the "egg hunt." During the course of the project, "some thousand eggs" were "harvested" from women surgical patients undergoing total hysterectomy.[78] In what "unquestionably was a delicate ethical maneuver," Hertig and Rock asked women awaiting hysterectomies to keep rhythm charts and also note times of intercourse, scheduling the surgeries shortly after ovulation.[79] Over the next fourteen years, Hertig and Rock collected thirty-four fertilized eggs, "gleaned from 211 unknowingly pregnant women," representing the first seventeen days of life.[80]

Not only were surgeries restructured but also new surgical methods were developed and applied in the quest for research materials.[81] On the whole, human embryo and ovum specimens were neither routinely available nor well-distributed. Scientists actively recruited such materials for many years to form complete collections. Even so, as Yerkes pointed out in 1925, "Generally, human subjects are more readily commanded for medical needs [than primates] because of clinical hospital resources and facilities."[82]

MUNDANE LIVE MATERIALS

Mundane live materials, though often acquired and used on an occasional basis, were also organized into colonies for routine research access. Here I examine cats, dogs, and other mundane species that became "laboratory" animals.

Cats, Dogs, and Surgical Precedents

Cats and dogs were routinely obtained for the animal houses of medical schools, where they were used in the teaching of anatomy, physiology, and especially surgery.[83] Local suppliers and other sources such as the ASPCA, dog catchers, pounds, and nearby farms associated with the laboratories were the usual providers.[84] The establishment of reproducing colonies of cats or dogs for research does not appear to have been pursued until the late 1930s.[85] Fresh supplies were therefore routinely needed.[86]

Reproductive scientists, especially those in medical school settings such as Allen and his associates, enjoyed established and very easy access to cats and dogs as mundane materials through the animal houses of their medical institutions.[87] The fact that the cat ovulates after coitus and the relatively long gestation periods of both cats and dogs made them less desirable than guinea pigs, rats, or mice as laboratory animals during the early twentieth century.

On the Making of Laboratory Animals

Several species of mundane materials were transformed into laboratory animals during the early twentieth century. Colonies of both research-only and research-and-production types were established, and their use became routine in the life sciences. Chief among the species that became laboratory animals in reproductive science were guinea pigs,[88] rats,[89] and mice.[90] All of these materials were relatively inexpensive to acquire (often from local pet shops and cottage industry suppliers) and to maintain (feed, house, and care for). Harry Steenbock provides an example from outside reproductive science; he established "the first rat colony in history for nutritional experiments" at the University of Wisconsin in 1908 "by purchasing at a cost of six dollars from his own pocket a dozen albino rats from a pet-stock dealer in Chicago."[91] Reproductive scientists could easily establish small rat colonies in a single room and fund them through normal departmental research budgets.[92]

One such modified animal material, the hypophysectomized rat, became "[o]ne of the most widely used tools of investigation in endocrinology and reproductive physiology."[93] Philip E. Smith, who developed and refined the technique of hypophysectomy, showed that removal of the anterior pituitary resulted in cessation of growth, loss of weight, and atrophy of the reproductive system, the thyroid gland, and the adrenal cortex. The injection of appropriately active hormones could "repair" these disabilities.[94] The hypophysectomized rat was, in many respects, a *tabula rasa* for the testing of a variety of hormone extracts. Smith's technique, along with the small size, rapid gestation, and relative inexpensiveness of rats for colonies, made them ideal "laboratory" animals.

EXOTIC LIVE MATERIALS

Exotic live materials were also organized into research colonies. Here I describe early opossum and primate colonies.

Opossum Colonies

Opossums were relatively mundane materials in terms of acquisition, since they were native to many parts of the United States and could be easily trapped. Yet they were exotic in terms of the difficulties they posed for the establishment and maintenance of living and reproducing colonies.

In 1912, Professor J. T. Patterson of the University of Texas offered Hartman an instructorship in his Department of Zoology along with the suggestion that he study the embryology of the opossum.[95] Why the opossum? Hartman answered:

> For the same reason that Patterson selected the armadillo: The prairie and the creek bottoms were full of both species and nobody had yet bothered to study the development of these, the most bizarre of all the American mammals. . . . [For thirteen years] I hunted the animals in the field, I also took care of them (at first at my home), operated on them, secured, fixed, and stained their eggs and photographed the sections myself.[96]

Hartman, building upon J. P. Hill's research on the kangaroo, ultimately studied nearly 1,000 female opossums, from which several thousand eggs were obtained. Hartman then described the main features of the anatomy of the sex organs, the estrus, ovulation, and fertilization, the early embryology, the birth, and postnatal development in the pouch of the opossum.[97]

The marsupial (pouched) opossum offered a particular opportunity or advantage for reproductive biologists, described as follows by Carl R. Moore and David Bodian of the University of Chicago:

> The American opossum (*Didelphys virginiana*), the only native marsupial on the North American Continent, presents decidedly unique mammalian material for many biological problems, particularly for those concerned with late embryonic development in which operative treatments are desirable. This is due to the fact

> that intra-uterine development is extremely short (12–13 days post-copulation) and later development occurs in the marsupial pouch where young are directly available. At birth posterior appendages are useless paddle-shaped structures but over-development of anterior appendages provided with claws permits the young to crawl among ventral hairs and reach the pouch. . . .The young find and attach to the nipples of the mammary glands and once attachment is gained fixation to it is uninterrupted until approximately day 60 to 70.[98]

Thus not only is the opossum relatively accessible, but so are its embryos for most of the gestational period—providing striking and unique research opportunities.

Hartman established and maintained an early colony of opossums while at the University of Texas from 1912 to 1925.[99] He caught the oppossums himself, and essentially created a minor and temporary colony while generating extensive specimens for study. Other zoologists and embryologists also attempted to colonize the opossum, but all met with failure until a major effort was launched during the experimental period of the Wistar Institute farm in Philadelphia.[100] Dr. M. J. Greenman, then director of the Institute, had an "unflagging interest in the opossum over a number of years." In fact, it took five years of effort (1930–1935) by Edward McCrady to develop the colony into a reliably [re]productive one.[101]

McCrady's main challenge in colonizing the opossum was the prevention of rickets. His colony was routinely devastated by the disease (despite a daily diet including milk, eggs, and even cod liver oil as sources of vitamin D) until he discovered, by opening the stomachs of fresh-caught animals, that they ate small vertebrates, bones and all. The addition of bone meal to their daily diet cured the rickets. Even then, however, the opossums failed to breed because of ovarian atrophy. A hint in Hartman's earlier work led the Wistar group to redesign the opossum cages to require exercise, which ultimately resolved this problem of ovarian atrophy.[102]

Moore and Bodian of the Departments of Zoology and Anatomy at the University of Chicago also established an opossum colony in the late 1930s, but in the city. An article of 1940 indicates that they accomplished this goal—at very low cost in an area not immediately surrounded by opossums—through a network of individual collectors and a biological supply house.[103] Their collectors, located in several southern and midwestern states, obtained special permits for trapping during the closed season to acquire females with pouch young. Moore and Bodian thus overcame the problems of acquiring this especially useful animal through careful organization of collectors and suppliers, as well as through their own breeding colony.[104] Opossum colonies thus represented a serious and often long-term investment of scientists' resources.

Primate Colonies

Primates became particularly attractive as research materials in the post-Darwin era, as their close relation to humans became increasingly clear.[105] While many scientists had kept a few live primates for observation and postmortem study, no serious research colonies existed until the 1920s.[106]

At that time, Yerkes and Corner initiated two major efforts in this direction.[107] According to the eminent primatologist Schultz, these efforts signaled the shift of primate study from "what had previously been regarded as merely the hobby of a few into a respected field of interest that began to receive adequate support and due academic recognition in some leading universities"—a new line of scientific work.[108]

Primates were required for the effective pursuit of certain problems in reproductive science. Corner, for example, required primates in order to study menstruation, which occurs only in such species.[109] Yerkes focused on problems of psychobiology, including reproductive function and behavioral studies.[110] For both Corner and Yerkes, expendable primate materials were essential.

Sociologically, primate colonies are of special interest because of the difficulty and expense of establishing and maintaining them. The expense of such colonies far exceeded normal university research budgets. The early colonies of Corner and Yerkes, like most later ones, drew on the fiscal resources of two or more major institutions, usually a university and a major foundation. If in the physical sciences, it was the need for elaborate and costly equipment that commonly pushed researchers to seek new sources of outside funding, it was the need for research colonies that commonly prompted such action in the life sciences.[111] The high costs of live materials drew biomedical scientists into new relations with external funding sources and into the arena we now call "Big Science."[112] Primate colonies were the extreme example of this process within American biomedical science between the two World Wars.[113]

The Hopkins/Carnegie Colony

Of the two initial efforts to establish primate colonies, the more famous was Yerkes' Yale Laboratories of Primate Biology, Inc., established at Orange Park, Florida in 1929.[114] As early as 1921, Corner, from the anatomy department at the Johns Hopkins Medical School of Baltimore, had established a colony of monkeys for reproductive scientific research. In his autobiography, Corner provided the following account:

> Obviously the best prospect was to use the one species, the Indian monkey *Macaca rhesus*, that could be purchased at moderate cost and was readily kept in American zoos . . . Dr. Weed found the money to buy eleven female monkeys. . . . This was the first monkey colony in the United States for long-term physiological experimentation.[115]

Zoo staff were often the only resource to which scientists could turn for information on maintaining live animals in captivity. Corner wrote as follows about his preparatory work for starting his colony in Baltimore:

> I learned that at the National Zoo in Washington and the Philadelphia Zoological Society's zoo, rhesus were being bred successfully. I went to Philadelphia to talk with the superintendent of the zoo, Mr. Ned Hollister, his head keeper, Mr. Blackburn, and the zoo's pathologist, Dr. Herbert Fox. They showed me a large colony of rhesus monkeys living to my surprise out of doors in a converted bandstand year around. With such animals I could work at my home laboratory with all the necessary facilities ready at hand.[116]

Significantly, Corner specified the zookeeper by name. If they lacked formal education, the keepers nonetheless had "tacit knowledge," extensive, lifelong, and even intergenerational experience of pivotal significance for scientists seeking to establish colonies.[117] Animal keeping had been a quasi-inherited position in many private collections developed by wealthy individuals whose servants' whole families often worked for them over several generations.[118]

Corner completed the first major study of the menstrual cycle with these rhesus monkeys,[119] stimulating parallel research on humans.[120] When Corner left Hopkins for the University of Rochester, Streeter, who became director of the Carnegie Institution's Department of Embryology at Hopkins after Mall's death, wanted to begin breeding rhesus for embryological and other reproductive investigations as part of a new and more experimental research thrust of the Department of Embryology.[121]

In 1925, Streeter invited Hartman, who was then planning an expedition to the Philippines to gather monkey embryos, "to join his laboratory and study the monkey embryology the safe and sane way, bringing the monkey to the laboratory."[122] Hartman inherited Corner's monkey paddocks and "got acquainted with the monkey."[123] Drawing upon his prior experience with opossums, Hartman succeeded in maintaining a large breeding rhesus colony in Baltimore.[124]

> [Hartman] provided the dated embryos that Heuser skillfully sectioned and Streeter analyzed and described. The result was one of the greatest achievements of American biology in its day, being the first complete description of the development of a primate from the one-cell stage through the whole embryonic period.[125]

Biomedical scientists used this colony in a wide variety of work, especially on reproduction and embryological development.[126]

YERKES' PROMOTION OF PRIMATE COLONIES

Although the Hopkins/Carnegie colony preceded the Yale colony, it was Yerkes who most actively promoted primate research and colonies in the United States. He conceived such work at the turn of the century, and in 1916 published a famous article in *Science* on "Provision for the Study of Monkeys and Apes."[127] Here Yerkes made a strong argument for primate research colonies, asserting first that the study of primates had been neglected:

> We have but fragmentary and unsatisfactory knowledge. . . . [Primates] are relatively difficult and expensive to obtain by importation or breeding, and to keep in normal condition. It is clear from an examination of the literature on these organisms and a survey of the present biological situation that the neglect by scientists of systematic study of all the primates excepting man is due, not to lack of appreciation of their scientific value but . . . to technical difficulties and the costliness of research.[128]

Such technical difficulties, Yerkes argued, could be addressed by careful planning and programming of research.

Yerkes was sensitive to the high costs involved in establishing and maintaining a primate research colony. He estimated the costs of a colony in southern California to be $50,000 per year, requiring an endowment of $1 million.[129] Asserting that his program would be cost-effective through the conservative use of expensive resources, Yerkes sought to justify major multi-institutional investments in primate research colonies on the grounds that they would ultimately help in solving practical human problems:

> I wish to emphasize the important relation of the plan I have outlined to strictly human interests and problems. It is eminently desirable that all studies of infrahuman organisms, and especially those of the primates which are most similar, structurally and functionally, to man, should be made to contribute to the solution of our own intensely practical, medical, social and psychological problems. . . . [Such problems] . . . may be solved, at least in large measure, most directly and economically through the use of monkeys and anthropoid apes.[130]

Finally, Yerkes called upon other biologists to support his plan, arguing that the task of securing adequate provisions for systematic and long-term study of primates "is by far the most important task for our generation of biologists, and the one which we shall therefore be most shamed by neglecting."[131]

Yerkes' promotion of primate research colonies was an ambitious undertaking. He initiated this effort in the scientific community in 1916, promoted it throughout his tenure at the National Research Council (ca. 1919–1924), garnered Carnegie funds for preliminary investigations (ca. 1924–1925), and in 1925 obtained a Rockefeller Foundation grant to establish a pilot laboratory and purchase several chimpanzees.[132]

Significantly, Yerkes drew special attention to international scientific competition in making his case for primate colonies in the United States. In his book, *Almost Human* (1925), Yerkes went to great lengths to compare the primate colonies and related research of W. Koehler in Germany, Nadie Kohts in the new Union of Soviet Socialist Republics, and the Pasteur Institute in French Guiana with the relative absence of such facilities in the United States.[113] These and other arguments were successful. In 1929, the Rockefeller Foundation, in cooperation with Yale University, funded the Yale laboratory of Primate Biology in Orange Park, Florida.[134]

CONCLUSION

The shift of emphasis to experimental physiological approaches in the life sciences at the turn of this century in the United States led to radically different needs for research materials. This chapter examines some of the constraints, opportunities, and resources involved in pursuit of such materials in the case of early twentieth-century reproductive science. Access to materials was initially achieved by scientists themselves for both exotic and mundane materials. Inexpensive and easily acquired materials predominated.

Physicians, biologists, and agricultural scientists often "inherited" particular research materials and established routes of access to them from their professors and/or at a given site. Staffs of zoos, museums, hospitals, and other personal and professional networks also helped the scientists achieve access to desired materials. Gradually, these scientists pioneered in establishing colonies for routine on-site access to materials. The acquisition of adequate supplies of materials represented a considerable investment of scientists' resources.

Once means of access were arranged for a particular material on an ongoing basis, scientists seem to have developed working commitments to continued use of these arrangements and research materials—continued cultivation not only of the actual materials but of access arrangements as well.[135] In other words, once successful access to given research materials was established, scientists seemed reluctant to abandon these materials unless and until their usefulness was exhausted through examination of a wide variety of reproductive biological problems. Certain materials, such as the rat (hypophysectomized or not), were conventionalized to such a degree that they became laboratory animals in reproductive and other life sciences.

Laboratory colonies were the major institutional solution to the problem of routine access to live and fresh materials. The importance of biological colonies lay not only in their immediate convenience in terms of access, but also in the broader scope of the research problems that could be addressed through standardized live materials maintained over time. The expense of such biological colonies drew life scientists into the arena of "Big Science" through sponsorship by major foundations.

The centrality of research materials to the production of biomedical knowledge may seem evident, but the topic has received almost no scholarly attention. Modern biomedical research requires a highly organized materials infrastructure, including animal colonies. Infrastructural organizations qua emergent social institutions deserve historical and sociological examination, especially in terms of their interaction with the social organization of scientific work and the development of scientific problem structures over time. To date, the historical sociology of research materials remains itself embryonic. The study of the relations among scientists, their research materials, their home institutions, their external funding sources, biomedical problems, emergent lines of scientific work, and even general political issues should prove to be a highly rewarding arena of investigation.

NOTES

1. An earlier version of this paper was presented at the meetings of the American Association for the History of Medicine, San Francisco, CA, 1984. I am grateful to Howard S. Becker, Kathy Charmaz, Nan P. Chico, Joan H. Fujimura, Elihu M. Gerson, Marilyn Little, Sheryl Ruzek, Leonard Schatzman, S. Leigh Star, and Anselm L. Strauss for their ongoing support of the project on which this paper is based. Merriley Borell, Jane Maienschein, and other participants in the conference from which this volume emerged also provided invaluable assistance. The Special Collections of the University of Chicago, the University of California, Davis, the Chesney Archives of the Johns Hopkins Medical Institutions, the Carnegie Institution of Washington, and the Rockefeller Archives graciously allowed me access to unpublished materials. The research has been supported by the Tremont Research Institute, the University of California, San Francisco, and the Rockfeller University.
2. This shift of emphasis took place in most areas of biological, medical, and agricultural research. See, for example, Garland Allen, *Life Sciences in the Twentieth Century* (Cambridge, UK: Cambridge Univ. Press, 1978); Adele E. Clarke, "Emergence of the Reproductive Research Enterprise: A Sociology of Biological, Medical and Agricultural Science in the United States, 1910–1940" (Ph.D. diss., Univ. of California, San Francisco, 1985); William Coleman, *Biology in the Nineteenth Century: Problems of Form, Function and Transformation* (New York: Cambridge Univ. Press, 1977); Coleman, "The Cognitive Basis of the Discipline: Claude Bernard on Physiology," *Isis* 76(1985): 49–70; Gerald L. Geison, "Divided We Stand: Physiologists and Clinicians in the American Context," in *The Therapeutic Revolution: Essays in the Social History of American Medicine*, ed. Morris J. Vogel and Charles Rosenberg (Philadelphia, PA: Univ. of Pennsylvania Press, 1979) pp. 67–90; Geison, *Michael Foster*

and the Cambridge School of Physiology: The Scientific Enterprise in Late Victorian Society (Princeton, NJ: Princeton Univ. Press, 1978); Gerson, "Styles of Scientific Work and the Population Realignment in Biology, 1880–1925" (paper presented at the Conference on History and Philosophy of Biology, Ohio, 1983); Gerson, "The Realignment of Population Biology, 1880–1925" (paper presented at the Society for Social Studies of Science, Philadelphia, PA, 1982); A. McGehee Harvey, *Adventures in Medical Research: A Century of Discovery at Johns Hopkins* (Baltimore, MD: Johns Hopkins Univ. Press, 1976); Rosenberg, "Rationalization and Reality in Shaping American Agricultural Research, 1875–1914," in *The Sciences in the American Context: New Perspectives*, ed. Nathan Reingold (Washington, DC: Smithsonian Institution, 1979) pp. 143–163; Rosenberg, *No Other Gods: On Science and American Social Thought* (Baltimore, MD: Johns Hopkins Univ. Press, 1976); and Margaret Rossiter, "The Organization of the Agricultural Sciences," in *The Organization of Knowledge in Modern America, 1869–1920*, ed. Alexandra Oleson and John Voss (Baltimore, MD: Johns Hopkins Univ. Press, 1979) pp. 279–298.

3. For some of the leading examples of such developments, see the other chapters in this book, especially those of Diana Long, Merriley Borell, and Jane Maienschein.
4. For analyses of problems of disciplinary emergence and substantive cases, see Adele E. Clarke, "Emergence"; Geison, ed., *Professions and Professional Ideology in America* (Chapel Hill: Univ. of North Carolina Press, 1983); Geison, "Scientific Change, Emerging Specialties and Research Schools," *Hist. Sci.* 19(1981): 20–40; Loren Graham, Wolf Lepenies, and Peter Weingart, eds., *Functions and Uses of Disciplinary Histories* (Boston: Kluwer, 1983); Robert Kohler, *From Medical Chemistry to Biochemistry: The Making of a Biomedical Discipline* (Cambridge, UK: Cambridge Univ. Press, 1982); Gerard Lemaine, Roy MacLeod, Michael Mulkay, and Weingart, eds., *Perspectives on the Emergence of Scientific Disciplines* (Chicago: Aldine, 1976); Philip J. Pauly, "The Appearance of Academic Biology in Late 19th Century America," *J. Hist. Biol.* 17(1984): 369–397; Rosenberg, "Toward an Ecology of Knowledge: On Discipline, Contexts and History," in *The Organization of Knowledge*, pp. 440–455; Rosenberg, "Rationalization and Reality"; and Rosenberg, *No Other Gods.*
5. See, for example, Maienschein, "Agassiz, Hyatt, Whitman and the Birth of the Marine Biological Laboratory," *Biol. Bull. Woods Hole* 168 Suppl. (1985): 26–34; Maienschein, "Early Struggles at the Marine Biological Laboratory," *Biol. Bull. Woods Hole* 168 Suppl. (1985) 192–196; Reingold, ed., *The Sciences in the American Context*; and Jeffrey Werdinger, "Embryology at Woods Hole: The Emergence of a New American Biology" (Ph.D. diss., Indiana Univ., 1980).
6. In terms of the history of materials, for example, articles have appeared on the history of cell lines and tissue cultures, but the question of biological materials is largely subordinated to the careers of the individuals who developed them and their institutional settings. See, for example, Frederick B. Bang, "History of Tissue Culture at Johns Hopkins," *Bull. Hist. Med.* 51(1977): 516–537; Harvey, *Adventures*; and Jan A. Witkowski, "Experimental Pathology and the Origins of Tissue Culture: Leo Loeb's Contribution," *Med. Hist.* 27(1983): 269–288. Works focused on other matters occasionally discuss biological materials tangentially; see, for example, Coleman, "The Cognitive Basis of the Discipline"; George W. Corner, *Seven Ages of a Medical Scientist: An Autobiography* (Philadelphia, PA: Univ. of Pennsylvania, 1981); Bruno Latour, *Les Microbes: Guerre et Paix Suivi de Irreductions* (Paris: Metaillie, 1984); Susan E. Lederer, "Hideyo Noguchi's Luetin Experiment and the Antivivisectionists," *Isis* 76(281) (1985): 31–48; and Lederer, "The Rights and Wrongs of Making Experiments on Human Beings," *Bull. Hist. Med.* 58(1984): 380–398. On instrumentation in physiology, see chapter XIII by Merrily Borell in this book.
7. See, for example, Kohler, *From Medical Chemistry*; and Geison, "Divided We Stand." The blurred boundaries between disciplines at the turn of the century were clear to the historical actors themselves. Ross Harrison, for example, once stated that departmental affiliations, "present fortuitous attachments," were often a matter of institutional contingency. See John B. Blake, "Anatomy," in *The Education of American Physicians: Historical Essays*, ed. Ronald L. Numbers (Berkeley: Univ. of California Press, 1980), p. 43.
8. Rosenberg, "Toward an Ecology of Knowledge."
9. See, for example, Becker, *Art Worlds* (Berkeley: Univ. of California, 1982); Rue Bucher and Anselm L. Strauss, "Professions in Process," *Am. J. Sociol.* 66(1961): 325–334; Bucher, "Pathology: A Study of Social Movements Within a Profession," *Soc. Prob.* 10(1962): 40–

51; Eliot Freidson, "The Division of Labor as Social Interaction," *Social Problems* 23(1976): 304–313; Everett C. Hughes, *The Sociological Eye* (Chicago, IL: Aldine Atherton, 1971); Strauss, "Work and the Division of Labor," *Sociological Q.* 26(1985): 1–19; Strauss, *Negotiations: Varieties, Contexts, Processes and Social Order* (San Francisco, CA: Bass, 1979); Strauss, *Professions, Work and Careers* (San Francisco, CA: Sociology, 1971). My own approach draws on the grounded theory research method as developed by Barney G. Glaser and Strauss. See Glaser and Strauss, *The Discovery of Grounded Theory: Strategies for Qualitative Research* (New York: Aldine, 1968); Glaser, *Theoretical Sensitivity: Advances in the Methodology of Grounded Theory* (Mill Valley, CA: Sociology, 1978); and Strauss, *Qualitative Analysis* [in German] (Munich, FRG: Wilhelm Fink, 1985), [in English] (Boston, MA: Cambridge Univ. Press, 1987).

10. See, for example, the varied perspectives of Gerson, "Scientific Work and Social Worlds," *Knowledge* 4(1983) pp. 357–377; Latour and Steve Wolgar, *Laboratory Life: The Social Construction of Scientific Facts* (Beverly Hills, CA: Sage, 1979); John Law, "Theories and Methods in the Sociology of Science: An Interpretative Approach," in *J. Soc. Sci. Inf.* 13 (1974): 163–172; Karen Knorr-Cetina and Michael Mulkay, eds., *Science Observed: Perspectives on the Social Study of Science* (Beverly Hills, CA: Sage, 1983); Star, "Simplification in Scientific Work: An Example from Neuro-science Research," *Soc. Stud. Sci.* 13(1983): 208–226; Star, "Scientific Work and Uncertainty," *Soc. Stud. Sci.* 15(1985): 391–427; and Rachel Volberg, "Constraints and Commitments in the Development of American Botany, 1880–1920" (Ph.D. diss., Univ. of California, San Francisco, 1983).

11. At the conference from which this book is derived, Robert Frank noted that materials, instruments, and techniques are commonly "packaged" together in scientific work. A classic example is the hypophysectomized rat, discussed later. For an analysis of the relation of such packaging to overall scientific research production, seen Fujimura, "Constructing Doable Problems in Cancer Research: Articulating Alignment," in *Soc. Stud. of Sci.*, in press.

12. There is considerable debate about whether this was really a shift, or whether it was instead an integration into biology of approaches previously reserved to medial domains, and about whether the changes were revolutionary or evolutionary in character. See Allen, "Morphology and Twentieth Century Biology: A Response," *J. Hist. Biol.* 14(1981): 159–176; Allen, "Naturalists and Experimentalists: The Genotype and the Phenotype," in *Studies in the History of Biology 3*, ed. Coleman and Camille Limoges (Baltimore, MD: Johns Hopkins Univ. Press, 1981) 179–209; Allen, "The Transformation of a Science: T. H. Morgan and the Emergence of a New American Biology," in *The Organization of Knowledge*, ed. Oleson and Voss; Allen, *Thomas Hunt Morgan: The Man and His Science* (Princeton, NJ: Princeton Univ. Press, 1978) 173–210; Allen, *Life Sciences*; Keith R. Benson, "American Morphology in the Late Nineteenth Century: The Biology Department at Johns Hopkins University," *J. Hist. Biol.* 18(1985): 163–205; Benson, "Problems of Individual Development: Descriptive Embryological Morphology in America at the Turn of the Century," *J. Hist. Biol.* 14(1981): 115–128; Frederick B. Churchill, "In Search of the New Biology: An Epilogue," *J. Hist. Biol.* 14(1981): 177–191; Coleman, "The Cognitive Basis of the Discipline"; Coleman, *Biology in the Nineteenth Century*; Paul L. Farber, *The Emergence of Ornithology as a Scientific Discipline* (Boston, MA: D. Reidel, 1982); Farber, "The Transformation of Natural History in the Nineteenth Century," *J. Hist. Biol.* 15(1982): 145–152; Maienschein, "Experimental Biology in Transition: Harrison's Embryology, 1895–1910," in *Studies in the History of Biology 6*, ed. Coleman and Limoges (Baltimore, MD: Johns Hopkins Univ. Press, 1983) 107–127; Maienschein, "Shifting Assumptions in American Biology: Embryology, 1890–1910," *J. Hist. Biol.* 14(1981): 89–113; Maienschein, Ronald Rainger, and Benson, "Introduction: Were American Morphologists in Revolt?" *J. Hist. Biol.* 14(1981): 83–87; and Werdinger, "Embryology." For analyses of the shift in population biology, see Gerson, "The Realignment."

13. My chief sources here include the Frank Rattray Lillie Papers, the Zoology Department Papers [hereafter, the Zoology Papers], University of Chicago Archives, and the President's Papers [hereafter, the President's Papers] from Special Collections, University of Chicago Archives; the Embryology Department Papers of the Carnegie Institution of Washington, DC (also held at the Johns Hopkins Medical Institution, Chesney Archives); the Animal Science Department Papers [hereafter, the Animal Science Papers], Univ. of California, Davis, Special Collections; the Rockefeller Foundation Papers, Rockefeller Archives Center,

North Tarrytown, NY; and the Hunterian Laboratory Papers, Johns Hopkins Medical Institution, Chesney Archives.

14. Lillie to President Judson, 10 December 1914, p. 3 and part of his proposal, "Department of Zoology: The Chief Need," p. 1. Zoology Papers, box IV, folder 2.
15. Lillie also noted that Princeton University was currently building such a vivarium at a cost of about $25,000: see E. G. Conklin to Lillie, 23 January 1915, Zoology Papers, box IV, folder 2. Such sharing of architectural plans and facilities proposals was common practice in scientific networks.
16. See, for example, the President's Reports to the Regents of the University of California, ca. 1910–1940, sections on Gifts to the University, especially to the Museum of Vertebrate Zoology.
17. See, for example, Corner, *Seven Ages*; and E. C. Amoroso and Corner, "Herbert McLean Evans, 1882–1971," *Biogr. Mem. Fellows R. Soc. Lond.* 21(1975): 83–186.
18. The application of statistical methods to materials and colonies is a problem worthy of investigation. See, for example, Daniel J. Kevles, *In the Name of Eugenics: Genetics and the Uses of Human Heredity* (New York: Knopf, 1985); and Donald MacKenzie, *Statistics in Britain, 1865–1930* (Edinburgh: Univ. of Edinburgh Press, 1981).
19. Elsewhere (Clarke, "Emergence") I discuss these assumptions of physiological parallelism and argue that they led to systematic research biases that routinely emphasized similarity and deemphasized diversity both within and across species. Cf. William Wimsatt, "Reductionist Research Strategies and Their Biases in the Units of Selection Controversy," in *Scientific Discovery: Case Studies*, ed. T Nickles (Dordrecht, The Netherlands: Reidel, 1980) pp. 213–259.
20. See, for example, Constance Holden, "An Omnifarious Data Bank for Biology?" *Science Wash. DC* 228(1985): 1412–1413; and W. Lane-Petter, "The Experimental Animal in Research," in *Techniques in Endocrine Research*, ed. Peter Eckstein and Francis Knowles (London: Academic, 1963), pp. 149–159.
21. Such collections are expensive to develop and maintain, whether of zoological or botanical specimens. See, for example, Miriam Rothschild, *Dear Lord Rothschild: Birds, Butterflies and History* (Philadelphia, PA: Balaban, 1983); and, in botany, Volberg, "Constraints."
22. The secondary literature has little to say about collecting expeditions for morphological specimens; in contrast, there is a growing scholarly and popular literature on collecting expeditions for live animals for zoos. On morphological expeditions, see Donna Haraway, "Signs of Dominance: From a Physiology to a Cybernetics of Primate Society: C. R. Carpenter, 1930–1970," *Studies in the History of Biology 6*, pp. 129–219. On zoo expeditions, see for example, David Attenborough, *The Zoo Quest Expeditions* (London: Lutterworth, 1980); Robert Bendiner, *The Fall of the Wild and the Rise of the Zoo* (New York: Dutton, 1981); William Bridges, *Gathering of Animals: An Unconventional History of the New York Zoological Society* (New York: Harper and Row, 1966); and more popular books such as Frank Buck, with Edward Anthony, *Bring 'Em Back Alive* (New York: Simon and Schuster, 1930); Gerald Durrell, *A Zoo in My Luggage* (Middlesex, UK: Penguin, 1964); and Cecil S. Webb, *A Wanderer in the Wild: The Odyssey of an Animal Collector* (London: Hutchinson, 1953).
23. I am indebted to S. Leigh Star for this insight. She is currently preparing a manuscript on the topic "preserving natural history" that will fill significant gaps in the secondary literature.
24. The issue of access to materials is, of course, not limited to mammalian reproductive science, but rather affects most if not all sciences. For example, one of the major features of both the Naples Zoological Station and the Woods Hole Marine Biological Laboratory was ease of access to plentiful and fresh materials, especially marine invertebrates. In these cases, the scientists went to the site of the materials to do research. See, for example, Allen, *Life Sciences*; Maienschein, "First Impressions: American Biologists at Naples," *Biol. Bull. Woods Hole 168* Suppl. (1985): 187–191; Maienschein, "Agassiz"; Maienschein, "Early Struggles"; and Werdinger, "Embryology."
25. Robert M. Yerkes, "Provision for the Study of Monkeys and Apes," *Science* 43(1916): 231.
26. See Becker, *Art Worlds*, for a discussion of the ways in which organizational commitments become conventionalized, and the ultimate consequences of such conventions.
27. See Clarke, "Emergence." I am indebted to Reuben Albaugh, extension agent emeritus,

University of California, Davis, for an interview that confirmed this aspect of his work. The Animal Science Papers record expenditures for herds and research. The [Annual] Reports of the Secretary of the University of California, ca. 1890–1940, document the not insignificant income (used for subsequent research) that various agricultural programs generated through services to farmers and ranchers and through sales of animals and products that grew out of such research.

28. Charles Otis Whitman, "A Biological Farm for the Experimental Investigation of Heredity, Variation and Evolution and for the Study of Life Histories, Habits, Instincts and Intelligence," *Science* 16(1902): 504–510. See also Whitman, "A Biological Farm," *Biol. Bull. Woods Hole* 3(1902): 214–224.
29. See chapter VII by Maienschein in this book and Pauly, "The Appearance."
30. Whitman, "A Biological Farm for the Experimental Investigation," p. 504, emphasis in original.
31. Whitman, "A Biological Farm for the Experimental Investigation," p. 505, emphasis in original. Lillie, quoted earlier along similar lines, was Whitman's student and successor as both Chairman of the Department of Zoology at the University of Chicago and as chief executive officer of the Woods Hole Marine Biological Laboratory.
32. See Rosenberg, *No Other Gods*; Rosenberg, "Rationalization and Reality"; and Rossiter, "The Organization."
33. Whitman, "A Biological Farm for the Investigation," p. 507. The history of materials, record keeping, and the breeding of colony animals evolved into the notion of controlled, standardized materials, perfected models with extensive control of genetic, environmental, and other sources of variation. An early criterion was known age; a later criterion was freedom from certain parasites or pathologies; recently, a host of new standards have been invoked. See, for an early example of an argument for standardization, the Hull Zoological Laboratory Report, President's Report, University of Chicago Archives, 1902, p. 438. See also Yerkes, "Provision"; Yerkes, "The Significance of Chimpanzee-Culture for Biological Research," Harvey Lect. 31(1935–1936): 57–73; Robert M. Saver ed., "Care and Diseases of the Research Monkey," *Ann. N. Y. Acad. Sci.* 85(1960) art. 3 pp. 735–992; and D. A. Valerio et al., *Macaca mulatta: Management of a Laboratory Breeding Colony* (New York: Academic, 1969). Also of interest are the Rockefeller Foundation Papers on the National Research Council's work on "Genetic Stocks, 1939–1941," Rockefeller Archives Center, RG1.1, S200D, box 51, folders 1855–1856.
34. Whitman, chair, and faculty members J. M. Coulter, T. R. Williston, Frank R. Lillie, Edwin O. Jordan, and William L. Tower for the biological departments, to Martin Ryerson, president of the board of trustees, 1 December 1906, President's Papers I, box 15, folder 12.
35. President's Report, University of Chicago Archives, 1902, p. 437. There seems to have been some misunderstanding of the meaning of the term "biological farm." Whitman ("A Biological Farm for the Investigation," p. 509) drew a careful distinction between a farm as a research endeavor and public zoological or botanical gardens. In the President's Papers at the University of Chicago, however, the biological farm proposal was filed under "Animal Husbandry," likely reflecting earlier misunderstanding. See President's Papers I, box 7, folder 1. The Rockefeller Institute purchased a New Jersey farm for laboratory animal production in 1908, subsequently called "Hell Farm" by antivivisectionists (Lederer, "Hideyo Noguchi," p. 37). Many years later, Geoffrey H. Bourne referred to primate research centers in their breeding capacity as "animal farms." See Bourne, ed., *Nonhuman Primates and Medical Research* (New York: Academic, 1973), p. 490.
36. Florence Rena Sabin, *Franklin Paine Mall* (Baltimore: Johns Hopkins, 1934), p. 213.
37. See Amoroso and Corner, "Herbert McLean Evans"; Sabin, *Franklin Paine Mall*, pp. 200–201.
38. Corner, *Seven Ages*, pp. 21, 43, 86, 102–106.
39. Ibid., p. 106.
40. Carl G. Hartman, "The Scientific Achievements of George W. Corner," *American Journal of Anatomy* 98(1956), p. 8. Hartman further noted that Corner later used materials from rabbits, monkeys, and occasionally dogs, guinea pigs, and humans in his research on the corpus luteum. Applying cytological criteria, Corner was able "as early as 1915 to distinguish

seven stages of pregnancy as reflected in the corpus luteum of the sow;" parallel research on the rat by Joseph Long and Herbert Evans followed six years later.

41. Corner, *Seven Ages*, p. 159.
42. The funds were provided by the Department of Anatomy, Ibid., p. 159.
43. Ibid., p. 156.
44. Wilson was then a member of Corner's Department of Anatomy at Johns Hopkins. Karl M. Wilson, "Histological Changes in the Vaginal Mucosa of the Sow," *Am. J. Anat.* 37(1926): 418–432.
45. Ibid., p. 419.
46. When abattoirs disappeared from New York City, as zookeeper Crandall noted, they could no longer keep certain animals such as vampire bats for which they had relied on abbattoirs for a regular supply of fresh blood. Lee S. Crandall in collaboration with William Bridges, *A Zoo Man's Notebook* (Chicago, IL: Univ. of Chicago, 1966), p. 21.
47. See Lillie to G. O. Fairweather, 3 October 1918, Frank R. Lillie Papers, University of Chicago Archives, box VII, folder 14 [hereafter, Lillie Papers]. There is routine mention of the stockyards as a source of materials in both the Lillie Papers and the Zoology Department Papers.
48. Lillie to Wallace Heckman, Faculty Exchange, 14 March 1919, Lillie Papers, box VII folder 15.
49. See Lillie, "The Free-Martin: A Study of the Action of Sex Hormones in the Foetal Life of Cattle," *Journal of Experimental Zoology* 23(1917): 371–452; and H. H. Newman, "History of the Department of Zoology in The University of Chicago," *Bios* 19(1948): 215–239.
50. See Clarke, "Emergence," chapt. 7.
51. Ibid.
52. See, for example, Bernhard Zondek and Michael Finkelstein, "Professor Bernhard Zondek: An Interview," *J. Reprod. Fertil.* 12(1966): 3–19. Zondek's comments on stallion urine as an excellent source of female estrogenic hormones are especially interesting.
53. I am indebted to Andrew Nalbandov, professor emeritus of the University of Illinois, for the interview on which this information is based.
54. This was considerably easier than collecting sperm from many domestic species for artificial insemination. See Harry A. Herman, *Improving Cattle by the Millions* (Columbia, MO: Univ. of Missouri, 1981).
55. See H. H. Cole and G. H. Hart, "The Potency of Blood Serum of Mares in Progressive Stages of Pregnancy in Effecting the Sexual Maturity of the Immature Rat," *American J. Physiol.* 94(1930): 57–68; and Zondek and Finklestein, "An Interview."
56. For this information, I am indebted to Perry T. Cupps and Hubert Heitman of the University of California, Davis.
57. I have not included some fascinating accounts of access to cadavers at Hopkins and Rochester due to limitations of space, and because of their tangential use in reproductive biological research. But see Sabin, *Franklin Paine Mall*; and Corner, *Seven Ages*.
58. Elemer Scipiades, "Young Human Ovum Detected in Uterine Scraping," *Carnegie Contrib. Embryol.* 163(1938): 97.
59. Letter from Mall to President Woodward, 21 March 1914, Department of Embryology drawers, folder "Embryology Miscellany I," Carnegie Institute of Washington (hereafter, Embryology file). The Department of Embryology, Carnegie Institute of Washington, established in 1913, was physically located at and was closely associated with Johns Hopkins University and its medical school. See Mall, "Note on the Collection of Human Embryos in the Anatomical Laboratory of Johns Hopkins University," *Bull. Johns Hopkins Hosp.* XIV 143(1903): 29–33; and "The Embryological the Carnegie Institution," undated but printed document, Embryology file.
60. Mall, "Early Human Embryos and the Mode of their Preservation," *Bull. Johns Hopkins Hosp.* IV 36 (1893): 115.
61. Mall, "The Embryological Collection," p. 29.
62. See "Data Bearing Upon the Cause of Abortion and Age of the Embryo," undated form bearing Mall's name and the Johns Hopkins Medical School address. Embryology file.
63. Mall, "Note On the Collection," from The Embryological Collection," pp. 29–30.
64. "Request . . ." from Mall, June 1913. Embryology file. This document appears to have been

mailed to departmental alumni of the medical school.

65. Letter "Dear Doctor," from Evans, undated. Embryology file.
66. For example, the Department of Embryology also collected alligator, monkey, and baboon embryos. For alligators, see letter to President Woodward from George L. Streeter, 4 February 1919, regarding a trip to British Guyana by Albert M. Reese of West Virginia University, Embryology file. Regarding monkey embryology, see Personnel file of Dr. Chester H. Heuser; and, regarding baboon embryology, see Personnel folders of George L. Streeter, letter to O. A. Scherer from Corner, 20 August 1947; Personnel drawers at the Carnegie Institute of Washington.
67. A. H. Schultz, "The Rise of Primatology in the Twentieth Century," in *Proceedings of the Third International Congress of Primatology*, Zurich, 1970, vol. 1, ed. J. Biegert and W. Leutenegger (New York: S. Karger, 1971) p. 4.
68. Corner, *Seven Ages*, p. 205.
69. G. W. D. Hamlett, "Primordial Germ Cells in a 4.5-M. M. Human Embryo," *Anat. Rec.* 61(1935): 273.
70. See Lillie's paper on the freemartin calf for the classic study of fetal endocrine interrelations. Lillie, "The Free-Martin."
71. A. W. Diddle and T. H. Burford, "A Study of a Set of Quadruplets," *Anat. Rec.* 61(1935): 282.
72. Edgar Allen, J. P. Pratt, Q. U. Newell and L. J. Bland, "Human Tubal Ova; Related Early Corpora Lutea and Uterine Tubes," *Carnegie Contrib. Embryol.* 27(1930): 45–75.
73. Allen and his associates reported that Corner had recovered the first primate ovum from rhesus (1923); Allen recovered three tubal primate ova (1927, 1928); and degenerating uterine ova were recovered by Corner (1923) and Hartman (1929). Ibid.
74. Ibid., p. 48. There is no mention in their article of informed consent processes in relation to the research.
75. Ibid.
76. Ibid., p. 55.
77. Arthur T. Hertwig and John Rock, "On a Complete Normal 12-day Human Ovum of the Pre-villous Stage." *Anat. Rec.* 73 Suppl. (1939): 26–27; and Carl G. Hartman, "Studies on Reproduction in the Monkey and their Bearing in Gynecology and Anthropology," *Endocrinology* 25(1939): 676. In their article Hertig and Rock do not, in fact, discuss their means of access to the materials; this was likely done in conversation with Hartman who reported it. See also, Loretta McLaughlin, *The Pill, John Rock, and the Church* (Boston, MA: Little, Brown, 1982), esp. chapt. 4, "The Human Egg Hunt." McLaughlin notes (p. 61) that Hertig had worked with Streeter at the Carnegie Department of Embryology in 1933 on primate embryology.
78. McLaughlin, *The Pill*, pp. 59–60.
79. Ibid, p. 63. McLaughlin pursues the informed consent and abortion issues; see pp. 63–64.
80. Ibid, p. 66.
81. The results of this research focused on (1) the timing of ovulation in terms of the extrusion of the first polar body; (2) the stage of development of corpora lutea in relation to ovulation; (3) the timing of ovulation in relation to the menstrual cycle and related epithelial changes; and (4) determination of whether ovulation always accompanies menstruation. Allen and associates, "Human Tubal Ova," pp. 73–4.
82. Yerkes, *Almost Human* (New York: Century, 1925), p. 269.
83. See the Hunterian Laboratory Papers in the Chesney Archives at Johns Hopkins.
84. Ibid. Antivivisectionists often influenced the access of scientists and faculty to these mundane materials, and animal house managers went further and further afield to achieve their quotas.
85. See the Rockefeller Foundation Papers on the National Research Council's work on "Genetic Stocks, 1939—1941," Rockefeller Archives Center, RF RG1.1, S200D, Box 51, folders 1855–6. The Rockefeller Foundation provided extensive support to Charles Stockard, Department of Anatomy, Cornell University Medical School, to develop a colony of genetically standardized dogs.
86. Aside from works on antivivisectionism, these issues have not been subjected to adequate historical research.

87. See, for example, Allen et al., "Human Tubal Ova"; and Alden B. Dawson and Harry B. Friedgood, "The Time and Sequence of Preovulatory Changes in the Cat Ovary After Mating or Mechanical Stimulation of the Cervix Uteri," *Anat. Rec.* 76(1940): 411–429.
88. Leo Loeb, "The Cyclic Changes in the Ovary of the Guinea Pig," *J. Morphol.* 23(1911): 37–70; Charles R. Stockard and George Papanicolaous, "The Existence of a Typical Oestrus Cycle in the Guinea-Pig, with a Study of its Histological and Physiological Changes," *Am. J. Anat.* 22(1917): 225–283; and Sewell Wright, *Evolution in Mendelian Populations* (Chicago, IL: Univ. of Chicago Press, 1931).
89. John Q. Griffith, *The Rat in Lab Investigations* (Philadelphia, PA: Lippincott, 1942); J. A. Long and H. M. Evans, "The Oestrus Cycle in the Rat and Its Associated Phenomena," *Mem. Univ. Calif.* 6(1922); and Philip E. Smith, "The Disabilities Caused by Hypophysectomy in the Rat and Their Repair," *J. Am. Med. Assoc.* 88(1927): 158–161.
90. Allen, "The Oestrus Cycle in the Mouse," *Am. J. Anat.* 30(1922): 297–372; and E. C. MacDowell and E. M. Lord, "The Relative Viability of Male and Female Mouse Embryos," *Am. J. Anat.* 37(1926): 127–140.
91. Howard A. Schneider, "Harry Steenbock (1886–1967): A Biographical Sketch," *J. Nutr.* 103(1973): 1238. I am indebted to Merriley Borell for locating this information. Fruit flies were also colonized with considerable success in genetics research. See Allen, "The Life Sciences"; and Allen, *Thomas Hunt Morgan.*
92. The development of colonies of smaller laboratory animals deserves study, especially in relation to the emergence of the biological supplies industry. See, for example, the Hunterian Laboratory Papers, Chesney Archives at Johns Hopkins.
93. John F. Fulton and Leonard G. Wilson, *Selected Readings in the History of Physiology*, 2nd ed. (Springfield, IL: Thomas, 1966), pp. 399–400.
94. For a detailed explanation of Smith's technique, see Frederick J. Agate, Jr., "Philip Edward Smith," in *Dictionary of Scientific Biography* XII, ed. Charles Coulston Gillispie (New York: Scribner, 1975), pp. 472–478.
95. Rudolph P. Vollman, "Fifty Years of Research on Mammalian Reproduction: Carl G. Hartman," USDHEW, PHS Publication no. 1281 (1965): 5.
96. Ibid., p. v.
97. Ibid., pp. 407, vi.
98. Carl R. Moore and David Bodian, "Opossum Pouch Young as Experimental Material," *Anat. Rec.* 76(1940): 319.
99. Vollman, "Fifty Years," p. 5.
100. Moore and Bodian, "Opossum," p. 320.
101. Edward McCrady, "The Embryology of the Opossum," *Am. Anat. Mem.* 16(1938): 9–14. McCrady notes (on p. 9) that in his own research he also utilized materials collected and left by Hartman at the Wistar Institute.
102. Ibid., pp. 11–14.
103. Moore and Bodian, "Opossum."
104. See also G. E. Coghill, "Studies on Rearing the Opossum (Delphys Virginiana)," *Ohio J. Sci.* 39(1939): 239–249.
105. See Schultz, "The Rise of Primatology," p. 2; and L. H. Schmidt, "Problems and Opportunities of Breeding Primates," in *Breeding Primates: Proceedings of the International Symposium on Breeding Non-Human Primates for Laboratory Use*, ed. W. I. B. Beveridge (New York: Karger, 1972) 1–22. I use the term "materials" throughout this chapter to refer to both living and dead animals, both as wholes or parts. Some contemporary scientists have objected to use of the term "materials" in reference to nonhuman primates, and specifically to the term "primate materials." They are accustomied to using the term "subjects" for the nonhuman primates they study. In contrast, my focus here is not on the animals per se but rather on their organization in the service of scientific research, as part of the requisite infrastructure. Hence the term "materials" seems appropriate, if not customary.
106. Corner, *Seven Ages*; and Solly Zuckerman, "The History of Apes and Monkeys as Objects of Enquiry," in *Beyond the Ivory Tower: Frontiers of Public and Private Science*, ed. Solly Zuckerman (New York: Taplinger, 1970), appendix I, pp. 205–214. Animals displayed in some zoological gardens were occasionally used for research purposes. See R. I. Pocock, "Notes Upon Menstruation, Gestation and Parturition of Some Monkeys that Have Lived in the Society Gardens," *Proc. Zool. Soc. Lond.* (1906): 558–570; and Zuckerman, "The Menstrual

Cycle of the Primates, Part I: General Nature and Homology," *Proc. Zool. Soc. Lond.* (22 October 1930): 691–754. They were not, however, intentionally organized for research but rather for public observation. Research was usually carried out on them only after their death. Circuses were another source of primate cadavers. See Schultz, "The Rise of Primatology."

107. Among other such efforts were those of Allen at Washington University, St. Louis, and Gertrude Van Wagenen and John Fulton at Yale. See Carl Backman, "Institutional History," (unpublished manuscript, 1982); Backman, "Resource Utilization in Biomedical Science: Patterns of Research on Nonhuman Primate Reproductive Physiology," (Ph.D. diss., Cornell University, 1983); Allen, "The Time of Ovulation in the Menstrual Cycle of the Monkey, M. Rhesus," *Proc. Soc. Exp. Biol. Med.* 23(1926): 281–383; and van Wagenen, "The Monkey," in *The Care and Brreding of Laboratory Animals*, ed. E. J. Farris (New York: Wiley, 1950), pp. 1–42.
108. Schultz, "The Rise of Primatology," p. 7.
109. Corner, "Ovulation and Menstruation in Macacus Rhesus," *Carnegie Contrib. Embryol.* 75(1923); 73–110.
110. Yerkes, *Chimpanzees: A Laboratory Colony* (New Haven, CT: Yale Univ. Press, 1943); and Yerkes, "Yale Laboratories of Primate Biology," *Science* 82(1935): 618–620.
111. Regarding the physical sciences, see Reingold, "The Sciences in the American Context." Regarding the life sciences, see, for example, the Grants-In-Aid files of the Rockefeller Foundation, RF RG1.1 S200 B38 F433 and passim, Rockefeller Archives Center. In 1939, requests to the Rockefeller Foundation for support of various colonies to produce standardized materials became so numerous that Warren Weaver of the Foundation requested the assistance of the National Research Council in developing an improved understanding of the needs in this area to which the Foundation could best respond. See the Rockefeller Foundation Papers on the National Research Council's work on "Genetic Stocks, 1939–1941," RF, RG1.1, S200D, Box 51, folders 1855–1856, Rockefeller Archives Center.
112. Foundations and other philanthropies had earlier sponsored buildings and expeditions to collect specimens. See, for example, Haraway, "Signs of Dominance."
113. By comparison, Sewall Wright's major colony of guinea pigs at the University of Chicago after 1924/1925 was costly for its day, with an annual feed budget of about $2500–$3000 (Zoology Department Papers, box 11, folder 3). However, such figures did not begin to approach primate colony costs.
114. Yerkes, "Yale Laboratories," p. 618.
115. Corner, *Seven Ages*, pp. 163–164.
116. Ibid., p. 163.
117. For example, the head zookeeper of the New York Zoological Park in 1908, Samuel Stacey, was himself the son of the Duke of Wellington's water bailiff—in charge of water fauna. As a youth, Sam Stacey had become the Dutchess' "bird boy" and eventually moved on to the London Zoological Gardens as a bird keeper, coming to New York for the same purpose and to train a staff of bird keepers. Crandall, *Zoo Man's Notebook*, pp. 3–4.
118. For an account of a private primate colony in Cuba, see Yerkes, *Almost Human.*
119. Corner, "Ovulation."
120. See, for example, George W. Bartelmez, "Menstruation," *Physiol. Rev.* 17(1937): 28–72; Bartelmez, "Histological Studies on the Menstruating Mucous Membrane of the Human Uterus," *Carnegie Contrib. Embryol.* 142(1933): 151–186; Hartman, "Studies on Reproduction in the Monkey and Their Bearing;" and Jessie L. King, "Menstrual Records and Vaginal Smears in a Selected Group of Normal Women," *Carnegie Contrib. Embryol.* 95(1926): 79–94.
121. Regarding this shift of focus, see Heuser and Streeter, "Development of the Macaque Embryo," *Carnegie Contrib. Embryol.* 181(1941): 15–55. The Department had been committed for many years to the development of a human embryo collection and related embryological work. Streeter argued that a monkey colony would allow conceptions to be controlled and fetuses to be dated (unlike those of human materials), while a host of other problems could also be pursued. For the full rationale for this colony, see Memorandum to Doctor Merriam from George L. Streeter, 20 April 1925, folder "Embryology-Director, 1913–1935," Department of Embryology Files, at the Carnegie Institution in Washington, D.C.
122. Hartman quoted in Vollman, "Fifty Years," p. vii.

123. Ibid., p. 7.
124. In Vollman's words: "Thus, for the first time, a primate became a laboratory animal." Somehow Corner's earlier work was ignored in this claim, which is odd in that (according to both Corner and Hartman) Hartman's breeding practices were based in part on Corner's research, which had yielded crucial data on the timing of fertility and the menstrual cycle that was helpful in breeding programs. Ibid., p. vii. Hartman also published on the development of primate colonies. See Hartman, "Bimanual Rectal Palpation as Applied to the Female Rhesus Monkey," *Anat. Rec.* 45(1930); 263; and Hartman, "The Mating of Mammals [including monkeys]," in Special Issue: Animal Colony Maintenance, *Ann. NY Acad. Sci.* 46(1945): 23–44.
125. Corner, *Seven Ages*, p. 284.
126. See, for example, Bartelmez, "The Circulation in the Intervillous Space of the Macaque Placenta," *Anat. Rec.* 61 Suppl. A (1935): 4; Hartman, "Studies in the Reproduction;" Hartman, "Ovulation and the Transport and Viability of Ova and Sperm in the Female Genital Tract," in *Sex and Internal Secretions*, ed. Edgar Allen (Baltimore: Williams and Wilkins, 1932); and Heuser and Streeter, "Development."
127. Yerkes, "Creating a Chimpanzee Community," [edited by Roberta W. Yerkes from archival materials] *Yale J. Biol. Med.* 36(1963): 205–223; and Yerkes, "Provision."
128. Yerkes, "Provision," p. 231.
129. Ibid, p. 233.
130. Ibid.
131. Ibid., p. 234.
132. See Yerkes, "Provision"; Yerkes, *Almost Human*; Yerkes, "Yale Laboratories,"; and Yerkes, "Creating a Chimpanzee Colony."
133. Quite similar arguments were made regarding federal assumption of sponsorship of major primate colonies. See Willard H. Eyestone, "Scientific and Administrative Concepts Behind the Establishment of the U.S. Primate Centers," in *Some Recent Developments in Comparative Medicine. Symposium of the Zoological Society of London* no. 17: 1–9, ed. R. N. Fiennes (London: Academic Press for Zoological Society, 1966).
134. Yerkes, "Yale Laboratories."
135. For a discussion of such commitments and their consequences, see Becker, *Art Worlds*; and Elihu Gerson, "On Quality of Life," *Am. Soc. Rev.* 41(1976): 793–806.

XV

Instruments, Techniques, and Social Units in American Neurophysiology, 1870–1950

LOUISE H. MARSHALL

A full history of neurophysiology in the United States has not yet been written, and this chapter seeks only to sketch in a few of its major features, with special attention to the role of experimental techniques and instruments in its maturation as a recognizable subdiscipline. The chapter is divided into two distinct parts, with some names and a few of the themes from the first part recurring in the second part. The first part begins by noting the extent of British influence on the history of American neurophysiology. However, its major aim is to identify the leading social units in the subdiscipline in the United States, including the American Physiological Society (APS), several major university research centers, and the informal network known as the "Axonologists."

In the second part, the focus shifts to leading concepts, experimental techniques, and laboratory instruments in the history of neurophysiology in the United States. Special attention is accorded American contributions to new methods of amplifying and recording biological signals, notably the cathode-ray oscilloscope, the electroencephalograph, and the microelectrode. The effort to localize centers of neurophysiologic activity utilizing the stereotaxic instrument is also described. A concluding section draws some connections between the first and second and reflects upon broader issues this chapter suggests.

MAJOR CENTERS OF NEUROPHYSIOLOGY IN THE UNITED STATES

American Neurophysiologists Abroad: British Influence

No historical account of American neurophysiology can ignore its European and especially British antecedents. By the time the subdiscipline took root in this country, the complex morphology of the nervous system had already begun to be sorted out through new staining technologies, and some functional characteristics of clusters of adjacent neurons such as those found in ganglia and in cortical areas, had been perceived. Those discoveries led to two fields, the spinal reflex physiology of Charles Scott Sherrington in England and the conditioned reflex physiology of Ivan Pavlov in Russia.[1] In the late nineteenth century those two "channels" received the major attention of European physiologists. Although the great German and other Continental laboratories attracted many young American physiologists for postdoctoral training (as Robert Frank has shown in chapter I in this book), British laboratories drew their fair share of those who later specialized in the nervous system.

Many of these young American scientists were drawn to the laboratory of Sherrington (1857–1952), where some of the most fundamental neurophysiologic concepts of the early twentieth century were developed. Among many important contributions, he and his collaborators identified the motor unit and the synapse, the excitatory and inhibitory states of the central nervous system, and reciprocal innervation. Sherrington's major contribution is considered to be the conceptualization of the "functional counterpart of the neuron theory . . . namely, integration."[2] Among the Americans who worked with Sherrington and whose own later research had significant impact on modern neurophysiology were Harvey Cushing and John Farquhar Fulton, both later associated with Yale, and Alexander Forbes of Harvard.

Cushing (1869–1939) visited Sherrington's laboratory at Liverpool in 1900. He was an imaginative surgeon who applied the precision of human neurosurgery to preparing the craniums of three higher apes (chimpanzee, orangutan, and gorilla) for Sherrington's determination of the inverted representation of the body in the precentral motor cortex. This was Cushing's introduction to neurophysiology. Later, he stimulated the exposed cortex of conscious patients and established that the reactions in higher apes (as in Sherrington's laboratory) were applicable to humans.[3] By 1912 Cushing had shown that meticulous removal of the pituitary in experimental animals produced adiposity and sexual immaturity,[4] thus emphasizing the interrelatedness of neuroendocrinology, neuroanatomy, and neurophysiology.

By the time Fulton (1899–1960) began working with Sherrington in the early 1920s, the English mentor had transferred his laboratory to Oxford. Despite Fulton's many important contributions to the field, his early research at Oxford also introduced a pseudophenomenon into the study of reflexes. Using the optical myograph, a technique of which the Oxford group was "most proud," Fulton thought he had discovered a novel detail in the record of the descending phase of muscular contraction: a steady tension at the peak was followed by rapid relaxation, the transition being called "the angle." When the optical myograph was redesigned to reduce the effects of friction and enhance its responsiveness, the angle no longer appeared and the muscle twitch was now recorded as a smooth rise and decline. John Eccles, also at Oxford, later engagingly recounted the resulting consternation when the imaginary angle was unmasked. With this work, the investigation of nerve-muscle physiology came full circle within the same laboratory at Oxford.[5]

Forbes (1882–1965), who visited Sherrington's laboratory during the academic year 1911–1912, was to become the brightest star on the neurophysiologic horizon at Harvard. Forbes's long career in the field began in 1915 with a series of electrical studies of mammalian reflexes, and he became the acknowledged American expert on the spinal reflex. We shall hear more of him later as one of the Axonologists and in connection with the microelectrode as it was utilized by his research group.

Sherrington was not the only influential English mentor of American neurophysiologists. Especially significant after World War I was Edgar Douglas Adrian (1889–1977) of Cambridge, who had been a student there of one of the great early "engineers" in neurophysiology, Keith Lucas. In 1912 Lucas and Adrian together proposed the "all-or-none law" for propagation of the nervous impulse, thus extending to nervous action a law already established for muscular contraction. Thereafter, Adrian's international prestige among neurophysiologists rose steadily, and he seemed to enjoy hosting visitors from abroad at least as much as those from his own country.

By 1930, Adrian's distinguished American students included Sherrington's earlier visitor, Forbes (with Adrian in 1921), as well as future Nobel laureate Herbert Spencer Gasser (with Adrian and A. V. Hill in 1923) and Detlev Wulf Bronk. While working with Adrian at Cambridge in 1928–1929, Bronk learned to cut down muscle tissue to single fibers in order to study their neurophysiological properties—a technical innovation that was soon also applied to the study of nerves and had major advantages compared with bundles of fibers that produced compound action potentials.

Neurophysiology at Home: The American Physiological Society

American neurophysiologists continued to look toward Europe and especially England well beyond 1930, but by then they could and often did choose to undertake training at home instead. One reason is suggested in a letter of 1923 that Gasser, then working with Adrian, sent back to Joseph Erlanger, his chief at Washington University in St. Louis. Gasser wrote as follows of the contrasts that could be drawn by then between physiology on the two sides of the Atlantic: "The laboratory here is not as good as ours, I mean yours. There are many beautiful instruments for electrical investigations but they seem to belong mostly to [A.V.] Hill as the result of purchases from grants."[6]

Perhaps Gasser's concern with instruments and techniques reflected a deeper and long-standing American style of approach to research problems in neurophysiology. In 1870, the notable American physician, teacher, and man of letters, Oliver Wendell Holmes (1809–1894) described his verification of European estimates of the speed of propagation of the impulse in motor and sensory nerves:

> I have myself instituted a good many experiments with a more extensive and expensive machinery than has ever been employed, namely, two classes each of ten intelligent students who with joined hands represented a nervous circle of about sixty-six feet: so that a hand-pressure transmitted ten times round the circle traversed six hundred and sixty feet, besides involving one hundred perceptions and volitions. . . . After some practice, my second class gradually reduced the time of transmission. . . which, like that of the first class, had stood at fourteen and fifteen seconds, down to ten seconds, that is, one-tenth of a

second for the passage through the nerves and brain of each individual—less than the least time I have ever seen assigned for the whole operation. . . .[7]

In any case, Americans who went abroad for advanced laboratory training in neurophysiology could also draw inspiration from a domestic tradition of special interest in problems of the nervous system. By the time the APS was founded in 1887, the nervous system constituted a major research interest among its members. Six of its first twenty-eight charter members were engaged solely in neurophysiological investigations,[8] and all five of the papers at the first annual meeting of the Society, in 1888, were on neural topics. In addition, the Society offered prizes (with money contributed by S. Weir Mitchell of Philadelphia) for research on nerve conduction and regeneration. For at least another quarter century, physiologists lacked equipment that could approach an accurate quantitative measurement of the nerve impulse, and only recently have there been any positive results in studies of mammalian nerve regeneration.[9] Yet from the outset, several of the pioneers in American physiology clearly believed that the important unresolved problems lay in the nervous system.

The specific questions and techniques that interested these pioneers—as recorded in the papers they presented at the initial meetings of the APS—included the following: the nature of reflexes (Warren P. Lombard, who devised the reflex harp that simultaneously recorded eleven to fifteen muscles in the frog leg), the vasomotor nerves of the limbs (H. P. Bowditch, using his plethysmograph), fever centers in the brain (I. Ott), cutaneous sensibility (Henry Herbert Donaldson), nerve transmission (E. T. Reichert), and sensory reinforcement of the knee jerk (J. W. Warren). The talk delivered at the first Society meeting by an English guest, Dr. Victor Horsley, was titled "Observations on Negative Variation in the Spinal Cord and on the Relation of Clonic Spasms to Cortical Stimulation," thus combining in a single stroke, so to speak, both of the main channels of modern neurophysiology previously mentioned—spinal reflexes and cerebral localization.

Although Mitchell (1829–1914) apparently presented no papers at the first few meetings of the APS, he made other important contributions to American neurophysiology during its infancy. The scope of his research in the field was wide: motor disturbances after cerebellar injury (1869), injuries of nerves (1872), and the knee-jerk reflex (1886). Mitchell's great reputation as a practicing neurologist and his encouragement of research through cash prizes conveyed a strong impetus to neurophysiology in Philadelphia and elsewhere. The proximity of the Wistar Institute and several medical schools contributed to the momentum.

By the eve of World War I, the annual meetings of the APS were no longer dominated by papers on the nervous system. In 1913, for example, neurophysiological topics represented only thirty-six percent of the twenty-nine papers read. By then, the founding of departments of physiology in American liberal arts colleges[10] and universities signified a widespread interest in the discipline of physiology as a whole, and neurophysiology became just one of several subdisciplines, albeit a major one among them. We now turn to the place of neurophysiology within the larger discipline as it became institutionalized in university departments.

Neurophysiology in American Universities

Vastly more important for the early development of American neurophysiology than Mitchell's individual influence or even the American Physiological Society was the founding of research laboratories at Harvard under Bowditch (1871) and at Johns Hopkins under H. Newell Martin (1876). The larger story of those laboratories is described elsewhere in this book; here attention is confined to the place of neurophysiology within them. The neurophysiologic direction of Bowditch's research on the vasomotor nerves and knee-jerk reflex is well known. One major result of his thirty-year stewardship of the Harvard department was the education and subsequent appointment to its faculty of one of the first completely American-trained physiologists and one of the most influential physiologists of this century, Walter Bradford Cannon (1871–1945). Even before he left to serve in World War I, Cannon had introduced neurophysiology into the study of the emotions, as summarized in his 1915 publication, *Bodily Changes in Pain, Hunger, Fear and Rage.* Noting the sensitivity of the mechanical activity of the digestive tract to emotion, Cannon and his collaborators showed that the effect was due to impulses from the sympathetic nervous system; "from the results emerged the concept that the sympathetic system acts to maintain stable states in the organism, to preserve *homeostasis.*"[11]

At Johns Hopkins the early situation for neurophysiology was very different. Because the new university lacked a medical school during its first seventeen years, research on the nervous system was initially confined to marine species (in conjunction with the Chesapeake Zoology Laboratory). The Hopkins biology department was by no means unfriendly to physiologists, however.[12] Among them was Donaldson (1857–1938), with interests "in the neurological side of physiology."[13] In 1906 Donaldson went to the Wistar Institute of Anatomy and Biology in Philadelphia, where his studies on the nervous system continued. When the Hopkins medical school finally opened in 1893, the anatomy department rather than physiology became the hub of nervous system studies because research in the physiology department was dominated by its chairman, W. H. Howell, who concentrated on studies of the blood and circulation.

Perhaps the most important Hopkins contribution to neurophysiology during this period was Ross Granville Harrison's invention of the hanging drop method of culturing embryonic tissue. Neurophysiology benefited spectacularly from the new technique, for Harrison observed the outgrowth of nerve fibers and thus laid to rest the amoeboid theory of the motility of nerve endings. The idea of nerve endings making and breaking contact with each other provided an appealingly simple explanation of rapid changes in mood states such as hysteria and depression that was receiving credence during that period, especially in Europe.[14] Harrison's hanging drop technique clearly had survival value, for it is still in use for culturing glial cells from the central nervous system.

The association formed at Hopkins during this period between Gasser (1888–1963), then a medical student, and Harry Sidney Newcomer also had far-reaching consequences for American neurophysiology. Even then, Gasser had a deep interest in the propagation of the nerve impulse, and Newcomer, with engineering expertise and friends at Western Electric, was well prepared to attempt the recording of electrical changes in muscle and nerve. Their teamwork at Hopkins and elsewhere after World War I culminated in a 1921 paper on action currents in the phrenic nerve amplified with vacuum tubes.[15]

In 1933, the Hopkins physiology department became a major center for neurophysiological research with the appointment of Philip Bard (1898–1977) as chairman. Bard had done his graduate work at Harvard with Cannon, participating in the "sham-rage" experiments that led to the activation theory of emotions, which melded neurophysiology and psychology together in what has been called the first real neurophysiological theory.[16] Using ablation techniques, Bard extended his studies to locating the brain structures involved in postural "hopping and placing" reactions and sexual behavior in cats.

Meanwhile, the rich Philadelphia tradition in neurophysiology found renewed expression at the University of Pennsylvania in 1929, when Bronk (1897–1975) organized the laboratories of the Eldridge Reeves Johnson Foundation for Medical Physics. Work there on the physiology of vision, transduction of sensory signals to neuronal activity, and spinal reflexes was augmented by the foundation's first two visiting scientists, William Rushton from England and Ragnar Granit from Sweden. Bronk's early training had been in physics, and he gathered together a group at Penn (and later at Hopkins) that gave close attention to the biophysical aspects of neurophysiology.

It should not be supposed that American neurophysiology was confined to Harvard, Hopkins, Penn, or eastern universities more generally. Even before World War I, and especially thereafter, the Midwest also was home to several important research units in neurophysiology. By the 1930s, as we shall see, Chicago could claim three such units and other significant centers formed at the University of Michigan and at Washington University in St. Louis.

At the University of Michigan, the early direction of neurophysiology was set mainly by Lombard, chairman of the physiology department there from 1892 until his retirement in 1923. His work provides evidence of the preoccupation in the later nineteenth century with spinal reflexology, as nerve-muscle physiology was then known. Under Lombard, medical and dental students at the University of Michigan were assigned laboratory exercises that focused on the knee-jerk reflex, Lombard's abiding passion. His interest in the knee jerk, first awakened in Carl Ludwig's laboratory at Leipzig, yielded a large amount of data, although it is said that Lombard failed to understand their implications.[17] In 1923, when Robert Gesell succeeded Lombard as professor of physiology, following a ten-year association with Joseph Erlanger at Washington University in St. Louis, the physiology program at Michigan was greatly expanded. Under Gesell, research focused on cardiovascular problems and culminated in the location of the respiratory center in the brain stem and descending pathways in the spinal cord.

At the University of Chicago, the appointment of German-born Jacques Loeb (1859–1924) as first professor of physiology in 1892 bespoke an early concern there with the biological as opposed to the medical side of physiology.[18] Loeb's demonstration of the response of animals to chemical stimuli (chemotropism) served to flame the debate between proponents of mechanism and free will, an issue to which neurophysiology was central. In the wake of Loeb's extravagant claims on behalf of mechanism, the perennial argument about whether or not organisms behave in response to processes within a substratum of brain and nerves was pursued diligently by psychologists, philosophers, and physiologists.

In the 1930s Chicago gained its full luster as a center of neurophysiological research. By then, three major units had formed: (1) the physiology department at the University of Chicago under the chairmanship of Ralph Waldo Gerard; (2) Stephen Walter Ranson's Institute of Neurology at Northwestern University Medical School, established in 1928, and (3) the group that gathered around Percival Bailey (1892–1973), a graduate of Northwestern Medical School who was then professor of surgery at the University of Chicago. Gerard's laboratory and Ranson's institute each became closely identified with a particular instrument—the microelectrode and the stereotaxic instrument, respectively—and we shall read more of them in the second part of this chapter. The Bailey group deserves a brief separate discussion here.

In the 1930s Bailey organized an informal neurology club, which brought together monthly some of the most outstanding local scientists with interests in the nervous system. Bailey had visions of creating a division of the neurosciences at the University of Chicago composed of such outstanding members of the faculty as C. Judson Herrick, Heinrich Klüver, A. J. Carlson, and Paul Weiss, in whose laboratory fundamental work on nerve regeneration and axoplasmic flow was under way; indeed, Weiss's student Roger Sperry had already discovered the specific patterned growth of nerves. After years of trying to bridge the barriers between three departments—surgery, medicine, and pediatrics—and an unsupportive administration, Bailey moved across town to the University of Illinois Neuropsychiatric Institute (INI) in 1939. He had earlier collaborated with J. G. Dusser de Barenne at Yale, and when Dusser died in 1940, his group was recruited for the biological laboratory in the basement of the newly constructed INI in Chicago.

The most colorful figure in the Bailey group, Warren Sturgis McCulloch (1898–1971), was called a "rebel genius." His training in biophysics and psychology made it easy for him to combine the real with the unreal in his "thinking of neuron systems in terms of topological rather than topographical space."[19] He interpreted neuronal behavior in terms of Boolean logic, an approach that contributed importantly to the development of cybernetics. Actually, Norbert Wiener dedicated his famous book, *Cybernetics,* to yet another "American" neurophysiologist, Arturo Rosenblueth from Mexico, who collaborated with Cannon at Harvard during the 1940s.[20] Earlier, Forbes (also of Harvard), in an authoritative 1922 review article, "demonstrated the possibility of an experimentally based, self-consistent system of interpretation"[21] of nervous system activity, a stimulating concept to incipient cyberneticists.

In St. Louis, even before the 1930s, Washington University had also achieved fame as a major center for neurophysiological work under Erlanger (1874–1965) and his associate Gasser. As we shall see in the second part of this chapter, their research program, like those of Gerard at Chicago and Ranson at Northwestern, also relied heavily on a single instrument, in their case the cathode-ray oscilloscope and vacuum tube amplifier.

The "Axonologists" and Other Aspects of the Social History of Neurophysiology in America

After this brief survey of the leading institutionalized centers for neurophysiology in the United States, we now turn to other aspects of its social history. One of the most significant was the informal network known as the "Axonologists," which can be credited largely to Gerard's organizational talents.

The evening before the 1930 annual meeting of the APS, held that year in Chicago, Gerard hosted "all those working on nerve," for dinner and "a prolonged chat about our common interests."[22] The name "Axonologists" was suggested in jest by Forbes and, to his dismay, could not be dislodged. Their common interests were, of course, the nerve action potential, including the very real possibility that synaptic transmission might be chemical in nature: "soup at the synapse" was Forbes's contribution to the fun. Although some sessions may have been more overtly convivial than scientific, a special topic and discussant were usually designated. This unusual series of meetings helped boost neurophysiology to the forefront of the subspecialties in the 1930s, when the "Axonologists were the important people, and almost strutted through the corridors, being very conscious that they alone were in the frontiers of physiological discovery."[23]

The instant popularity of the gatherings made them unwieldly and less intimate, and they were abandoned a decade later, although the "superaxonologists" continued to dine together annually if that could be arranged. But the impact had already been felt. Gerard had facilitated the most effective of promotional processes—interpersonal communication. Those "prolonged chats" among men of recognized stature were conducive to the rapid spread of ideas, a crucial feature of all scientific conferences.[24] In 1952, ten years after the axonology meetings were abandoned, the custom of an informal assembly of neurophysiologists held the day before the annual meeting of the American Physiological Society was revived by a new generation of investigators. A modest contribution was extracted to keep one's name on the mailing list, and a planned program was followed by dinner at an announced restaurant. This custom flourished until the first meeting of the Society for Neuroscience in 1971.

The Axonologists and other such informal networks were not the only sign of the emerging independence and status of the subdiscipline of neurophysiology between the two World Wars. New centers of neurophysiological research—independent of the other subdisciplines of physiology—were established and advanced courses in neurophysiology were offered in an increasing number of graduate schools. At the 1930 meeting of the American Physiological Society, papers on some aspect of the nervous system constituted 42 percent of the total presented. The subdiscipline also acquired its own specialty publications during this period. Fulton of Yale was responsible for two of them, the *Journal of Neurophysiology* in 1937, and a textbook,[25] the first edition of which was published in 1938 and the third in 1949. After Fulton's death in 1960, the American Physiological Society added the journal to its stable of distinguished serials. When the publication committee decided to sponsor a handbook series, the first section undertaken was *Neurophysiology*, published in 1958.[26]

By the 1950s the number of major themes of neurophysiological inquiry had multiplied and the number of investigators had more than kept pace. The formation of the Society for Neuroscience in 1969 led to a good deal of handwringing by some members of the American Physiological Society, who feared there would be a stampede to join the new interdisciplinary society. The effects on overall membership were not immediately discernible; the ultimate outcome may depend on a self-sorting process and should be of interest to sociologists of science. The older, more conservative organization (i.e., APS), with a tradition of safe leadership in the biomedical community, enjoys the loyalty of neurophysiologists of the same stamp, whereas the newer Society for Neuroscience, whose governance was dominated by "young Turks," appeals to the younger and more venturesome types. For some women neurophysiologists the choice of membership may be easily made on the basis of the absence of their gender from the seats of power in the American Physiological Society.

MAJOR CONCEPTS, INSTRUMENTS, AND TECHNIQUES IN AMERICAN NEUROPHYSIOLOGY

Any list of the major themes and concepts in twentieth-century neurophysiology would include (1) the central nervous system localization for control of hormonal secretion and body homeostasis, (2) the identification of control of movement at several levels of the higher brain centers, (3) the characterization of the elements of the compound action potential, (4) the forces influencing neuronal regeneration, and (5) the electrochemical theory of nervous transmission. American neurophysiologists were involved in all of them, and it is difficult to distinguish whether they were involved more or less importantly than their English colleagues.

Immediately after Otto Loewi's demonstration in 1921 of adrenaline production when the vagus nerve was stimulated, fundamental research on chemical transmission of the nerve impulse across synapses was centered in England, especially around Sir Henry Dale and his associates. The work of Cannon's Harvard laboratory on sympathin (noradrenaline) also contributed to the chemical theory in a major way by strengthening the key role of neurotransmitters in the autonomic nervous system. In the late 1930s, David Nachmansohn and his associates at Columbia University found a relatively high concentration of an acetylcholine splitting enzyme in all conducting tissue. This discovery made neurochemists overnight of some of the more flexible neurophysiologists and helped to produce general agreement that chemical transmission at nerve endings was a fact.[27] Agreement was hardly universal, however, and the concept of nerve conduction as a purely physical process based on ionic movement continued to be supported by other neurophysiologists. Not until 1951, for example, did John Eccles of Oxford relinquish his belief that the fast phase of synaptic transmission was electrical in nature.[28] As late as 1953, in summarizing a symposium, Gerard of Chicago "spoke at length in favour of 'electrical' transmission using many arguments . . . which at the time still carried some conviction."[29]

The most striking feature of the work of American neurophysiologists in these areas was their quickness at adopting and improving technically ingenious novelties—so much so as to suggest that this feature might help to define a distinctively American style in neurophysiology. This characteristic becomes apparent in a survey of American contributions to some of the leading research problems in neurophysiology.

Amplification of Biological Currents and the Cathode-Ray Oscilloscope

During World War I, many American laboratories of physiology were diverted into research that had battle-related clinical implications—circulation, respiration, fatigue, and above all traumatic shock, which became a major focus of concern at Cannon's Harvard laboratory and Erlanger's at Washington University in St. Louis, among others. Although neurophysiology seemed less immediately applicable to the war effort, it received a major indirect benefit: the experience gained in developing electronic surveillance equipment. Forbes is a prime example. Among several other assignments during his service with the United States Navy, Forbes took charge of installing the new radio compass in destroyers and supply ships. That experience honed his already extensive knowledge of electronics, and after returning to teaching and research at Harvard, he published with Catherine Thacher in 1920 the first electron tube–amplified action currents of nerve, using a significantly improved Einthoven string galvanometer as the recording device.[30] When the armistice was signed and research on shock was no longer imperative, Gasser of Washington University in St. Louis interested Erlanger in pursuing experiments similar to Forbes's on amplification and recording of weak electrical signals from nerve. As mentioned earlier, Gasser (with Newcomer) had already published the results of the application of the vacuum tube to action currents in the phrenic nerve.[31] Erlanger had never worked on nerve, but he was highly accomplished in electrical measurements of cardiac muscle action. Gasser had learned of the possibilities of the new Braun tube, and he and Erlanger began to build a cathode-ray oscillograph from distilling flasks. Their series of papers commenced in 1922 with a description of recording without distortion the rapid nerve action potential by the cathode-ray tube, its first application to a biological phenomenon.[32] The impetus this new instrument gave to American neurophysiology soon spread to laboratories abroad. When Ali Monnier returned to Paris in 1938 after a year in Erlanger's laboratory, he took with him a complete "plant" (the contemporary term for a collection of apparatus). Canadian physiologist Herbert Jasper in turn arranged for time in Monnier's laboratory to learn the state of the art.

The award of the 1944 Nobel Prize in physiology or medicine to Gasser and Erlanger for their analysis of conduction in peripheral nerves marked the success of a single-minded attack on a problem by a dominant leader. Erlanger ran a very taut ship, and his small unit broke up (after Gasser had left for Cornell) when George Bishop and Peter Heinbecker temporarily forgot the lines of command and informed the chief of new findings[33] only after the results were written up. Erlanger was furious that he had not been informed while the work was in progress and demanded the raw data; when they were cheerfully provided, he incorporated the information into his next publication.[34]

Brain Waves and the Electroencephalograph

In 1929 the German psychiatrist Hans Berger reported that he had detected electrical discharges from the human scalp. His reports met a skeptical response from research physiologists. As Berger worked in secrecy, it is not surprising that his "brain waves" were not accepted at once. After Adrian and Brian Matthews in England confirmed Berger's findings,[35] brain waves were considered real rather than artifactual, and many neurophysiologists diverted their time, talent, and resources to the new applied field of electroencephalography, which in fact proved to be unproductive in yielding fundamental information about the activity of single nerve cells. Hallowell Davis at Harvard was among the first Americans to demonstrate human brain waves.[36] In addition, he conducted EEG studies at the private laboratory of Alfred Lee Loomis at Tuxedo Park, New Jersey, where a monstrous kymograph was available that could record multiple physiological changes during a 24-hour period.[37] Davis soon left those studies to return to his important work on hearing. Forbes, on the contrary, was not diverted, for he had been advised for the sake of his reputation not to engage in such improbable studies.

Although the EEG was limited from the point of view of basic research in neurophysiology, the intense investigation of the electroencephalogram of normal subjects of all ages, asleep and awake, and of epileptic patients soon established its utility as a diagnostic tool comparable with the electrocardiogram. The demand for the new apparatus placed a premium on electrical engineers who appreciated the possibilities and limitations of quantifying biological variables. A notable example was the team of Ellen and Albert Grass, associated with the physiology department at Harvard in the 1930s. When the value of the EEG in diagnostic localization of brain dysfunction became clear, and as the need for it peaked with World War II casualties, the Grass Instrument Company manufactured and shipped the first commercially produced electroencephalographs, thus inaugurating the American biomedical polygraph industry.

Localization of Control Mechanisms and the Stereotaxic Instrument

From its first use by C. Matteucci in 1844,[38] the term "electrophysiology" had been synonymous with "neurophysiology." That changed in the United States during the 1930s, when investigations of nerve impulse conduction and reflex action acquired an added dimension. Two independent ideas that had been gathering force now came into focus. One has already been mentioned: the view that transmission of impulses at the synapse was not merely electrical (the chemical theory of nervous transmission). The second is our concern here: that control of movement was localized at several levels of the higher brain centers. The stereotaxic instrument was crucial to the development of this new conception.

The attempt to demonstrate a hierarchy of supraspinal control was a prime concern to Ranson's Institute of Neurology at Northwestern University Medical School, where the Horsley-Clarke stereotaxic instrument was revived soon after the Institute was established in 1928.[39] The "stereotack" had been invented in England twenty years earlier by Robert H. Clarke, a physician and experimentalist, who applied geometric principles to positioning a probe within brain structures accurately and reproducibly. Clarke and Horsley, the most noted neurosurgeon of the time in London, published a full description of the instrument and their findings on cerebellar function in the cat in 1908.[40] The merit of the novel apparatus was not at first widely recognized in England, but Clarke continued to work with his invention, and in 1920 published in the United States an atlas with the first tridimensional brain maps.[41] A young American neurosurgeon, Ernest Sachs, who was then studying with Horsley in London, had the instrument replicated, and applied it to an extensive study of the optic thalamus.[42] Sachs used the stereotaxic instrument again at Washington University in St. Louis, where in 1919 he became the first neurosurgeon appointed to a professorship in the world.

Ranson saw Sachs's instrument in St. Louis, and had one built in Chicago by a local machinist. Thereafter, it "was kept humming in a range of projects on the role of the hypothalamus and lower brain stem in visceral integration, emotional expression, and the regulation of feeding, fighting, mating, and other vital behaviors."[43] During the fourteen years of Ranson's directorship of the Institute of Neurology at Northwestern, a steady flow of graduate students and visiting research fellows learned to use the new instrument. It was decisive in turning the attention of many neurophysiologists from the peripheral to the central nervous system. Ranson set the tone of his group by his own example: diligent, serious, and directed toward publishable experimental results. To make sure these results and his group became widely known, Ranson each year sent 500 bound volumes of reprints of Institute papers gratis to an international distribution list.

THE MICROELECTRODE

By the 1930s, as noted earlier Gerard had begun to develop a major center for neurophysiological work in the department of physiology at the University of Chicago. With a good eye for promising ideas and people, Gerard published with his students a succession of papers on new topics that he did not usually follow through to completion. If that style tended to work against sound competence in any one area, it also led to several important innovations. These included a primitive "electrical atlas" of the cat brain constructed from serial sections borrowed from Ranson, and the application of the evoked-potential technique to the somatic sensory areas of the central nervous system.[44]

Another significant work of Gerard's group at Chicago was the perfection of the capillary glass electrode, which opened up an entirely new domain in neurophysiological research: single-unit recording from within the nerve cell. "[In] view of the discrepancy which is possible between unit and aggregate behavior,"[45] it was a great advantage to be able to record from a single nerve fiber rather than from bundles as in a nerve trunk. Unlike the sudden appearance of the stereotaxic instrument, full-blown and successful in its original configuration, the microelectrode underwent continual refinement for thirty years before yielding satisfactory results. Also called capillary electrodes, the devices were fashioned from a segment of glass tubing held at each end and twirled while heated, then pulled apart at just the right moment of malleableness. Ideally, this maneuver produced two capillaries whose walls tapered to an open end at an angle such that the tube was neither too brittle nor too flexible. Later, mechanical electrode pullers were marketed by apparatus companies alert to the microelectrode's extensive application in biological research.

Immediately after World War I, microelectrodes were put to use by physiologists to stimulate muscle cells or unicellular organisms.[46] These early electrodes had tips of less than four to eight micrometers in diameter and were filled with an electrolyte solution or mercury. In 1927, in an obscure journal, Samuel Gelfan, a young Russian-American at Berkeley, published a detailed description of glass microelectrodes one-fourth the size he used to record protoplasmic conductivity from inside the unicellular *Paramecium*.[47]

In the United States, the earliest reports of recording intracellularly from mammalian cells came from the University of Chicago laboratory of biophysicist Kenneth S. "Kacy" Cole (1900–1984), who in 1935 used the microelectrode to study embryonic rat heart cells.[48] With the introduction from Great Britain of the giant squid axon by J. Z. Young at a Cold Spring Harbor, New York Symposium in 1936,[49] American neurophysiologists hastened to catch up with their British peers in recording electrical activity with microelectrodes inserted inside the axon and parallel to its length. Taking advantage of this combination of microelectrodes and macroneurons, British neurophysiologists charted the pathways of the ionic activities creating action currents in nerve,[50] and in the United States the resting membrane potential came under systematic scrutiny, especially by Cole.[51]

Cole's invention of the voltage clamp[52] to stabilize the voltage of the membrane was "an indispensible tool for exploring the mechanisms underlying excitability" and greatly facilitated the study of ionic transfer in neuronal membranes.[53] However, the electrodes were still too large and could not be pushed through the cell membrane at right angles without injuring or breaking the cell. In Gerard's laboratory at the University of Chicago, Judith Graham further miniaturized the diameter of the electrode.[54] In 1949, a definitive description of the microelectrode from the same laboratory[55] includes a drawing of a properly configurated glass tube having a tip of one micrometer or less, with which "highly constant" membrane potentials in frog muscle single fibers were obtained.

In 1939, Forbes and his Harvard collaborators reported that they had introduced the microelectrode into brain tissue.[56] Its positioning was controlled by a simple arrangement of a microscope rack and pinion for vertical penetration in small increments. The resulting records of potentials from nerve cell surfaces were compared with those from axons, although the Harvard group was forced to concede that "Just how closely the recorded responses parallel the sequence of changes in a single neuron is not clear."[57] This important paper addressed in detail most of the major concerns of neurophysiologists during the 1930s: verification of electrode tip placement, the effect of various anesthetic agents and depth of anesthesia on nerve cell activity, the configuration and electrical characteristics of microelectrodes, the species differences in the results, and their possible meaning and significance. Within months, Forbes and his collaborators became convinced that the action potentials they were recording with microelectrodes and the cathode-ray oscillograph were derived from single neurons.[58]

World War II, Radioisotopes, and the Evoked-Potential Technique

During World War II, neurophysiological techniques became part of the war effort because of the threat of enemy war gases. Studies of the mechanism of "nerve gas" and the testing of antidotes were carried out in United States Army laboratories at Edgewood Arsenal, Maryland, among other places. The United States Navy, in its facility at Bethesda, conducted tests on the effectiveness of analgesic agents, using the evoked-potential technique, but these neurophysiological and neuropharmacological studies were transient and did not enter into the mainstream of basic research.

Yet World War II did have major and still lasting consequences for research methodology in neurophysiology. One was the new availability of radioactive tracers, which allowed ionic and molecular movements to be measured quickly and accurately, and thus opened the way to metabolic and ion transfer studies in the nervous system as well as other tissues. Even more important perhaps was the experience gained during the war with the evoked potential. Already a recognized technique, evoked potentials now became perhaps the dominant tool in neurophysiology and did much to shift attention from the peripheral axon to activity in the central nervous system.

In the United States, the evoked-potential technique was carried by Wade Hampton Marshall (1907–1972) from Gerard's laboratory at the University of Chicago, where the term was coined, to Bard's department at Hopkins in 1937. Using state-of-the-art equipment constructed in consultation with Albert Grass at Harvard and the movement of a single hair on a cat's paw to evoke an action potential recorded on the brain surface, Marshall inaugurated at Hopkins a systematic mapping of the sensory cortex of cat and monkey,[59] which was subsequently completed in careful detail by Clinton N. Woolsey.[60]

By 1951, the nature of the stimulus had become important. In that year George Bishop at Washington University wrote to Fulton of Yale[61] that it was preferable to show the cat a mouse rather than a discrete spot of light, thus introducing a behavioral dimension to "tunnel vision" neurophysiology. However, the artificial stimulus yielded informative results in the hands of Torsten Weisel and David Hubel, who in 1959[62] identified single cells in the cat's brain that responded only to stimuli in the visual field that were moving in specific directions. Their work was part of the most active area of sensory research at that time in neurophysiology.

After 1948, the earlier interest in conduction of the nerve impulse by the axon shifted to axonal transport of materials in nerve. The discovery of axoplasmic flow in Weiss's laboratory at Rockefeller University,[63] coupled with the application of radioisotopes as labels of transported material, stimulated new studies on regeneration of peripheral nerves, a problem whose attack had been encouraged sixty years before by a prize offered by Mitchell through the American Physiological Society.[64]

CONCLUDING REFLECTIONS

It is clear that technical innovations and their rapid utilization, including especially new electronic amplifying and recording devices, played a crucial role in the history of American neurophysiology. Between the two World Wars, the major research themes, and the leading concepts that emerged from them, owed much to such innovations as the stereotaxic instrument and the cathode-ray oscilloscope. The question of whether or not there is reason to think that such technical ingenuity can be seen as part of a distinctively "American style" in the history of neurophysiology should be a productive line of research.

Until the 1930s, many leading American neurophysiologists—among them, Forbes, Gasser, and Bishop, who had been trained in physics as well as biology—built and maintained their own electronic equipment. However, the increasing complexity and growing demand for amplifying and recording apparatus gave rise to a new cadre of assistants in the physiology laboratory: the "electronikers," electrical engineers who knew the capabilities of such apparatus. The notable example of the Grasses was mentioned in connection with the electroencephalograph. Until manufactured "black boxes" became widely available, neurophysiologists depended heavily on the expertise of such "electronikers."

Unorganized networks such as the Axonologists, or Bailey's informal neurology club at Chicago deserve a place in the history of American neurophysiology. Such groups are too often lost from sight in the tendency to focus on individual "stars" and formal institutionalized centers of research in universities. Those groups can be very consequential, even if they occur outside the category of "research schools" or other formal, organized units of analysis that are now attracting so much attention from historians and sociologists of science.

The Axonologists, for example, were a major force in American neurophysiology without being a locally cohesive or otherwise traditional social unit. They met only annually (in connection with the American Physiological Society) and were spread among the physiology departments at Washington University, Harvard, Pennsylvania, Chicago, and Rochester. Except for Erlanger's associates at St. Louis, each member of the initial group of axonologists worked independently and only rarely collaborated on a publication. Yet they clearly did constitute a group in some important sense, united by their common interest in neural research, their eminence, and perhaps even a form of "male bonding." In the case of at least one absolutely central issue—the idea that the transmission of the nerve impulse across the synapse might be chemical in nature—they exerted a direct and telling influence on one another. Thus the Axonologists were usually receptive to the chemical theory of nervous transmission, while many other eminent neurophysiologists (notably Eccles) continued to cling for a time to a strictly physical (i.e., electrical) interpretation of the phenomenon.

Even within the formally institutionalized centers of neurophysiology in the United States, we need to know much more about the variety of conditions under which research was carried out. We have described how, as neurophysiology emerged as a recognized subdiscipline in the United States, some local groups or research schools coalesced around a single leader, a distinctive research program, or—most emphasized here—a particular technique or instrument. Indeed, a technique or instrument could itself be a major force in the formation of a coherent research program or group. Examples include the close association between the cathode-ray oscilloscope and Erlanger's group at Washington University, the stereotaxic instrument and Ranson's group at Northwestern, and the microelectrode and Gerard's group at the University of Chicago.

In addition, the more narrowly social features of these and other research units in American neurophysiology varied widely. In St. Louis, Erlanger led by fiat and without charisma; it was only the utilization of the new amplification and recording techniques that bound the team together. Gerard's group at the University of Chicago had no conceptually coherent leadership; rather, it constituted an exploitation of new ideas from many directions. The most personally charismatic leader on our list was Bailey. Although he failed initially to secure the institute for the neurosciences at the University of Chicago he sought, in the end the small group that gathered around him at the INI enjoyed their creative intellectual work together and contributed substantively to neurophysiology's progress.

Perhaps only Ranson's Institute of Neurology at Northwestern can be regarded as something like a complete research school in terms of the ideal criteria set forth by Gerald L. Geison.[65] Ranson's Institute had a new technique (stereotaxis) in an emerging field (the central nervous system control of body functions), a steady supply of young recruits, and a do-it-yourself publication outlet in the form of an international distribution of reprints. Although Ranson was not himself particularly sociable, his group clearly pursued a coherent and innovative intellectual program.

In short, the research units in American neurophysiology that we have surveyed varied widely in their social styles, ranging from the very tight control of apparatus and research in Erlanger's group in St. Louis to the much freer situation in Bailey's group at Chicago. The latter style was more prevalent in the United States than abroad, and senior American figures in the subdiscipline generally encouraged the participation of younger investigators in both formal and informal settings. Indeed, perhaps the relatively democratic tone of American neurophysiology during this period, resulting in a pluralism among research groups, was as much a part of any distinctive style it may have had as was its technical ingenuity.

NOTES

1. For a discussion of the earlier work see E. Clarke and L. S. Jacyna, *Nineteenth Century Origins of Neuroscientific Concepts* (Los Angeles: Univ. of California Press, 1987), chapt. 4.
2. V. B. Mountcastle, ed., *Medical Physiology*, 14th ed. (St. Louis, MO: Mosby, 1980), p. 185.
3. These spectacular experiments on humans were described by H. Cushing, "A Note Upon the

Faradic Stimulation of the Postcentral Gyrus in Conscious Patients," *Brain* 32 (1909): 44–53.

4. See Cushing, *The Pituitary Body and Its Disorders* (Philadelphia, PA: Lippincott, 1912).
5. J. F. Fulton devoted twenty pages in the first two editions of his famous textbook, *Physiology of the Nervous System* (London: Oxford Univ. Press, 1938, 1943) to discussion of the meaning of the angle. A full account of this amusing incident is found in J. C. Eccles and W. C. Gibson, *Sherrington, His Life and Thought* (Berlin: Springer International, 1979).
6. H. S. Gasser, Letter to J. Erlanger, 9 October 1923, Washington University School of Medicine Library, St. Louis, MO, by permission.
7. This description of Holmes's straightforward experiment is quoted in J. C. Burnham, *Science in America: Historical Selections* (New York: Holt, Rinehart, Winston, 1971), p. 191.
8. The six were H. P. Bowditch (1840–1911), H. H. Donaldson (1857–1938), W. P. Lombard (1885–1939), T. W. Mills (1847–1915), S. W. Mitchell (1829–1914), and Isaac Ott (1847–1916).
9. Degeneration studies on nerve are summarized by D. M. Barnes, in "Research News: What Makes Nerves Regenerate?" *Science* 230 (1985): 1024–1025.
10. An extraordinarily early example was one of those "peculiar" institutions, then named Vassar Female College, which opened in 1861 with a department of physiology as noted in an anonymous editorial in *Harper's Weekly* (9 February 1861).
11. W. B. Cannon, *The Way of an Investigator: A Scientist's Experiences in Medical Research* (New York: Hafner, 1945), p. 92.
12. The reasons for the chairman, Henry Newell Martin, looking favorably on students who wished to pursue physiological problems are discussed by Jane Maienschein in chapter VII in this book.
13. Donaldson's postgraduate research was described thus by W. H. Howell, "The American Physiological Society During its First Twenty-Five Years," in *History of the American Physiological Society Semicentennial 1887–1937* (Baltimore, MD: American Physiological Society, 1938), pp. 1–89, esp. p. 25.
14. See S. Black, "Pseudopods and Synapses: The Amoeboid Theories of Neuronal Mobility and the Early Foundation of the Synapse Concept, 1894–1900," *Bull. Hist. Med.* 55 (1981): 34–58.
15. Gasser, and H. S. Newcomer, "Physiological Action Currents in the Phrenic Nerve: An Application of the Thermionic Vacuum Tube to Nerve Physiology," *Am. J. Physiol.* 57 (1921): 1–26.
16. See K. Lissak and P. Molnar, "Emotional Behavior: The Control of the Emotions and Their Expression," in *The Life and Contributions of Walter Bradford Cannon 1871–1945*, ed. C. Mc. Brooks, K. Koizumi, and J. O. Pinkston (Brooklyn: State Univ. of New York Downstate Medical Center, 1975), p. 122.
17. Lombard's role at Michigan is described in the recent work of Horace W. Davenport, who points out that among the early members of the American Physiological Society, several more eminent than Lombard spent various periods there, for example, Howell, J. Vaughn, D. W. Bronk, and H. Sewall.
18. Chapter VII by Maienschein and chapter VIII by Philip Pauly in this book detail the influence of Jacques Loeb on the transition from a disease-oriented discipline to one encompassing dynamic life processes.
19. R. W. Gerard, "Warren Sturgis McCulloch: Rebel Genius," *Trans. Am. Neurol. Assoc.* 95 (1970): 344–345, esp. 344.
20. I am indebted to E. Clifford Barger for bringing this dedication to my attention.
21. Quoted from W. O. Fenn, "Alexander Forbes, May 14, 1882–May 27, 1965," *Natl. Acad. Sci. Biogr. Mem.* 40 (1969): 113–141, esp. 118–119. The review is A. Forbes, "The Interpretation of Spinal Reflexes in Terms of Present Knowledge of Nerve Conduction," *Physiol. Rev.* 2 (1922): 361–414.
22. From Gerard's letter to Bronk, 17 March 1930, Rockefeller University Archives, with permission. In additioin to Bronk, copies of the letter were sent to W. R. Amberson, *G. H. Bishop, *H. Davis, *J. Erlanger, *W. O. Fenn, *A. Forbes, *H. S. Gasser, A. B. Wilson, P. Heinbecker, *R. S. Lillie, *G. P. McCouch, *A. M. Monnier, *W. A. H. Rushton, and *F. O. Schmitt. Asterisks indicate those who are known to have attended.
23. From a letter of Alan C. Burton quoted in Fenn, *History of the American Physiological Society:*

The Third Quarter Century, 1937–1962 (Washington, DC: American Physiological Society, 1963), p. 92–93.

24. As one of the original group, Hallowell Davis wrote a quarter century later, "Only after 'The Axonologists' had dissolved. . . did the members realize how important it had been in stimulating and shaping the development of neuro-electrophysiology"; Davis, "Joseph Erlanger," *Natl. Acad. Sci. Biogr. Mem.* 41(1970): 110–139, quote on 123.
25. Fulton, *Physiology of the Nervous System* (New York: Oxford Univ. Press, 1938).
26. For a definitive history of the *Handbook of Physiology* see F. K. O'Malley and Horace W. Magoun, "The First American-Based *Handbook of Physiology*," *Physiologist* 28 (1985): 35–39.
27. David Nachmansohn wrote that Fulton, in whose laboratory he spent three years, was the first neurophysiologist "to integrate ACh into the metabolic pathways of the nerve cell" in Nachmansohn, *Chemical and Molecular Basis of Nerve Activity* (New York: Academic, 1959), p. vi.
28. L. G. Brock, J. S. Coombs, and Eccles. "The Recording of Potentials From Motoneurones With an Intracellular Electrode," *J. Physiol. Lond.* 117 (1952): 431–460.
29. Z. M. Bacq, *Chemical Transmission of Nerve Impulses. A Historical Sketch* (Oxford, UK: Pergamon, 1975), p. 93.
30. Forbes and C. Thacher, "Amplification of Action Currents with the Electron Tube in Recording With the String Galvanometer," *Am. J. Physiol.* 52 (1920): 409–471.
31. See Gasser and Newcomer, "Physiological Action Currents."
32. Gasser and Erlanger, "The Cathode Ray Oscillograph as a Means of Recording Nerve Action Currents and Induction Shocks," *Am. J. Physiol.* 59 (1922): 473–475.
33. Heinbecker and Bishop, "Differentiation Between Types of Fibers in Certain Components of Involuntary Nervous System," *Proc. Soc. Exp. Biol. Med.* 26(1929): 645–647.
34. George H. Bishop's gracious assessment of the incident is quoted in L. H. Marshall, "The Fecundity of Aggregates: The Axonologists at Washington University, 1922-1931," *Perspect. Biol. Med.* 26 (1983): 613–636.
35. E. D. Adrian and B. H. C. Matthews, "The Berger Rhythm: Potential Changes From the Occipital Lobes of Man," *Brain* 57 (1934): 355–385.
36. The early days of the EEG were described from first-hand knowledge in J. L. O'Leary and S. Goldring, *Science and Epilepsy: Neuroscience Gains in Epilepsy Research* (New York: Raven, 1976), esp. p. 136.
37. Alfred L. Loomis was a wealthy engineer who occupied himself and his well-equipped laboratory with, in addition to electroencephalography, the same types of biophysical problems as the Eldridge Reeves Johnson Foundation for Medical Physics at the University of Pennsylvania, with which he had contacts.
38. C. Matteucci, *Traité des Phénomeñes Electro-physiologiques des Animaux* (Paris: Fortin, Masson, 1844).
39. W. R. Ingram et al., "Results of Stimulation of the Tegmentum With the Horsley-Clarke Stereotaxic Apparatus," *Arch. Neurol. Psychiatry* 25 (1932): 513–541.
40. The first report of the instrument was made by Horsley at the meeting of the British Medical Association, in Toronto that year, 1906. The published description was clear and detailed, so much so that replicas or improvements could be made from it, as Ranson and others twenty years later did. The illustrated paper was V. Horsley and R. H. Clarke, "The Structure and Function of the Cerebellum Examined by a New Method," *Brain* 31 (1908): 45–124.
41. Clarke and E. E. Henderson, *Atlas of Photographs of the Frontal Sections of the Cranium and Brain of the Rhesus Monkey (Macacus rhesus)*, special vol., part II (Baltimore, MD: Johns Hopkins Hospital Reports, 1920).
42. E. Sachs, "On the Structure and Functional Relations of the Optic Thalamus," *Brain* 32 (1909): 95–186. Years later Sachs presented this instrument to Magoun, "the new torch-bearer."
43. Magoun, "The Northwestern Connection with the Reticular Formation," *Surg. Neurol.* 24 (1985): 250–252, esp. 250.
44. Both these innovative contributions were published in Gerard, W. H. Marshall, and L. J. Saul, "Electrical Activity of the Cat's Brain," *Arch. Neurol. Psychiatry* 36 (1936): 675–738. According to Magoun (personal communication), Ranson was not aware that his sections were to be published.

45. F. H. Pratt, "Excitation of Microscopic Areas: A Non-Polarizable Capillary Electrode," *Am. J. Physiol.* 43 (1917): 159–168.
46. Four years after Pratt's description of his method (loc. cit.), Ida H. Hyde at the University of Kansas, the first woman to become a member of APS, in 1902, published her work, "A Micro-Electrode and Unicellular Stimulation," *Biol. Bull.* 40 (1921): 130–133.
47. S. Gelfan, "The Electrical Conductivity of Protoplasm and a New Method of its Determination," *Univ. Calif. Publ. Zool.* 29 (1927): 453–465. A year later Gelfan was a postdoctoral student of Ralph S. Lillie in the physiology department at the University of Chicago.
48. Graham Hoyle's letter correcting some misconceptions about the development of the microelectrode [Hoyle, "Origins of Intracellular Microelectrodes," *Trends Neurosci.* 6 (1983): 163] accorded first place to K. S. Cole's laboratory with the following publication: B. M. Hogg, C. M. Goss, and Cole, "Potentials in Embryo Rat Heart Muscle Cultures," *Proc. Soc. Exp. Med. Biol.* 32 (1934–1935): 304–307.
49. J. Z. Young, "Structure of Nerve Fibres and Synapses in Some Invertebrates," *Cold Spring Harbor Symp. Quant. Biol.* 4 (1936): 1–6.
50. A. L. Hodgkin and B. Katz, "The Effect of Sodium Ions on the Electrical Activity of the Giant Axon of the Squid," *J. Physiol. Lond.* 108 (1949): 37–77.
51. Cole, and H. J. Curtis, "Electric Impedance of Nerve During Activity," *Nature Lond.* 152 (1938): 209.
52. Cole, "Dynamic Electrical Characteristics of the Squid Axon Membrane," *Arch. Sci. Physiol.* 3 (1949): 253–258.
53. E. Jakobsson and R. Guttman, "Continuous Stimulation and Threshold of Axons: The Other Legacy of Kenneth Cole," in *The Biophysical Approach to Excitable Systems*, ed. W. J. Adelman and D. E. Goldman (New York: Plenum, 1981), pp. 197–211, esp. p. 198.
54. Graham and Gerard, "Membrane Potentials and Excitation of Impaled Single Muscle Fibers," *J. Cell. Comp. Physiol.* 28 (1946): 99–117.
55. G. Ling and Gerard, "The Normal Membrane Potential of Frog Sartorius Fibers," *J. Cell. Comp. Physiol.* 34 (1949): 383–396.
56. B. Renshaw, Forbes, and B. R. Morison, "Activity of Isocortex and Hippocampus: Electrical Studies with Micro-Electrodes," *J. Neurophysiol.* 3 (1940): 74–105. An abstract was published in 1938.
57. Ibid., p. 99.
58. P. O. Therman, Forbes, and R. Galambos, "Electric Responses Derived From the Superior Cervical Ganglion With Micro-Electrodes," *J. Neurophysiol.* 3 (1940): 191–200.
59. Marshall, C. N. Woolsey, and P. Bard, "Cortical Representation of Tactile Sensibility as Indicated by Cortical Potentials," *Science* 85 (1937): 388–396.
60. Woolsey, Marshall, and Bard, "Representation of Cutaneous Tactile Sensibility in Cerebral Cortex of Monkey as Indicated by Evoked Potentials," *Bull. Johns Hopkins Hosp.* 70 (1942): 399–441.
61. Bishop to Fulton, 5 June 1951, Yale Historical Library, New Haven, CT, with permission.
62. D. H. Hubel and T. N. Weisel, "Receptive Fields of Single Neurones in the Cat's Striate Cortex," *J. Physiol. Lond.* 148 (1959): 574–591.
63. P. Weiss and H. B. Hisco, "Experiments on the Mechanism of Nerve Growth," *J. Exp. Zool.* 107 (1948): 315–395.
64. Sidney Ochs, in addition to contributing importantly to the subject, has written on the history of nerve repair and regeneration in "A Brief History of Nerve Repair and Regeneration," in *Nerve Repair and Regeneration: Its Clinical and Experimental Basis*, ed. D. L. Jewett and H. R. McCarroll (St. Louis, MO: Mosby, 1980), pp. 1–8.
65. G. L. Geison, searching for a universally valid conception of scientific change, enumerated the characteristics of an ideal group in "Scientific Change, Emerging Specialties, and Research Schools," *Hist. Sci.* 19 (1981): 20–40.

Author Index

This book has no bibliography in the usual sense. The author index can be used as a substitute, since full citations can be found in the pages listed after each author's name.

Subject Index